T0310077

EGG BIOSCIENCE AND BIOTECHNOLOGY

EGG BIOSCIENCE AND BIOTECHNOLOGY

Edited by

YOSHINORI MINE
Department of Food Science
University of Guelph

WILEY-INTERSCIENCE
A JOHN WILEY & SONS, INC., PUBLICATION

Published by John Wiley & Sons, Inc., Hoboken, New Jersey
Published simultaneously in Canada

Library of Congress Cataloging-in-Publication Data:

Egg bioscience and biotechnology / [edited by] Yoshinori Mine.
p. ; cm.
Includes bibliographical references.
ISBN 978-0-470-03998-4 (cloth)
1. Eggs—Biotechnology. I. Mine, Yoshinori.
[DNLM: 1. Biotechnology. 2. Eggs—analysis. 3. Birds. 4. Egg Shell—physiology.
TP 248.65.E35 E29 2008]
TP248.65.E35E32 2008
660.6—dc22

2007015573

Printed in the United States of America
10 9 8 7 6 5 4 3 2 1

CONTENTS

PREFACE

Avian eggs have long been recognized as an excellent source of nutrients and foods. The egg is the largest biological cell known to originate from one cell division and is composed of various important chemical substances that form the basis of life. The egg is a complete set of biological substances containing nutrients such as proteins, lipids, inhibitors, enzymes, and various biologically active substances, including growth promoting factors as well as defense factors against bacterial and viral invasion. Since the mid-1990s, numerous extensive studies characterizing biophysiological functions of egg components and seeking novel biologically active substances in hen eggs have been conducted. These applications are being developed to utilize the nutritional and functional contributions of eggs not only in food products but also as biologically active components that may be used as nutraceutical and functional food ingredients with the potential to reduce the risk of disease and enhance human health and also as materials for drug and cosmetic applications. This book focuses mainly on the most recent advances in biologically active (bioactive) components such as nutraceuticals, pharmaceuticals, and cosmetics derived from egg components.

Chapter 1 defines the basic fundamental structural and chemical characteristics of eggs. Key elements of the physical structure, chemical composition, and properties of the eggshell, albumen, and yolk are described, including highlights of the most recent literature on factors that may affect structure and chemical composition, as well as novel functional or bioactive properties that are attributed to specific chemical components of eggs. The avian egg is a reproductive structure that has been shaped through evolution to resist physical, microbial, and thermal attack from an external and possibly aggressive environment, while satisfying the needs of the developing embryo. Chapter 2 focuses on recent proteomic and genomic analyses of the eggshell and draws attention to the impact of this information on the current understanding of eggshell function. Chapter 3 deals with utilization of the refined eggshell as a source of dietary calcium and the eggshell membrane as a biological dressing for the treatment of skin injures. Chapters 4 and 5 describe the recent scientific body of literature on the numerous biological activities of egg white and egg yolk components. These include novel antimicrobial activities; antiadhesive and antioxidant properties; hypercholesterolemic, immunomodulatory, anticancer, and antihypertensive activities; protease inhibitory function; nutrient bioavailability; and antibody and functional lipids activities, highlighting the importance of egg and egg components in human health and disease prevention and treatment. The hen's egg is one of the most common sources of food allergens, especially in children. Chapter 6 focuses on providing the reader with up-to-date knowledge about the biochemical and molecular characteristics of egg allergens, as well as prospects for novel immunotherapeutic and preventive strategies against egg allergy. The potential to produce novel proteins in chicken eggs was identified in 1990, when the introduction of genetic modifications into the chicken genome was first suggested. Chapter 7 presents the background and most recent developments in the production of novel proteins in eggs. Along with changes in egg-processing technology, there has been a continuing growth of further processed egg products. Today, approximately 30% of the total consumption of eggs is in the form of further processed egg products. Chapter 8 discusses future perspectives of egg science and technology to increase the value of eggs. Continued research to identify new and existing biological functions of hen egg components will help define new methods to further improve the value of eggs, as a source of numerous biologically active compounds with specific benefits for human and animal health.

The editor has succeeded in bringing together many renowned international egg experts to review current egg bioscience and biotechnology and is grateful to all the authors for their state-of-the-art compilation of the most recent developments in this field. The editor believes that this book will be the first to unlock the secrets of eggs as an excellent source of biologically active substances for various applications. It certainly warrants a broad readership in the disciplines of nutrition, pharmacology, nutraceutical/functional foods, poultry science, food science, biology, biochemistry, biotechnology, and life science. This book could also be used as a reference by senior undergraduate and graduate students.

YOSHINORI MINE

University of Guelph
Ontario, Canada, 2007

CONTRIBUTORS

Icy D'Silva Department of Food Science, University of Guelph, Guelph, Ontario, Canada N1G 2W1

*__Robert J. Etches__ Vice President Research, Origen Therapeutics, 1450 Rollins Road, Burlingame, CA 94010 [Email: REtches@OrigenTherapeutics.com]

*__Glenn W. Froning__ Food Science and Technology, Institute of Agriculture and Natural Resources, University of Nebraska–Lincoln, 357 Food Industry Complex, Lincoln, NE 68583 [Email: gfroning@unlnotes.unl.edu]

Joël Gautron INRA, UR83 Recherches Avicoles, F-37380 Nouzilly, France

*__Hajime Hatta__ Department of Food and Nutrition, Kyoto Women's University, Higashiyama-ku, Kyoto 605-8501, Japan [Email: hatta@kyoto-wu.ac.jp]

*__Maxwell T. Hincke__ Department of Cellular and Molecular Medicine, Faculty of Medicine, University of Ottawa, 451 Smyth Road, Ottawa, Ontario, Canada K1H 8M5 [Email: mhincke@uottawa.ca]

*__Hajime Hiramatsu__ R&D Division, Q.P. Corporation, 5-13-1 Sumiyoshi-Cho, Fuchu-shi, Tokyo, Japan 183-0034 [Email: hajime_hiramatsu@kewpie.co.jp]

Lekh Raj Juneja Taiyo Kagaku Co., Ltd, Yokkaichi, Mie, Japan

Mahendra P. Kapoor Taiyo Kagaku Co., Ltd, Yokkaichi, Mie, Japan

*Corresponding author.

Hyun-Ock Kim The University of British Columbia, Faculty of Land and Food Systems, Food Nutrition and Health Building, 2205 East Mall, Vancouver, BC, Canada V6T 1Z4

*__Eunice C. Y. Li-Chan__ The University of British Columbia, Faculty of Land and Food Systems, Food Nutrition and Health Building, 2205 East Mall, Vancouver, BC, Canada V6T 1Z4 [Email: eunice.li-chan@ubc.ca]

Marc D. McKee Faculty of Dentistry and Department of Anatomy and Cell Biology, McGill University, Montreal, Quebec, Canada H3A 2B2

Karlheinz Mann Max-Planck-Institut fur Biochemie, D-82152 Martinsried, Germany

Yasunobu Masuda R&D Division, Q.P. Corporation, 5-13-1 Sumiyoshi-Cho, Fuchu-shi, Tokyo, Japan 183-0034

*__Yoshinori Mine__ Department of Food Science, University of Guelph, Guelph, Ontario, Canada N1G 2W1 [Email: ymine@uoguelph.ca]

Yves Nys INRA, UR83 Recherches Avicoles, F-37380 Nouzilly, France

Oliver Wellman-Labadie Department of Cellular and Molecular Medicine, Faculty of Medicine, University of Ottawa, 451 Smyth Road, Ottawa, Ontario, Canada K1H 8M5

Marie Yang Department of Food Science, University of Guelph, Guelph, Ontario, Canada N1G 2W1

1

STRUCTURE AND CHEMICAL COMPOSITION OF EGGS

Eunice C. Y. Li-Chan and Hyun-Ock Kim
The University of British Columbia, Faculty of Land and Food Systems, Vancouver, BC, Canada

1.1. INTRODUCTION

The eggs of avian species have long been recognized as an excellent source of nutrients for humans. In recent years, innovative research revealing the diversity of structure and function of components in eggs has fueled increasing demand to more fully utilize this bioresource. These applications are being developed to take advantage not only of the nutritional and functional contributions of eggs in food products but also of the bioactive components that may be used as nutraceutical and functional food ingredients with potential to reduce risk of disease and enhance human health.

An understanding of the egg's structural and chemical characteristics provides the fundamental basis for investigation of its bioscience and biotechnology. Earlier references on this topic are reviewed in several monographs and reviews, including those by Burley and Vadehra (1989), Li-Chan and Nakai (1989), Stadelman and Cotterill (1995), and Yamamoto et al. (1997). The objective of this chapter is to build on that knowledge base of information on eggs from domestic hens. Key elements of the physical structure, chemical composition, and properties of the eggshell, albumen, and yolk are described, including highlights of the most recent literature on factors that may affect structure and chemical composition, as well as novel functional or bioactive properties that are attributed to specific chemical components of eggs. The details of biosynthesis, bioactivity, and applications of these components are described in subsequent chapters of this book.

Egg Bioscience and Biotechnology Edited by Yoshinori Mine

1.2. STRUCTURE OF EGGS

Eggs are composed of three main parts; the eggshell with the eggshell membrane, the albumen or white, and the yolk. The yolk is surrounded by albumen, which in turn is enveloped by eggshell membranes and finally a hard eggshell (USDA 2000) (Fig. 1.1).

1.2.1. Eggshell

The eggshell is composed of a foamy layer of cuticle, a calcite or calcium carbonate layer, and two shell membranes. As illustrated in Figures 1.2 and 1.3, ultrastructurally the eggshell includes shell membranes, mammillary zone, calcium reserve assembly, palisades, and cuticle, with 7,000–17,000 funnel-shaped pore canals distributed unevenly on the shell surface for exchange for water and gases (Dennis et al. 1996).

Eggshell formation takes place on the eggshell membrane in an acellular medium, the uterine fluid that contains the inorganic minerals and precursors of the organic matrix (Gautron et al. 2001). The high degree of eggshell structure results from deposition of calcium carbonate concomitantly with an organic matrix on the eggshell membranes, which are composed of two nets of type X collagen-containing fibrils (Dominguez-Vera et al. 2000). Onto these membranes, the mammillary knobs, specifically, the crystal nucleation sites, are deposited. The eggshell matrix proteins influence the process of crystal growth by controlling size, shape, and orientation of calcite crystals (Nys et al. 2001), thereby affecting texture and biomechanical properties of eggshell (Panheleux et al. 1999).

On the basis of observations by transmission electron microscopy (TEM), Fraser et al. (1999) hypothesized that vertical orientation of calcite crystals in

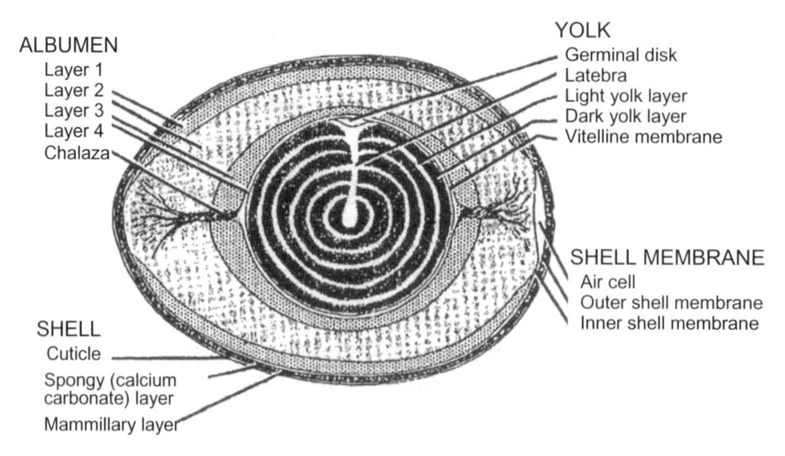

Figure 1.1. Structural components of the eggshell, eggshell membrane, albumen, and yolk. [*Source*: Adapted from USDA (2000) Egg Grading Manual.]

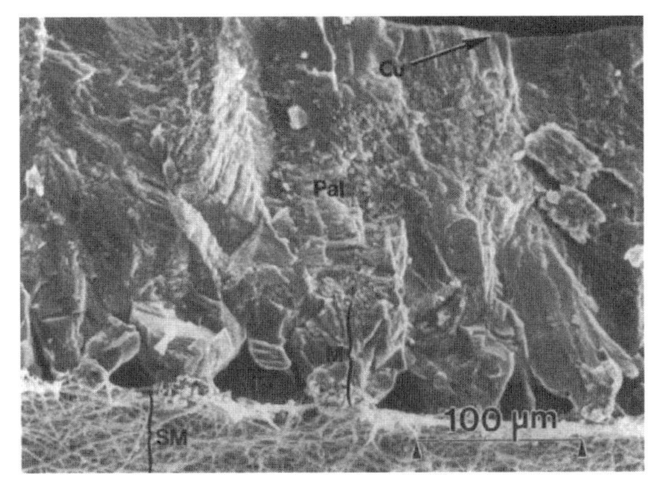

Figure 1.2. Scanning electron micrograph of a fractured eggshell showing the shell membranes (SM), the mammillary zone (M), the palisade region (Pal), and the cuticle (Cu). [*Source*: Dennis et al. (1996).]

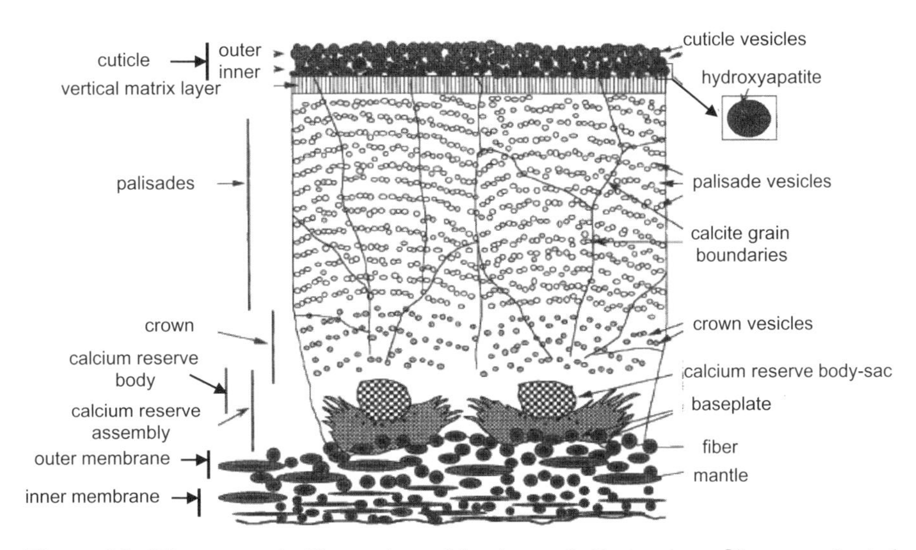

Figure 1.3. Diagrammatic illustration of hen's eggshell structure. [*Source*: adapted from Dennis et al. (1996).]

the vertical crystal layer is closely linked to a vertical matrix. At the outer palisade region of the eggshell TEM also revealed the presence of a two-layered cuticle, the inner layer containing vesicles that are absent in the outer. Cuticular vesicles contain hydroxyapatite and are thought to play a role in the termination of shell formation. At the outer palisade region of the eggshell the matrix fibers are perpendicularly oriented, becoming increasingly vertically aligned and more tightly packed toward the outer surface in the region of the vertical crystal layer.

Gautron et al. (1996) suggested that constituents of the eggshell matrix are involved in the control of calcite growth and crystallographic structure of the hen's eggshell. Using two-dimensional polyacrylamide gel electrophoresis (2D-PAGE) and sodium dodecyl sulfate PAGE (SDS-PAGE), Samata and Kubota (1997) elucidated the structure of the organic matrix in chicken eggshell, and also suggested that the eggshell formation may be controlled by several components contained in the organic matrix. Tsai et al. (2006) characterized eggshell and eggshell membrane particles prepared from the hen eggshell waste, and found typical type II pore structures for both biomaterials, characteristic of nonporous materials or materials with macropores or open voids. Fourier transform infrared (FTIR) spectra indicated that the chemical composition of the eggshell particles was strongly associated with the presence of carbonate minerals, in contrast to the eggshell membrane particles in which the presence of functional groups of amines and amides were observed, related to their chemical composition of fibrous proteins.

Many treatments have been reported to induce changes in eggshell quality and properties, as reviewed by Burley and Vadehra (1989). More recent studies include observations following heat exposure that resulted in a decrease in zootechnical performance and eggshell thickness, increase in egg breakage, and unchanged egg shape index (Lin et al. 2004). Vitamin E supplements in the feed led to significant differences in eggshell elasticity and shell thickness, although there was no obvious correlation to the amounts of vitamin E administered (Engelmann et al. 2001). Dietary supplementation with Zn, Cu, and Mn improved breaking strength and fracture toughness (resistance to fracture) in a study by Mabe et al. (2003), while Lien et al. (2004) reported that egg production, egg weight, eggshell strength, and eggshell thickness were not influenced by Cu or Cr supplementation. More recently Roberts and Choct (2006) studied effects on egg and eggshell quality in laying hens of commercial enzymes such as amylase, xylanase, cellulose, pectinase, and β-glucanase, added to standard commercial layer diets based on barley, wheat, or triticale. The positive effects of enzyme addition included improvements in eggshell breaking strength, shell weight, percentage shell and shell thickness for barley-based diet, and eggshell breaking strength for wheat-based diets. Negative effects of enzyme addition were slightly lighter-colored eggshells and reduced albumen quality.

1.2.1.1. Cuticle

The water-insoluble cuticle is the most external layer of eggs; it is about 10–30 μm thick and covers the pore canals. It helps protect the egg from moisture and microbial invasion. The cuticle is composed of an inner mineralized layer and an outer layer consisting of only organic matrix (Dennis et al. 1996). The layer adjacent to the shell has a foamy appearance, whereas the outer layer is more compact. The inner cuticular layer, referred to as the *vesicular cuticle*

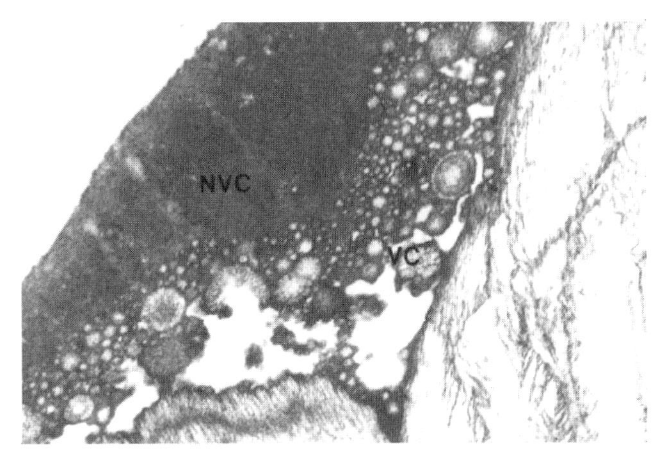

Figure 1.4. Transmission electron micrograph of vesicular cuticle (VC) and nonvesicular cuticle (NVC) layers. The inner VC shows large vesicles containing granular material which are absent in the outer NVC (×40,000). [*Source*: Fraser et al. (1999). © Taylor & Francis Ltd. `http://www.tandf.co.uk/journals.`]

(VC), is composed of a matrix-like material containing vesicles of various sizes that contain distinctly granular material with an electron-lucent core (C) and an electron-dense mantle (M). The outer layer of cuticle is much more compact and homogeneous, does not appear to contain any matrix vesicles, and is referred to as the *nonvesicular cuticle* (NVC). In all the eggshells examined by TEM, two layers were evident, but both cuticular layers were of variable thickness (Fraser et al. 1999) (Fig. 1.4).

Organic acids (e.g., 3% acetic or lactic acid) cause detachment of the surface cuticle of the eggshell (Favier et al. 2000). The effects of egg-washing chemicals, such as solutions of cetylpyridinium chloride or trisodium phosphate, on microstructural changes of the eggshell for fresh-laid eggs were examined by using scanning electron microscopy (Kim and Slavik 1996). Depending on the types of chemicals used in the washwater, different microstructural changes occurred in eggshell surfaces, and the more damaged eggshell surfaces allowed greater bacterial penetration (Kim and Slavik 1996).

Wong et al. (1996) evaluated the interior quality and shell properties of uncoated eggs and eggs coated with mineral oil or with solutions of egg albumen, soy protein isolate, wheat gluten, or corn zein. Eggshells coated with corn zein solution exhibited the least moisture loss and the strongest shell strength. Light microscopy depicted a more compact structure of eggshells coated with corn zein solution. Shell eggs coated with corn zein exhibited the lowest moisture loss and maintained a higher Haugh unit value (representing interior egg quality, based on the relationship of egg weight and thick albumen height) than did eggs coated with other treatments. Results also indicated that the protein-based coatings, corn zein and wheat gluten, added strength to the shell, which, in turn, served as a protective barrier.

More recently, egg weight, shell thickness, number of pores, cuticle deposition, and ability of *Salmonella enterica*, serovar *Enteritidis* (SE) to penetrate the shell were determined for eggs from one layer flock through the entire production period. No correlations were observed between any of the shell characteristics studied and the ability of SE to penetrate the shell (Messens et al. 2005).

1.2.1.2. Shell Matrix (Calcium Carbonate Layer)

The eggshell matrix consists of the vertical crystal layer, palisade layer (sponge layer), and mammillary knob layer. Calcium carbonate is the major component of inorganic substances. This calcite (the most stable form of calcium carbonate) forms elongated structures termed *columns*, *palisades*, or *crystallites*. Eggshell calcification is one of the most rapid biomineralization processes known, occurring in the oviduct of the hen, over a predetermined period 5–22 h after ovulation. Size, shape, and orientation of the calcite crystal in hen eggshell are affected by the interaction of calcium carbonate minerals with the organic matrix molecules (Arias et al. 1993; Nys et al. 1999).

Dennis et al. (1996) reported that the mammillary layer consists of the calcium reserve assembly (CRA) and crown region, each with a unique substructure. TEM images show that the matrix of the CRA consists of a dense, flocculent material partially embedded within the outer shell membrane, a mostly noncalcified region of the shell. The mantle of the collagen fibers of the shell membranes is rich in polyanions (cuprolinic blue–positive), as is the CRA matrix. The CRA is capped by a centrally located calcium reserve body sac (CRB sac) that contains numerous 300–400 nm, electron-dense, spherical vesicles. Directly above the CRB sac is a zone of matrix consisting of stacks of interconnected vesicles (similar in morphology to CRA vesicles) that are interspersed with a granular material.

The palisade region, the largest of the mineralized zones, contains hollow vesicles about 450 nm in diameter, with a crescent-shaped, electron-dense fringe. An interconnecting matrix material is also found between the vesicles in the palisades region. The bulk of the mineral within the eggshell is calcite, with small amounts of needle-like hydroxyapatite in the inner cuticle and occasionally, vaterite microcrystals found at the base of the palisade (cone) region. The well-crystallized calcite crystals within the palisade are columnar, typically about 20 μm wide by 100–200 μm long; aside from numerous entrapped vesicles and occasional dislocations, they are relatively defect-free. The bulk of the matrix found in the palisade and crown regions are thought to be residual components of the rapid mineralization process. The unique matrix structure within the CRB corresponds to the region of preferentially solubilized calcite used by the developing embryo and the hydroxyapatite found in the inner cuticle may play a role in the cessation of mineral growth.

The vertical crystal layer consists of short, thin crystals within the calcified spongy matrix (palisade layer), with their long axes oriented toward the shell surface. The palisade layer is very dense and hard. Its crystalline structure is formed by calcification of calcium carbonate containing a small amount of magnesium. The mammillary knob layer has a core and is interconnected to protein fibers of the outer shell membrane. Fraser et al. (1999) hypothesized that the vertical orientation of calcite crystals in the vertical crystal layer is closely linked to this vertical matrix.

Changes in eggshell crystal size could be due to changes in organic matrix composition (Ahmed et al. 2005). Dominguez-Vera et al. (2000) reported that the interaction of the uterine fluid with calcite contributed to eggshell structure. In young hens, the breaking strength of the eggshell is inversely related to the degree of calcite orientation. Conversely, reduced strength in eggshell from aged hens coincides with a high variability in crystallographic texture (Nys et al. 2001). Kim (2002) reported that specific gravity and breaking strength of eggshell are significantly increased by adding ascidian tunic shell to the diet.

This calcitic biomaterial forms in a uterine fluid where the protein composition varies during the initial, calcification, and terminal phases of eggshell deposition. Proteins are thought to influence shell formation and calcification and, thus, modify the resulting properties of the shell. Panheleux et al. (2000) investigated the potential of some of these proteins as biomarkers of eggshell quality by developing a competitive indirect enzyme-linked immunosorbent assay (ELISA) and identified ovotransferrin, ovalbumin, and ovocleidin-17 in eggshell extract.

Ruiz and Lunam (2000) used scanning election microscopy (SEM) to assess the relationship between the layers of the eggshell and egg viability. The relative thickness (absolute and percent) of the mammillary, palisade, vertical crystal, and cuticle layers relative to the total eggshell were measured over a 30-week laying period, and the percent contribution of each calcified layer was found to significantly differ over the egg production period. The cuticle was significantly thinner at the beginning and at the end of the laying period compared to 38 weeks of age. No significant correlation was observed between egg viability and the relative thicknesses of the mamillary, palisade, or cuticle layers, but there was a positive relationship between egg viability and thickness of the vertical crystal layer. The percent contribution of the mammillary and palisade layers was similar in the thinnest and thickest eggshells, suggesting conservation of the proportions of these layers independent of eggshell thickness.

Ahmed et al. (2005) investigated the effect of molt on eggshell mechanical properties, on composition and concentrations of organic matrix components, and on eggshell microstructure. After molt, bands associated with main proteins specific to eggshell formation (OC116 and OC17) showed higher staining intensity, while the intensity of the egg white proteins (ovotransferrin, ovalbumin, and lysozyme) decreased. Changes in eggshell crystal size could be due to changes in organic matrix composition.

1.2.1.3. Shell Membrane

The eggshell membrane structure is composed of inner and outer membranes that reside between the albumen and the inner surface of the shell. The inner membrane, also referred to as the *egg membrane*, has three layers of fibers that are parallel to the shell and at right angles to each other. On the other hand, the outer membrane, also referred to as the *shell membrane*, has six layers of fiber oriented alternately in different directions.

Using confocal scanning laser microscopy, Liong et al. (1997) reported that outer eggshell membrane fibers 1–7 μm in thickness could be seen emerging from the calcified layers of the eggshell. Inner membrane fibers 0.1–3 μm in thickness were interlaced with the outer membrane. The limiting membrane, when stained with fluorescein isothiocyanate (FITC), appeared as particles that filled spaces between the inner membrane fibers. The outer membrane layer (~50–70 μm thick) and the inner membrane (~15–26 μm thick) consisted of several discontinuous layers that were discernible as shifts in fiber position or orientation and changes in fiber size.

Eggshell membranes contribute to shell strength, probably by serving as a reinforcement of the crystalline part of the shell. Eggshell formation takes place on the eggshell membrane in an acellular medium, the uterine fluid that contains the inorganic minerals and precursors of the organic matrix. It is considered that the formation of β-aminoproprionitrile-sensitive crosslinks among the type X collagen molecules of the shell membranes play an essential role in normal eggshell formation (Arias et al. 1997). Type X collagen is highly crosslinked and insoluble. It was reported that the insoluble protein of hen eggshell membrane was stable and capable of binding heavy-metal ions from aqueous solution (Ishikawa et al. 1999).

1.2.2. Albumen

Albumen or egg white is composed of four distinct layers: an outer thin white next to the shell membrane, a viscous or outer thick white layer, an inner thin white, and a chalaziferous or inner thick layer. The contents of each layer are about 23.3%, 57.3%, 16.8%, and 2.7%, respectively (Burley and Vadehra 1989). The proportions may vary, however, depending on breed of the hen, environmental conditions, size of the egg, and rate of production (Li-Chan et al. 1995).

1.2.2.1. Thick and Thin Albumen

In fresh eggs, thick albumen covers the inner thin albumen and the chalaziferous layer, keeping the egg yolk in the center of the egg.

The viscosity of thick albumen is much higher than that of thin albumen because of its high content of ovomucin. The total amount of ovomucin isolated from the thick albumen of the eggs with high Haugh units (HU) was much higher than the amount isolated from the low-HU thick albumen, whereas the yield of ovomucin from thin albumen did not differ (Toussant and Latshaw 1999).

Thinning of egg white that occurs naturally during storage is a well-known phenomenon. A loss of thick albumen and a corresponding increase in free sulfhydryl groups in the albumen have also been observed after electron beam irradiation, resulting in a loss of foam volume and gel hardness (Wong and Kitts 2003).

1.2.2.2. Chalaziferous Layers and Chalazae Cord

The chalaziferous layer is a gelatinous layer that directly covers the entire egg yolk. In the long axis of the egg, the chalaziferous layer is twisted at both sides of the yolk membrane, forming a thick rope-like structure termed the *chalazae cord*. This cord is twisted clockwise at the sharp end of the egg and counter-clockwise at the opposite end. The chalazae cord stretches into the thick albumen layer to both sides; thus the egg yolk is suspended in the center of egg (Okubo et al. 1997). The chalazae are slightly elastic and permit limited rotation of the yolk (Romanoff and Romanoff 1949).

1.2.3. Egg Yolk

Yolk is a complex system containing a variety of particles suspended in a protein solution and encircled by a vitelline membrane. The particles include yolk spheres, free-floating granules, low-density lipoprotein globules, and myelin figures. Most of the yolk consists of yellow yolk, which is composed of layers of alternate light and deep yellow yolks. Less than 2% of the total egg yolk is white yolk, which originates from the white follicle maturing in the ovary.

Mineki and Kobayashi (1997) elucidated the fine structure of fresh yolk by using SEM in conjunction with a freeze–cutting fixation method with liquid nitrogen. Yolk spheres show polyhedral structures, but with varying shape and size in each layer and latebra part (Fig. 1.5). The spheres contain protein granules with high electron density, which also vary in distribution and shape between the outer and inner layers.

The microstructure of egg yolk granules has been reported to be affected by the presence of mineral cations (Causeret et al. 1992). Dialysis causes partial release of Ca^{2+} and breaks the granular structure. Na^{+} present in the form of NaCl can replace Ca^{2+} bound to the proteins and also causes destabilization of the granule structure. In contrast, Fe^{3+} binding is not modified by the presence of Na^{+}. The chelating agent EDTA binds all of the cations, breaking the phosphocalcic bridges and destabilizing the structure. On the other hand, the addition of bivalent or trivalent cations at low concentration strengthens the structure, enabling it to resist changes in pH or ionic strength.

1.2.3.1. Vitelline Membrane

Mineki and Kobayashi (1997) reported that in fresh yolk, the yolk vitelline membrane has a thickness of about 10 μm and is composed of three layers (Fig. 1.6a) The outer and inner layers are each about 5 μm thick, but the former

Figure 1.5. Shape of the yolk spheres in fresh native yolk by light microscopy (LM) or scanning electron microscopy (SEM): (a) yolk sphere in outer layer (LM); (b) yolk sphere in outside yolk (SEM); (c) yolk sphere in inside yolk (SEM); (d) yolk sphere in latebra (LM). [*Source*: Mineki and Kobayashi (1997).]

has a coarse structure with fiber layers, while the latter has a closer structure with dense net-like fibers. Diameters of the fibers are 200–600 and 15 nm in the outer and inner layers, respectively. Mineki and Kobayashi (1997) suggested that the slender grains observed in the inner layer are glycoproteins because of their low staining characteristics. Closely adhering to the inner layer is a thin film 0.1–0.2 μm in thickness, referred to as the *continuous layer*, which consists of a piled, sheet-like structure of granules with an estimated diameter of 7 nm.

The permeability of the vitelline membrane of eggs during cold storage may be changed by conjugated linoleic acid (CLA) supplement in the diet (Aydin 2006). Dietary CLA caused higher levels of C16:0 and C18:0 and lower levels of C16:1(*n*-7) and C18:1(*n*-9) fatty acids compared to the control group. The results indicated that dietary CLA influenced fatty acid composition of eggs and had negative effects on the quality of eggs stored at 4°C or 15°C, but not at room temperature (21–24°C).

Eggs from young hens were found to have yolks with a vitelline membrane breaking strength significantly higher than that noted from older birds (Ngoka et al. 1983). Kirunda et al. (2001) investigated supplementation of hen diets with vitamin E as a means to alleviate egg quality deterioration associated with high-temperature exposure (HTE). Supplementation of HTE hen diets with 60 international units (IU) vitamin E/kg feed improved egg pro-

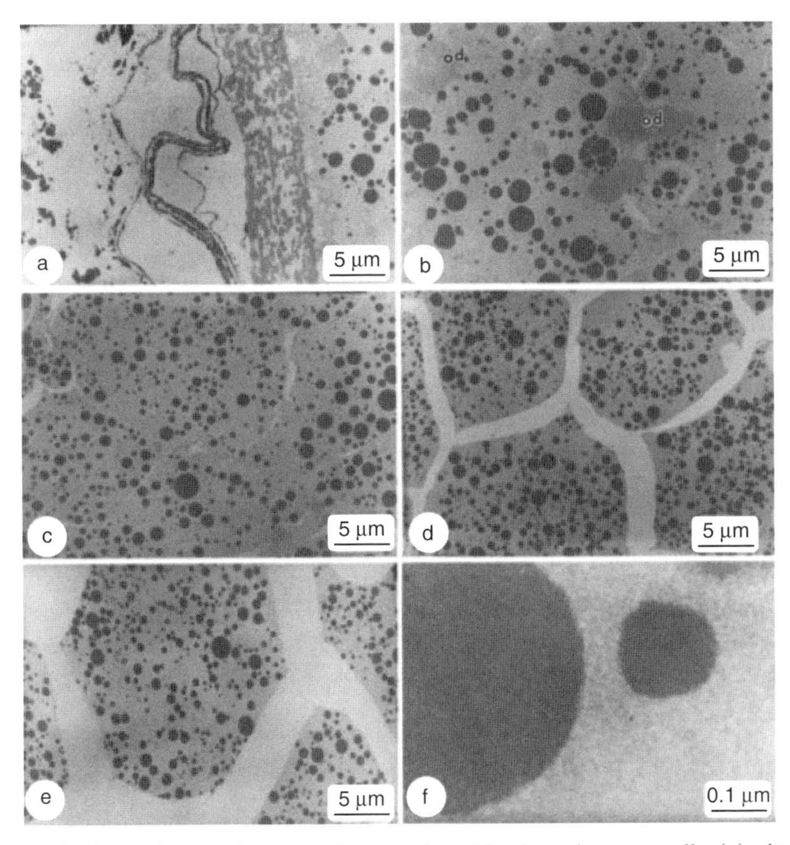

Figure 1.6. Transmission electron micrographs of fresh native egg yolk: (a) vitelline membrane; (b) cortical layer of yolk (od = oil droplet); (c) yolk sphere bordering cortical layer; (d) yolk sphere in outer layer; (e) yolk sphere in inner layer; (f) protein granules in yolk sphere. [*Source*: Mineki and Kobayashi (1997).]

duction, vitelline membrane strength, and yolk and white solids. Novak and Scheideler (2001) reported that Ca supplementation significantly ($p < .03$) increased yolk solids.

1.2.3.2. Yellow Yolk

Yellow yolk consists of two types of lipoprotein emulsion: the deep yellow yolk and the light yellow yolk. The deep yellow yolk is formed in the daytime, and the light yellow yolk is formed at night, leading to the alternate and circular appearance of these yolk layers (Romanoff and Romanoff 1949).

Mineki and Kobayashi (1997) examined the structure of fresh egg yolk using a combination of light, scanning electron and transmission electron microscopy (LM, SEM, and TEM). These studies revealed that the cortical layer of the yolk immediately inside the vitelline membrane (Figs 1.6b and 1.6c) has a structure different from the other parts, with undeveloped yolk spheres, a shapeless membrane structure, small granules regarded as protein, and larger granules regarded as oil spheres. The yolk spheres in the outer layer (Fig. 1.6d) are rounded and smaller than the polyhedral spheres observed in the inner layer (Fig. 1.6e). In the yolk spheres, protein granules with high electron density are dispersed (Fig. 1.6f). The yolk spheres in the middle layer are slender polyhedrons with a relatively uniform size (Fig. 1.5c). The yolk spheres in the latebra have an undefined boundary, and many of them are rounded in shape (Fig. 1.5d). The ground substrate of the yolk shows a homogenous structure consisting of fibrous materials under high magnification. No difference in properties of the ground substrate was observed among the yolk parts.

Microstructural changes were observed in hen egg yolk stored at room temperature (20 ± 1°C) for 0, 5, 10, 20, and 30 days (Mineki and Kobayashi 1998). Yolks at day 0 had polyhedral spheres that were closely packed at the vitelline membrane, uniform yolk spheres, and evenly distributed protein granules. Vesicles appeared in the ground substrate of eggs stored for 5 days. After storage, interstitial spaces between yolk spheres in outer layers were broadened and had oval-shaped particles released from the yolk spheres. The protein granules within these spheres were unevenly distributed, and many of them were fused. Fine spots appeared in the granules. These changes were observed throughout every layer from the eggs stored for 20 days. These results clearly show that the microstructure in the yolk spheres changed with storage.

1.3. CHEMICAL COMPOSITION OF EGGS

Table 1.1 shows the approximate composition of the edible portion of whole egg, and the contents of the main constituents in terms of minerals, vitamins, fatty acids, amino acid, and pigments (USDA 2006). Water constitutes approximately 75% of the total weight of the edible portion of eggs, while lipids and proteins are the major contributors to the nutrient value. Minor amounts of carbohydrate (present in the form of simple sugars, including glucose, sucrose, fructose, lactose, maltose, and galactose) and minerals or "ash" are also present.

Whole egg consists of 9–11% eggshell, 60–63% albumen, and 28–29% yolk. Table 1.2 shows the typical range of approximate composition for whole egg, compared to the proximate composition in the eggshell, albumen, and yolk. Proteins are present primarily in egg yolk and egg white, while lipids are almost exclusively in the yolk. Minerals are the primary component of the eggshell.

In addition to the nutrients listed in Table 1.1, eggs contain many other components. While these may be present in minor quantities, they could be significant when considering bioscience and biotechnological applications. For example, sialic acid is an important component in eggs, which is found at higher concentrations in yolk membrane (13.7 ± 0.5 μg/mg), chalaza (5.1 ± 0.2 μg/mg), and yolk (5.2 ± 0.4 μg/mg) than in egg white (2.2 ± 0.3 μg/mg), and lowest in

TABLE 1.1. Chemical Composition for Edible Portion of Whole Egg[a]

Nutrient	Value per 100 g
(a) Approximate Composition	
Water	75.84 g
Protein	12.58 g
Total lipid (fat)	9.94 g
Ash	0.86 g
Carbohydrate, by difference	0.77 g
(b) Minerals	
Calcium, Ca	53 mg
Iron, Fe	1.83 mg
Magnesium, Mg	12 mg
Phosphorus, P	191 mg
Potassium, K	134 mg
Sodium, Na	140 mg
Zinc, Zn	1.11 mg
Copper, Cu	0.102 mg
Manganese, Mn	0.038 mg
Fluoride, F	1.1 μg
Selenium, Se	31.7 μg
(c) Vitamins	
Vitamin C (total ascorbic acid)	0.0 mg
Thiamin	0.069 mg
Riboflavin	0.478 mg
Niacin	0.070 mg
Pantothenic acid	1.438 mg
Vitamin B_6	0.143 mg
Folate, total	47 μg
Choline, total	251.1 mg
Betaine	0.6 mg
Vitamin B_{12}	1.29 μg
Vitamin A, IU	487 IU
Retinol	139 μg
Vitamin E (α-tocopherol)	0.97 mg
β-Tocopherol	0.02 mg
γ-Tocopherol	0.50 mg
δ-Tocopherol	0.02 mg
Vitamin D	35 IU
Vitamin K (phylloquinone)	0.3 μg
(d) Lipids	
Fatty acids, total saturated	3.099 g
8:0	0.003 g
10:0	0.003 g
12:0	0.003 g

TABLE 1.1. (*Continued*)

Nutrient	Value per 100 g
14:0	0.034 g
15:0	0.004 g
16:0	2.226 g
17:0	0.017 g
18:0	0.784 g
20:0	0.010 g
22:0	0.012 g
24:0	0.003 g
Fatty acids, total monounsaturated	3.810 g
14:1	0.008 g
16:1 undifferentiated	0.298 g
18:1 undifferentiated	3.473 g
20:1	0.028 g
22:1 undifferentiated	0.003 g
Fatty acids, total polyunsaturated	1.364 g
18:2 undifferentiated	1.148 g
18:3 undifferentiated	0.033 g
20:4 undifferentiated	0.142 g
20:5 *n*-3	0.004 g
22:6 *n*-3	0.037 g
Cholesterol	423 mg
(e) Amino Acids	
Tryptophan	0.167 g
Threonine	0.556 g
Isoleucine	0.672 g
Leucine	1.088 g
Lysine	0.914 g
Methionine	0.380 g
Cystine	0.272 g
Phenylalanine	0.681 g
Tyrosine	0.500 g
Valine	0.859 g
Arginine	0.821 g
Histidine	0.309 g
Alanine	0.736 g
Asparticacid	1.330 g
Glutamicacid	1.676 g
Glycine	0.432 g
Proline	0.513 g
Serine	0.973 g
(f) Other Components	
β-Carotene	10 μg
β-Cryptoxanthin	9 μg
Lutein + zeaxanthin	331 μg

[a]Categories (b)–(f) listed above are important constituents.

Source: Adapted from the USDA (2006) nutrient database for edible portions of whole egg (raw, fresh, excluding eggshell).

TABLE 1.2. Approximate Composition of Whole Egg, Eggshell, Albumen, and Yolk

Egg Component (% of total)	Approximate Composition, % (w/w)				
	Moisture	Protein	Lipid	Carbohydrate	Ash (Minerals)
Whole egg (100%)	66.1	12.8–13.4	10.5–11.8	0.3–1.0	0.8–1.0
Eggshell (9–11%)	1.6	6.2–6.4	0.03	Trace	91–92
Albumen (60–63%)	87.6	9.7–10.6	0.03	0.4–0.9	0.5–0.6
Yolk (28–29%)	48.7	15.7–16.6	31.8–35.5	0.2–1.0	1.1

Sources: Adapted from Burley and Vadehra (1989), Li-Chan et al. (1995), and Sugino et al. (1997b).

shell membrane (0.6 ± 0.1 μg/mg) (Nakano et al. 1994). Hen egg yolk and white contain high amounts of lysophosphatidic acid (acyl LPA) in addition to small amounts of lysoplasmanic acid (alkyl LPA) (Nakane et al. 2001). Egg yolk acyl LPA contains predominantly saturated fatty acids as the acyl moiety, whereas egg white acyl LPA contains primarily polyunsaturated fatty acids, and the level of acyl LPA, especially polyunsaturated fatty acid–containing acyl LPA, is augmented markedly in egg white during incubation at 37°C, while there is little change in egg yolk. Nakane et al. (2001) suggested that egg yolk LPA and egg white LPA may play separate physiological roles in the development, differentiation, and growth of embryos.

The weight and composition of each structural part of hen eggs may vary depending on factors such as species, feeding, and age of the hen (Sugino et al. 1997b). Shafey (1996) reported that older hens lay larger eggs, with a greater proportion of albumen, a higher concentration of the polyunsaturated fatty acid linoleic acid, but less saturated fatty acids and total lipid than do younger hens. However, Souza et al. (1997) reported that composition of the eggs is unaffected by age of the hens, and Poser et al. (2003) observed little effect of hen age or storage on the lysozyme or conalbumin content of the egg white or protein composition of vitelline membranes.

Changes in the physical and chemical properties of eggs may be induced by altering the feed. The inclusion of organic selenium in the diet of laying hens could be used to maintain egg quality during storage (Surai 2002). Yalcin et al. (2006) reported that supplementing feeds with garlic powder to 10 g/kg significantly increased egg weight and significantly decreased concentration of egg yolk cholesterol compared with the control. Naulia and Singh (1998) found significant differences in the thickness of eggshell, albumen index, and phosphorus retention between eggs from hens on various diets containing fishmeal and phosphorus. Novak and Scheideler (2001) reported that flaxseed supplementation significantly increased albumen percentage and decreased yolk percent when compared to controls. Japanese green tea powder (GTP) supplementation to commercial diet of laying hens modified components in the edible part of the egg, leading to the characteristics favorable to consumers such as high durability of thick albumen and less cholesterol in yolk, without altering general performance of the layers (Biswas et al. 2000). An increase in egg weight associated with soy oil supplementation was accompanied by a simultaneous increase in the proportion of albumen and a decrease in yolk and shell percentages (Daenicke et al. 2000). Xiaoting et al. (1998) reported that addition of diludine (used to control *Eimeria coccidia*) increased egg weight, egg yolk weight, albumen height, HU value, shell thickness, and shell weight, and reduced egg breakage, abnormality, and total cholesterol.

Aydin et al. (2001) reported that dietary conjugated linoleic acids (CLA) caused abnormal pH changes of albumen and yolk when eggs were stored at 4°C. The pH of yolk and albumen from corn oil–fed hens after 10 weeks of storage was 6.12 ± 0.12 and 9.06 ± 0.03, respectively, versus 7.89 ± 0.25 and 8.32 ± 0.16, respectively, in eggs from CLA-fed hens. Olive oil prevented CLA-induced abnormal changes in the pH of albumen and yolks. The addition of activated charcoal with wood vinegar in layer diet resulted in improving egg quality by enhancing eggshell breaking strength, HU value, yolk index, and yolk color (Kim et al. 2006). Schaefer et al. (2001) reported that inclusion of CLA in layer diets altered the shape of the yolk and various egg parameters (albumen height, foam index, and yolk index).

1.3.1. Eggshell and Egg Membrane

Avian eggshell is sometimes referred to as a "natural composite bioceramic" containing organic (3.5%) and inorganic (95%) phases (Arias et al. 1993). It is composed of 1.0% (w/w) matrix proteins in addition to calcium carbonate (95%, w/w) (Daengprok et al. 2003). Fatty acid content in the insoluble eggshell layers (after decalcification) is in the range of 2–4%, which reflects both lipid and lipoprotein bound fatty acids (Miksik et al. 2003). Generally, there is little difference in fatty acid composition between the egghell layers, with the exception of behenic acid (a 22-carbon saturated fatty acid), which is found at much higher concentration in the cuticle layer (Table 1.3). Porphyrin pigments are also dominant in the cuticle layer (Miksik et al. 2003). Biliverdin is an important pigment in the eggshell of chickens and other avian species (Zhao et al. 2006).

The Ca, Mg, and P contents of eggshell are 2.21, 0.02 and 0.02 g/egg, respectively (Sugino et al. 1997b). Calcite crystals are composed of calcium carbonate, while dolomite crystals, which are mechanically harder than the calcite crystals, are composed of calcium and magnesium carbonate (Sugino et al. 1997b). The Mg/Ca ratio has been used as a paleothermometer in a range of calcite biominerals, including eggshells (Cusack et al. 2003). Magnesium distribution is not constant throughout the shell, decreasing from nucleation until after fusion of the mammillary caps and then increasing to termination. There

TABLE 1.3. Fatty Acid Composition of Eggshell Layers (ng/μg)

Fatty Acid	Cuticle Layer	Palisade Layer I	Palisade Layer II	Mammillary Layer
Decanoic acid (10:0)	0.5	0.8	0.3	0.4
Lauric acid (12:0)	0.5	0.4	0.5	0.3
Eicosapentaenoic acid (20:5)	0.1	0.1	0.1	0.2
Linolenic acid (18:3)	1.1	0.8	0.9	1.1
Myristic acid (14:0)	1.0	0.9	0.9	1.3
Docosahexaenoic acid (24:6)	1.2	1.2	1.2	1.0
Palmitoleic acid (16:1)	0.5	0.8	0.9	1.5
Arachidonic acid (20:4)	1.0	0.9	0.8	0.9
Linoleic acid (18:2)	1.2	1.1	1.2	1.7
Palmitic acid (16:0)	6.8	7.0	5.6	10.4
Oleic acid (18:1, c9)	1.9	1.8	1.5	2.7
Petroselinic acid (18:1, c6)	0.7	0.4	0.3	0.4
Elaidic acid (18:1, c9)	0.1	0.1	0.1	0.1
Stearic acid (18:0)	3.4	3.4	2.3	4.5
Arachidic acid (20:0)	4.9	1.4	1.9	1.3
Behenic acid (22:0)	14.8	4.3	2.5	3.1
Total saturated fatty acids	31.9	18.2	14.0	21.3
Total unsaturated fatty acids	7.8	7.2	7.0	9.6
Total fatty acids	39.7	25.4	21.0	30.9
Ratio saturated/unsaturated	4.09	2.53	2.00	2.22

Source: Adapted from Miksik et al. (2003).

is a greater increase in Mg in the outer region of eggshells from older birds, and the variation in Mg concentration does not appear to correlate with organic content. Phosphorus occurs in the outer quarter of the eggshell and rises to termination (Cusack et al. 2003).

The eggshell contains an average of 0.024% of its dry weight as uronic acid, a carbohydrate moiety of glycosaminoglycans. The eggshell glycosaminoglycans consist of approximately 48% hyaluronic acid and 52% galactosaminoglycan. In the latter, chondroitin sulfate–dermatan sulfate copolymers are the major galactosaminoglycans with dermatan sulphate disaccharide as a relatively minor component. The inorganic material recovered after decalcification accounts for ~14% of the dry weight of the eggshell, and contains 24.11% Ca, 0.04% P, and 0.23% Mg, with an undetectable amount of N (Nakano et al. 2001).

The uronic acid, sialic acid, and nitrogen contents of chicken eggshell and shell membranes are shown in Table 1.4 (Nakano et al. 2003). Uronic acid concentrations are similar ($p > .05$) between the inner shell membrane and outer shell membrane but approximately five-fold higher ($p < .05$) in the organic matter of eggshell. Sialic acid concentrations are highest ($p < .05$) in the organic matter of eggshell and higher ($p < .05$) in the inner than in the

TABLE 1.4. Analysis of Decalcified Eggshell and Eggshell Membranes[a]

Constituent	Decalcified Eggshell	Inner Shell Membrane	Outer Shell Membrane
Uronic acid	6.34 ± 0.20[b]	1.30 ± 0.10[c]	1.15 ± 0.18[c]
Sialic acid	4.83 ± 0.56[b]	1.70 ± 0.07[c]	0.48 ± 0.04[d]
Nitrogen	127.1 ± 6.9[b]	150.4 ± 2.4[c]	149.8 ± 1.4[c]

[a]Values are expressed as μg per mg of organic matter.
[b–d]Means in the same row with different superscripts are significantly ($p < .05$) different.
Source: Adapted from Nakano et al. (2003).

TABLE 1.5. Amino Acid Composition of Decalcified Eggshell and Eggshell Membranes[a]

Amino Acid	Decalcified Eggshell	Inner Shell Membrane	Outer Shell Membrane
Asx	8.1 ± 0.1[b]	8.4 ± 0.4[b]	8.8 ± 0.1[b]
Thr	6.2 ± 0.1[b]	6.9 ± 0.0[b]	6.9 ± 0.2[b]
Ser	9.7 ± 0.1[b]	9.2 ± 0.2[b]	9.2 ± 0.0[b]
Glx	11.8 ± 0.2[b]	11.1 ± 0.4[b]	11.9 ± 0.3[b]
Gly	13.0 ± 0.3[b]	11.1 ± 0.2[c]	10.6 ± 0.2[c]
Ala	6.9 ± 0.4[b]	4.6 ± 0.2[c]	4.1 ± 0.2[c]
Val	7.3 ± 0.1[b]	7.2 ± 0.2[b]	7.9 ± 0.1[b]
Met	2.0 ± 0.2[b]	2.3 ± 1.0[b]	2.3 ± 1.0[b]
Ile	2.6 ± 0.1[b]	3.3 ± 0.4[b]	3.4 ± 0.1[b]
Leu	6.1 ± 0.2[b]	5.6 ± 0.5[b]	4.8 ± 0.2[c]
Tyr	1.8 ± 0.1[b]	2.2 ± 0.1[b]	1.7 ± 0.3[b]
Phe	2.1 ± 0.1[b]	1.6 ± 0.1[b]	1.5 ± 0.1[b]
His	4.2 ± 0.1[b]	4.1 ± 0.4[b]	4.3 ± 0.4[b]
Lys	3.6 ± 0.1[b]	3.6 ± 0.2[b]	3.4 ± 0.2[b]
Arg	5.9 ± 0.1[b]	5.7 ± 0.3[b]	5.8 ± 0.2[b]
Pro	8.3 ± 0.5[c]	11.6 ± 0.7[b]	12.0 ± 0.9[b]
Hyp[d]	0.3 ± 0.1[c]	1.5 ± 0.3[b]	1.4 ± 0.4[b]

[a]Values are expressed as mol% in organic matter.
[b,c]Means in the same row with different superscripts are significantly ($p < .05$) different.
[d]Hydroxyproline.
Source: Adapted from Nakano et al. (2003).

outer shell membrane. Nitrogen concentrations are lowest ($p < .05$) in the organic matter of eggshell but relatively constant between the two shell membranes. Amino acid analysis showed that the contents of glycine and alanine are higher ($p < .05$) and those of proline and hydroxyproline are lower ($p < .05$) in the eggshell layers compared to shell membranes (Table 1.5) (Nakano et al. 2003). Differences in amino acid composition are also observed between the cuticle, palisade layers I and II, and mammillary layers (Miksik et al. 2003).

Rubilar et al. (2006) investigated lactic fermentation for the processing of whole eggshells (WES) and eggshell membranes (ESM). After treatment of WES, the liquid phase contained the highest amounts of protein (85.4% and 73.2% in the presence and absence of citric acid, respectively). After fermentation and dialysis of ESM, the high-solubility liquid phase contained 88.3% soluble protein and 4.3% exopolysaccharides, whereas the low-solubility phase contained 60.2% insoluble protein. Further analysis of the liquid phase of fermented WES revealed the presence of chicken heatshock protein, glucose-regulated protein precursor, calcyclin, and 60*s* acidic ribosomal protein p2. The foamy phase of fermented WES was found to contain a protein homologous with calmodulin.

1.3.1.1. Cuticle

The solid matter in cuticle is composed of about 3% ash, 5% carbohydrate, and nearly 90% protein (mostly in the form of insoluble proteins) and glycoproteins. The cuticle also contains a large part of the pigments (protoporphyrin, biliverdin IXa, and its zinc chelate) in colored eggs and hydroxyapatite crystals (Burley and Vadehra 1989; Dennis et al. 1996; Fraser et al. 1999).

1.3.1.2. Shell Matrix

Eggshell matrix consists primarily (~97%) of calcium carbonate in the form of calcite crystals, together with an organic matrix (~2% of the eggshell) of proteins and polysaccharides rich in sulfated molecules (Nys et al. 1999). Chicken eggshell powder has been proposed as an attractive source of Ca for human health to increase bone mineral density in an elderly population with osteoporosis.

More recent analysis of the soluble components of the eggshell after demineralization has shown the identification and localization of numerous proteins in the eggshell matrix. Mammillae, especially the 100–300 vesicles of the calcium reserve body (CRB), contain a calcium binding molecule known as *mammillan*, a keratan sulfate proteoglycan (Fernandez et al. 1997; Fernandez and Arias 1999), which is involved in nucleation of the first calcite crystals of the shell (Arias and Fernandez 2001). The palisade is the thickest layer of the eggshell and is composed of integrated inorganic and organic components (shell matrix). It contains hyaluronic acid (Nakano et al. 2001) and ovoglycan, a unique 200-kDa dermatan sulfate proteoglycan (Fernandez et al. 1997; Fernandez and Arias 1999) with a core protein, ovocleidin-116 (Nimtz et al. 2004) and glycosaminoglycan chains containing dermatan sulfate (Dennis et al. 2000). Ovoglycan was considered as a copolymeric proteoglycan containing nonsulfated chondroitin and dermatan sulfate glycosaminoglycans (Nakano et al. 2001). Ovoglycan was involved in growth of the crystalline palisade (Arias and Fernandez 2001).

The eggshell matrix proteins that may be involved in the regulation of calcite growth during eggshell calcification include lysozyme (Dominguez-Vera et al. 2000), ovalbumin (Panheleux et al. 2000; Dominguez-Vera et al.

2000), ovotransferrin (Panheleux et al. 2000; Dominguez-Vera et al. 2000), clusterin (Mann et al. 2003), osteopontin (Fernandez et al. 2003; Lavelin et al. 2000), ovocleidin-17 (Reyes-Grajeda et al. 2002, 2004; Mann and Siedler 1999; Panheleux et al. 2000; Dominguez-Vera et al. 2000; Lakshminarayanan et al. 2005), ovocleidin-116 (Nimtz et al. 2004; Hincke et al. 1999; Dominguez-Vera et al. 2000), ovocalyxin-32 (Gautron et al. 2001; Dominguez-Vera et al. 2000), and ovoclyxin-36 (Dominguez-Vera et al. 2000).

Dominguez-Vera et al. (2000) examined the interaction between calcium carbonate and organic matrix by measuring the *in vitro* effect of uterine fluid collected at various phases of shell formation on precipitation kinetics, size, and morphology of calcite crystals. The SDS-PAGE profiles of the organic constituents differed between the different phases of eggshell formation. The predominant constituents were ovalbumin and ovotransferrin at the initial phase and lysozyme, ovocleidin-17, ovocalyxin-32, 36- and 21-kDa bands, and ovocleidin-116 at the growth phase. These proteins were numerous in the terminal phase, which also showed an increased staining of the 32- and 66-kDa bands and appearance of very low-molecular-weight bands. Components from the initial phase induced the formation of twinned crystals and of rounded corners in the rhombohedric crystals. The presence of components from the growth and terminal phases strongly modified the morphology of the calcite crystals.

The extracellular clusterin originates in the uterine fluid, where it is a disulfide-bonded heterodimer derived from the precursor polypeptide by proteolytic cleavage at the same site as in mammals. In the decalcified eggshell, immunofluorescence and colloidal gold immunocytochemistry revealed that clusterin was localized predominantly in the palisade and mammillary layers and also in the mantle and core of the inner and outer shell membranes. It has been suggested that clusterin acts as an extracellular chaperone. Thus clusterin could function in the uterine fluid to prevent the premature aggregation and precipitation of eggshell matrix components before and during their assembly into the rigid protein scaffold necessary for ordered mineralization (Mann et al. 2003).

Ovocleidin, a major protein of the avian eggshell calcified layer, occurs in the eggshell soluble organic matrix in at least two forms. The major form is a phosphoprotein with two phosphorylated serines (OC17), which has now been sequenced. A minor form is a glycosylated protein with identical sequence and only one phosphorylated serine (OC23). The site of glycosylation is Asn59, the only asparagine in the amino acid sequence contained in the *N*-glycosylation site consensus sequence (NAS). Ser61, which is part of this site, is phosphorylated in OC17 but not in OC23, indicating that the two modifications are mutually exclusive (Mann 1999). This is the first example of alternative glycosylation/phosphorylation occurring at an *N*-glycosylation site. The amino acid sequence of ovocleidin-17 contains 142 amino acids, including two phosphorylated serines (Mann and Siedler 1999). Reyes-Grajeda et al. (2004) determined the crystal structure of monomeric OC17 at 1.5 Å resolution, and showed that it has a mixed α/β structure containing a single *C*-type lectin-like domain.

However, although OC17 shares the conserved scaffold of the *C*-type lectins, it does not bind carbohydrates.

Daengprok et al. (2003) found that the total Ca transport across Caco-2 monolayers grown on a permeable support was increased 64% in the presence of soluble eggshell matrix proteins. The active enhancer had a molecular mass of 21 kDa and *N*-terminal sequence of Met–Ala–Val–Pro–Gln–Thr–Met–Val–Gln; it did not correspond to any previously identified protein.

The predicted sequence of ovocleidin-116 contains two consensus *N*-glycosylation sites, only one of which (Asn62) was found to be fully modified (Mann et al. 2002; Nimtz et al. 2004). Nimtz et al. (2004) identified 17 different oligosaccharide structures in the eggshell protein. Four of them were of the high-mannose type, eight were hybrid, and five were complex-type structures. Both hybrid and complex-type glycans were found in core-fucosylated and peripherally fucosylated structures. Most of the antennas contained the relatively rare lacdiNAc (GalNAcβ1-4GlcNAc) motif, which was fucosylated in 9 out of 15 structures. The lacNAc (Galβ1-4GlcNAc) motif, which is the more frequent motif in mammals, occurred in only 3 of the 17 glycoforms.

Osteopontin is localized in the core of the nonmineralized shell membrane fibers, in the base of the mammillae, and in the outermost part of the palisade (Fernandez et al. 2003), while ovocalyxin-32 localizes to the outer palisade layer, the vertical crystal layer, and the cuticle of the eggshell, at high levels in the uterine fluid during the termination phase of eggshell formation (Gautron et al. 2001).

The soluble organic matrix of chicken eggshell was extracted after demineralization with acetic acid (Borelli et al. 2003). Partial biochemical characterization suggested that the anticalcifying factor was a polyanionic and water-soluble molecule, which could be proteoglycans. Borelli et al. (2003) suggested that both carbonic anhydrase and anticalcifying activities are widespread and play a significant role in the regulation of biomineral formation.

Miksik et al. (2003) studied avian eggshell matrix proteins by two analytical approaches. Peptide mapping was done by trypsin and pepsin followed by collagenase cleavage; analyses were carried out by capillary electrophoresis and reverse-phase high-performance liquid chromatography (HPLC). The evidence indicated that the eggshell insoluble proteins were susceptible to cleavage by collagenase; however, the sequences split were not those typical for the main triple-helix core of collagenous proteins. It was proposed that the action of collagenase on eggshell proteins is caused by the side effect of collagenase described previously with synthetic peptides. Some of the proteins present are probably glycosylated.

Hincke et al. (2000) reported that lysozyme is present in the shell membranes as well as in the matrix of the calcified shell. Calcite crystals grown in the presence of purified hen lysozyme exhibited altered crystal morphology. Therefore, in addition to its well-known antimicrobial properties that could add to the protective function of the eggshell during embryonic development, shell matrix lysozyme may also be a structural protein that in soluble form influences calcium carbonate deposition during calcification.

1.3.1.3. Shell Membranes

The inner and outer shell membranes are composed of approximately 2% ash, 2% glucose, and 90% protein (dry weight basis) (Sugino et al. 1997b). Although early studies suggested that eggshell membranes are composed of keratin, it has more recently been reported that type X collagen is the main constituent of the shell membrane fibers (Fernandez et al. 2001), and this functions to inhibit mineralization (Arias and Fernandez 2001). The core of the shell membrane fibers also contains keratin sulfate, a glycosaminoglycan polyanionic molecule that is thought to be part of the glycosylated group of type X collagen (Fernandez et al. 2001). Mammillae, especially the 100–300 vesicles of the calcium reserve body, contain a calcium binding molecule known as *mammillan*, which is keratin oversulfated proteoglycan (Fernandez et al. 1997; Fernandez and Arias 1999).

Eggshell membranes (ESM) have been shown to exhibit antibacterial activity. ESM enzyme and biological activities are relatively constant across layer breeds and over extended storage (Ahlborn and Sheldon 2005). Ahlborn and Sheldon (2006) reported that ovotransferrin, lysozyme, and P-NAGase are the primary components responsible for ESM antibacterial activity. The combination of these proteins and perhaps other ESM components may interfere with interactions between bacterial lipopolysaccharides, sensitizing the outer bacterial membrane to the lethal affects of heat, pressure, and osmotic stressors. Lysyl oxidase activity has also been reported in hen ESM, supporting previous reports indicating lysine-derived crosslinks in ESM and the necessity of lysyl oxidase located in the isthmus of the hen oviduct for the biosynthesis of ESM (Akagawa et al. 1999). The coupling enzyme system of lysyl oxidase with catalase was considered to be involved in the biosynthesis of ESM and to protect the embryo against H_2O_2 (Akagawa et al. 1999).

Soluble eggshell membrane protein (SEP) was prepared from eggshell membrane by combined treatment with performic acid oxidation and pepsin digestion. SEP had high contents of acidic amino acids (320–340 residues/1000 residues), including cysteic acid (101–108 residues) converted from cystine, a small amount of saccharides, a main molecular weight of 12,000–22,000, and pI 4.2–4.8 (Takahashi et al. 1996). Improved physical and biochemical features of a collagen membrane were obtained by conjugating with soluble eggshell membrane protein (Ino et al. 2006).

1.3.2. Egg Albumen

Water is the major constituent of egg albumen, ranging from 84% to 89% from the outermost to innermost layers. Proteins are the major component of albumen solids, accounting for about 10–11% of the albumen weight, while carbohydrates (primarily glucose), lipids, and minerals are minor components (Li-Chan and Nakai 1989). Tables 1.6–1.8 show the composition of amino acids, fatty acids, and minerals and vitamins, respectively, in egg albumen compared to egg yolk.

TABLE 1.6. Amino Acid Composition[a] of Egg Albumen and Yolk

Amino Acid	Content (g/100g Albumen)	Content (g/100g Yolk)
Tryptophan	0.125	0.177
Threonine	0.449	0.687
Isoleucine	0.661	0.866
Leucine	1.016	1.399
Lysine	0.806	1.217
Methionine	0.399	0.378
Cystine	0.287	0.264
Phenylalanine	0.686	0.681
Tyrosine	0.457	0.678
Valine	0.809	0.949
Arginine	0.648	1.099
Histidine	0.290	0.416
Alanine	0.704	0.836
Aspartic acid	1.220	1.550
Glutamic acid	1.550	1.970
Glycine	0.413	0.488
Proline	0.435	0.646
Serine	0.798	1.326

[a]Values are expressed as g/100g edible portion of albumen or yolk.

Source: Adapted from USDA (2006) nutrient database for raw, fresh egg yolk and egg white.

1.3.2.1. Proteins in Albumen

Specific protein components of the hen's egg are associated with diverse biological properties, which include antimicrobial activity; protease inhibitory action; immunomodulatory, anticancer, and antihypertensive activities; vitamin binding properties; and antigenic or immunologic characteristics (Li-Chan et al. 1995; Mine and Kovacs-Nolan, 2006). Knowledge of the chemical properties of the individual components can enhance understanding of their biological and functional properties and thereby facilitate potential applications in the food industry. However, despite many studies to separate and identify the proteins located in hen's egg, many proteins remain uncharacterized or even unknown (Raikos et al. 2006).

Table 1.9 lists the content, isoelectric points, and molecular weights reported for some proteins in egg albumen. Amino acid composition of some of these proteins has been compiled by Li-Chan and Nakai (1989), and more detailed or recent information can be accessed through databases such as the UniProt knowledgebase (Swiss-Prot and TrEMBL). Variations in experimentally measured and theoretical molecular weights and isoelectric points accompany the polymorphism observed for many of these proteins (Guérin-Dubiard et al. 2005, 2006). Furthermore, the glycosylated isoforms and aglycoproteins may differ significantly in molecular weight (Raikos et al. 2006).

Desert et al. (2001) analyzed the protein composition of hen egg white by SDS-PAGE, native-PAGE, isoelectric focusing (IEF), and two-dimensional

TABLE 1.7. Fatty Acid Composition (g/100 g Total Lipids) of Egg Albumen, Yolk Plasma, and Yolk Granules

Fatty Acid[a]	Egg Albumen	Yolk Plasma	Yolk Granules
16:0	18.30	22.49	24.24
16:1 *n*-7	0.81	1.32	1.18
17:0	0.27	0.22	0.16
18:0	7.76	11.22	12.40
18:1 *n*-9	16.66	31.28	28.48
18:2 *n*-6	20.92	27.60	26.81
18:3 *n*-6	0.23	0.18	ND[b]
18:3 *n*-3	0.15	0.43	0.26
20:1 *n*-9	0.06	0.18	ND
20:2 *n*-6	0.23	0.31	0.23
20:3 *n*-6	0.40	0.27	0.27
20:4 *n*-6	9.92	2.36	3.53
22:4 *n*-6	1.14	0.16	ND
22:5 *n*-3	0.37	ND	ND
22:5 *n*-6	0.90	0.55	0.83
22:6 *n*-3	2.07	0.60	0.93
SFA	26.34	33.92	36.80
MUFA	17.53	32.78	29.66
PUFA	36.33	32.48	32.86
n-6	33.74	31.44	31.67
n-3	2.59	1.03	1.19
n-6/*n*-3	13.84	30.49	26.67
SFA/PUFA	0.73	1.05	1.12

[a]*Key*: SFA—total saturated fatty acids; MUFA—total monounsaturated fatty acids; PUFA—total polyunsaturated fatty acids; *n*-6—total omega-6 fatty acids; *n*-3—total omega-3 fatty acids.
[b]Not detected.

Source: Adapted from Watkins et al. (2003).

electrophoresis (2DE). Seven of the major known proteins were identified in at least one electrophoretic system. SDS-PAGE resulted in clearly identifiable bands for ovotransferrin, ovalbumin, ovomucoid, flavoprotein, ovostatin, ovoinhibitor, globulins G2 and G3, and Ch21 protein (quiescence-specific protein). The IEF procedure was able to separate isoforms of ovotransferrin, ovalbumin, and ovomucoid. Using native-PAGE, 6 or 12 protein bands were distinguished after staining with Coomassie brilliant blue or silver, respectively, and 2DE allowed separation of a relatively large number of spots. Each of the four systems revealed many unidentified minor proteins.

Raikos et al. (2006) employed high-resolution techniques for proteome analysis, including SDS-PAGE and 2D gel electrophoresis, combined with matrix-assisted laser desorption–ionization/time-of-flight (MALDI-TOF) mass spectrometry, to separate and identify several protein components in hen egg white and yolk. In addition to isoforms of ovalbumin and conalbumin, the

TABLE 1.8. Minerals and Vitamins in Whole Egg, Albumen, or Yolk of Shell Eggs

Constituent (Units)	Whole Egg	Egg Albumen	Egg Yolk
	Minerals (mg)		
Ca	29.2	3.8	25.2
Cl	96.0	66.1	29.9
Cu	0.033	0.009	0.024
I	0.026	0.001	0.024
Fe	1.08	0.053	1.02
Mg	6.33	4.15	2.15
Mn	0.021	0.002	0.019
P	111	8	102
K	74	57	17
Na	71	63	9
S	90	62	28
Zn	0.72	0.05	0.66
	Vitamins		
Vitamin A (IU)	264	—	260
D (IU)	27	—	27
E (mg)	0.88	—	0.87
B_{12} (μg)	0.48	—	0.48
Choline (mg)	11.0	2.58	8.35
Folic acid (mg)	237	0.46	238
Inositol (mg)	0.023	0.006	0.026
Niacin (mg)	5.94	1.52	4.35
Pantothenic acid (mg)	0.045	0.035	0.010
Pyridoxine (mg)	0.83	0.09	0.73
Riboflavin (mg)	0.065	0.008	0.057
Thiamine (mg)	0.18	0.11	0.07
	0.05	0.004	0.048

Source: Adapted from Watkins (1995).

presence of FLJ10305 and Fatso proteins in the proteome of *Gallus domesticus* was confirmed. Guérin-Dubiard et al. (2006) also performed proteomic analysis of hen egg white using 2D electrophoresis and mass spectrometry, separating 69 protein spots and identifying 16 of the proteins. Wide polymorphism was observed for 13 proteins, including nine isoforms of ovoinhibitor.

Some details of the chemical composition and structure of the main albumen proteins are described in the following sections.

1.3.2.1.1. Ovalbumin

Ovalbumin is the major protein in albumen, constituting about 54% of the total egg white protein (Li-Chan et al. 1995). It belongs to the serpin superfamily that includes more than 300 homologous proteins with diverse functions including the major serine protein inhibitors of human plasma (Huntington and Stein 2001). Interestingly, ovalbumin lacks this inhibitory activity even though it seems to have retained a putative reactive center at

Ala353–Ser 354 for proteolytic cleavage by elastase; in inhibitory serpins, such cleavage leads to a dramatic conformational change thought to be related to the inhibitory activity.

Ovalbumin is a glycoprotein with a single carbohydrate chain linked to Asn293. It is composed of 386 amino acid residues, and two genetic polymorphisms have been reported at residues 290 (Glu → Gln) and 312 (Asn → Asp). Among egg white proteins, ovalbumin is the only protein that has free SH groups, containing six cysteine residues and a single disulfide between Cys74 and Cys121. Ovalbumin molecules with two, one, and no phosphate groups are present in the ratio of 8:2:1 (Li-Chan and Nakai 1989; Huntington and Stein 2001).

Ovalbumin Y is a unique chimeric glycoprotein having an amino acid sequence similar to that of ovalbumin but a carbohydrate moiety similar to that of ovomucoid (Hirose et al. 2006). Nau et al. (2005) identified the ovalbumin gene Y protein in egg whites, after isolation by 2D PAGE and characterization with peptide mass fingerprinting. The ratio of ovalbumin gene Y to ovalbumin was found to be ~13–100. Unlike ovalbumin, ovalbumin gene Y is not phosphorylated; however, like ovalbumin, it is glycosylated. Three and five isoforms of ovalbumin gene Y and ovalbumin-related Y protein, respectively, have been revealed by electrophoretic analysis, each differing in their isoelectric points, but the polymorphism of this protein cannot be explained by various glycosylation or phosphorylation levels or by genetic variations (Guérin-Dubiard et al. 2006).

Ovalbumin may be transformed via an intermediate state into a noncleaved, thermally stable form (*S*-ovalbumin) during storage of unfertilized eggs or development of fertilized eggs. Storage temperature has a greater impact than does storage time on conversion of *N*-ovalbumin to *S*-ovalbumin. Huntington and Stein (2001) have suggested that the conversion is related to changing needs of the chick embryo at different developmental stages. On the basis of calorimetric analyses, Hatta et al. (2001) found that conversion into *S*-ovalbumin was slower in fertile eggs than in unfertile eggs under the same incubation conditions. The conversion also occurs on incubation of isolated ovalbumin under alkaline conditions.

During storage, loss of egg mass is associated with the type of ovalbumin present (*N*-, intermediate-, or *S*-ovalbumin). The isoelectric point of ovalbumin becomes slightly more acidic during storage (Schaefer et al. 1999), in accordance with the formation of *S*-ovalbumin. A partial insertion of the α-helical serpin loop into β-sheet A has been proposed as a mechanism for this transformation. However, Yamamoto et al. (2003) found that native ovalbumin and *S*-ovalbumin were similar with respect to the rates of loop cleavage by elastase and subtilisin, suggesting that *S*-ovalbumin is produced by a mechanism other than partial loop insertion.

Egg white containing *S*-ovalbumin forms a heat-induced gel of poor strength. In general, heat treatment at 90°C increased textural parameters favorably compared to 85°C, where only *N*-ovalbumin denaturation contributed to

TABLE 1.9. Content, Isoelectric Point, and Molecular Weight of Some Proteins Found in Egg Albumen

Protein	Percent of Albumen Proteins	Isoelectric Point	Molecular Weight, kDa
Ovalbumin	54	4.5 (5.1–5.3)	45 (42.4)
Ovalbumin Y		(5.3–5.5)	(53.4–54.3)
Ovotransferrin	12	6.1 (6.2–6.7)	76 (85–75)
Ovomucoid	11	4.1 (5.0–5.3)	28 (37.2–43.1)
Ovomucin	3.5	4.5–5.0	5500–8300
Lysozyme	3.4	10.7	14.3 (15)
Ovoglobulin		(6.1–5.3)	
G2 globulin	4.0	5.5	30–45
G3 globulin	4.0	4.8	N/D
Ovoinhibitor	1.5	5.1 (6.2–6.4)	49 (69.5–63.6)
Ovoglycoprotein	1.0	3.9 (5.0–5.4)	24.4 (37.2–43.1)
Ovoflavoprotein	0.8	4 (5.0–5.2)	32 (37.4–40)
Ovomacroglobulin	0.5	4.5	769
Cystatin	0.05	5.1 (6.1)	12.7(17)
Avidin	0.05	10	68.3
Tenp[a]	N/D[b]	(5.9–6.3)	(48.9–50.2)
Clusterin	N/D	(6.1–6.6)	(33–32.4)
Ch21[c]	N/D	(5.7)	(21)
VMO-1[d]	N/D	N/D	(17.6)

[a]Protein with strong homology with bacterial permeability increasing protein.
[b]Not determined.
[c]Quiescence-specific protein or extra fatty acid binding protein.
[d]Vitelline membrane outer layer protein 1.

Source: Data were compiled from Li-Chan et al. (1995), except for those shown in parentheses, which are from Guérin-Dubiard et al. (2006).

gelling (Hammershoj et al. 2002). Sugimoto et al. (1999) found that 83% and 90% of the ovalbumin population was in a heat-stable form in day 14 or stage 40 amniotic fluid and day 18 or stage 44 egg yolk, respectively, whereas ovalbumin in newly deposited eggs was in the heat-unstable, native form. Purified preparations of stable ovalbumin from egg white and amniotic fluid showed a less ordered configuration than did native ovalbumin, as analyzed by circular dichroism and differential scanning calorimetry. McKenzie and Frier (2003a, 2003b) isolated *S*-, S_1-, and S_2-ovalbumin from domestic hen egg *R*-ovalbumin, and showed that the *S*-ovalbumins are much more resistant to urea than are *R*-ovalbumin and that unfolding of S_1- and S_2-ovalbumin is an order of magnitude slower than that of *R*-ovalbumin.

Ovalbumin is a major allergen in hen egg white that causes IgE-mediated food-allergic reactions in children. Mapping of the immunodominant IgE binding epitopes of ovalbumin indicate that the critical amino acids involved in IgE antibody binding are primarily hydrophobic amino acids, followed by polar and charged residues, and comprised of β-sheet and β-turn structures.

One of the epitopes, Asp95–Ala102, consists of a single α-helix. IgE binding was reduced by heating of ovalbumin or by irradiation followed by heating (Kim et al. 2002), and irradiation has been suggested as a method to reduce allergenicity (Lee et al. 2001; Seo et al. 2004).

Bioactive peptides have been reported from ovalbumin, including angiotensin I-converting enzyme (ACE)-inhibitory peptides such as ovokinin and ovokinin-2–7 (Matoba et al. 2003; Scruggs et al. 2004; Miguel and Aleixandre 2006; Miguel et al. 2006), antimicrobial peptides (Kimura and Obi 2005; Pellegrini et al. 2004), and antioxidant peptides (Davalos et al. 2004).

1.3.2.1.2. Ovotransferrin

Ovotransferrin (OTf), also known as *conalbumin*, is similar to serum transferrin in animals. The transferrins are a family of bilobal glycoproteins that tightly bind ferric iron. Ovotranferrin has the ability to bind various metal ions in the ratio of 2 mol of ion per mole of proteins. For serum transferrin, ovotransferrin, and lactoferrin, each of the duplicate lobes binds one atom of Fe^{3+} and one carbonate anion (Lambert et al. 2005). Formation of an iron complex by ovotranferrin inhibits the growth of microorganisms that require iron. It was reported that chicken heatshock protein 108 (HSP108), the avian homolog of GRP94, was originally isolated from hen oviduct and binds Fe-ovotransferrin (Fe-OTf) (Weiner et al. 1997).

Each of the homologous *N*- and *C*-lobes in this glycoprotein contains a single iron-binding site situated in a deep cleft, and 15 disulfide bridges assist in maintining the bilobal structure. The *N*-terminal segment, Ala1–Tyr72, assumes a local native-like conformation in the two-disulfide form of the ovotransferrin *N*-lobe (Mizutani et al. 1997). Interaction with both lobes is considered to be necessary for efficient iron acquisition (Alcantara and Schryvers 1996). Nadeau et al. (1996) indicated that the presence or absence of basic residues in the metal-binding lobes of transferrins may account for the different anion- and metal-binding characteristics observed for the iron-binding sites of these proteins.

Yajima et al. (1998) examined the structural characteristics of the isolated fragments corresponding to the *N*- and *C*-terminal halves of ovotransferrin (OTf/2N and OTf/2C) with and without iron by means of small-angle neutron scattering, and inferred that OTf/2C tends to become more compact than OTf/2N on iron binding. Mizutani et al. (2000) reported the location of sulfate anion binding sites of the ovotransferrin *N*-lobe in the 1.90-Å-resolution apo crystal structure. Anion binding to the interdomain cleft, especially to sites 1 and 2, plays a crucial role in the domain opening and synergistic carbonate anion release, influencing iron release of the ovotransferrin *N*-lobe. The interdomain disulfide bond has no effect on the iron uptake and release function but significantly decreases the conformational stability in the *C*-lobe (Muralidhara and Hirose 2000).

Abdallah and Chahine (1998, 1999) investigated *in vitro* Fe^{3+} uptake and release by hen ovotransferrin, and reported similar although not identical mechanisms and lower affinities for Fe^{3+} compared to those of serum transferrin and lactoferrin. Transferrins bind Fe^{3+} very tightly in a closed interdomain cleft by the coordination of four protein ligands (Asp60, Tyr92, Tyr191, and His250 in ovotransferrin *N*-lobe) and of a synergistic anion, physiologically bidentate CO_3^{2-}. On Fe^{3+} uptake, transferrins undergo a large-scale conformational transition; the apo structure with an opening of the interdomain cleft is transformed into the closed holo structure, implying initial Fe^{3+} binding in the open form. The Fe^{3+}-soaked form shows almost exactly the same overall open structure as the iron-free apo form (Mizutani et al. 1999). Mizutani et al. (2000) demonstrated the quantitative Fe^{3+} release kinetics and the anion-mediated Fe^{3+} release mechanism in the ovotransferrin *C*-lobe.

Kurokawa et al. (1999) reported the three-dimensional crystal structure of hen apo-ovotransferrin by molecular replacement and refined by simulated annealing and restrained least squares to a 3.0 Å resolution. Both empty iron binding clefts are in the open conformation, supporting the theory that Fe^{3+} binding or release in transferrin proceeds via a mechanism that involves domain opening and closure. On opening, the domains rotate essentially as rigid bodies about an axis that passes through the two β-strands, linking the domains. The domains of each lobe contact one another at different sites in the open and closed forms, and domain opening or closing produces a see-saw motion between the two alternative close-packed interfaces. An interdomain disulfide bridge (Cys478–Cys671) is found only in the *C*-lobe, and may restrict domain opening but does not completely prevent it.

In addition to investigating the iron binding domain, several more recent studies have examined other structural features of the ovotransferrin molecule that may be associated with various bioactive properties. Embryos of avian eggs and mammals are highly sensitive to oxidative stress, and hence maintaining a steady reducing environment during the embryonic development is known to confer protection. Although information is lacking, proteins of avian egg albumen have been suggested as possible targets for this function. Ibrahim et al. (2006) found that ovotransferrin undergoes autocleavage at distinct sites on reduction with thiol reducing agent or thioredoxin-reducing system, through the unique chemical reactivity of four tripeptide motifs, which were located upstream and downstream of the two disulfide kringle domains (residues 115–211 and 454–544) of OTf. These reduction–scissile sequences, His/Cys–X–Thr, are evolutionary conserved self-cleavage motifs found in several autoprocessing proteins.

Ibrahim et al. (1998) identified a distinct antibacterial domain within the *N*-lobe of ovotransferrin, which retained the bactericidal activity independently of iron deprivation. A cationic antimicrobial fragment (OTAP92) was found within the 109–200 sequence of the *N*-lobe of ovotransferrin (Ibrahim et al. 2000), possessing a unique structural motif similar to that of insect defensins and capable of killing Gram-negative bacteria by crossing the outer membrane by a self-promoted uptake and causing damage to the biological function of the cytoplasmic membrane. Antiviral activity of several fragments from hen ovotransferrin [DQKDEYELL (residues 219–227), KDLLFK (residues 269–361 and residues 633–638)] toward Marck's disease virus infection of chicken embryo fibroblasts was reported by Giansanti et al. (2005). An angiotensin I–converting enzyme (ACE) inhibitory peptide derived from hen ovotransferrin and identified as Lys–Val–Arg–Glu–Gly–Thr–Thr–Tyr, was shown to act as a prodrug inhibitor with *in vivo* activity exerted in the form of Lys–Val–Arg–Glu–Gly–Thr (Lee et al. 2006).

1.3.2.1.3. Ovomucoid

Ovomucoid, which constitutes about 11% of egg white proteins, is a thermally stable glycoprotein belonging to the Kazal family of protein inhibitors (Li-Chan and Nakai 1989). The ovomucoid molecule is composed of three distinct domains crosslinked only by intradomain disulfide bonds. In chicken egg ovomucoid, domain II contains the active site for trypsin inhibitory activity, but wide variability in specificity and inhibitory activity of the domains has been reported for ovomucoids from different avian species (Li-Chan and Nakai 1989).

Chicken ovomucoid has an important role in the pathogenesis of IgE-mediated allergic reactions to hen egg white (Mine and Zhang 2001, 2002a; Mine and Rupa 2004a, 2004b). The allergenic potential of chicken egg white ovomucoid is thought to be related to its stability to heat treatment and digestion, necessitating development of methods of detection even after denaturation (Hirose et al. 2004, 2005). It was reported that only epitopes on the ovomucoid protein backbone and not the carbohydrate residues are responsible for IgE binding (Besler et al. 1997). Mine and Zhang (2002b) identified and determined eight IgG epitopes, 5–11 amino acids in length, and nine IgE epitopes, 5–16 amino acids in length, within the primary sequence in ovomucoid using arrays of overlapping peptides synthesized on cellulose membranes. Mutational analysis of the epitopes indicated that charged amino acids (aspartic acid, glutamic acid, and lysine) and some hydrophobic (leucine, phenylalanine, and glycine) and polar (serine, threonine, tyrosine, and cysteine) amino acids are important for antibody binding.

Van Der Plancken et al. (2004) conducted a kinetic study on the effect of heating at 75–110°C on the trypsin inhibition activity of ovomucoid, showing a time-dependent decrease in trypsin inhibitory activity that could be described accurately by a first-order kinetic model. Heat stability of ovomucoid is lowest at pH 7.6, and the presence of other egg white constituents, particularly the formation of ovomucoid–lysozyme complexes, decreases heat stability compared to the model system of ovomucoid in buffer.

Yoshino et al. (2004) reported peptic digestibility of ovomucoid in raw egg white over the pH range 1.5–2.5, but poor digestibitility at pH 3.0 or higher even after heat treatment. Pepsin-digested ovomucoid retains its trypsin (proteinase) inhibitor activities, and disulfide bonds play an important role in the digestive resistance and allergenicity of ovomuoid (Kovacs-Nolan et al. 2000). Chicken egg white ovomucoid was digested in simulated gastric fluid to examine the reactivity of the resulting fragments to IgE in sera from allergic patients; 21% of the examined patients retained IgE binding to a small (4.5-kDa) fragment (Takagi et al. 2005).

There have been many studies to modify the structure and composition of ovomucoid in an attempt to reduce its allergenicity, including modification of the ovomucoid third domain (Mine et al. 2003; Rupa and Mine 2006), gamma irradiation with heating (Lee et al. 2002a), heating in the presence of wheat flour (Kato et al. 2001), reduction of ovomucoid (Kovacs-Nolan et al. 2000), genetic modification (Rupa and Mine 2006), and deglycosylation by *endo*-β-*N*-acetylglucosaminidases (Yamamoto et al. 1998). The deglycosylated ovomucoid did not differ from native ovomucoid in trypsin inhibitory activity, but was very unstable to heat treatment, suggesting the importance of oligosaccharides of ovomucoid in stabilizing the protein structure against heat denaturation but not in its allergenicity (Yamamoto et al. 1998). Decrease in antigenic and allergenic potentials of ovomucoid by heating in the presence of wheat flour depended on wheat variety and intermolecular disulfide bridges, as it was greater in the durum than were other varieties (Kato et al. 2001). Considerable loss of human serum IgG and IgE binding activity occurred when phenylalanine at position 37 was replaced with methionine, which disrupted the α-helical structure forming part of the IgE and IgG epitopes (Mine et al. 2003; Mine and Rupa 2003). In addition, decreasing antigenicity resulted from substitution of glycine at position 32. Therefore, Gly32 and Phe37 were found to have an important role on its antigenicity and allergenicity as well as structural integrity of the third domain of ovomucoid.

1.3.2.1.4. Ovomucin

Ovomucin is a sulfated glycoprotein that confers the jelly-like structure to egg albumen. Insoluble ovomucin is the major component of the gel-like insoluble fraction of the thick albumen of egg white, whereas soluble ovomucin is the main component of the outer and inner albumen. Both insoluble and soluble fractions consist of two subunits with different carbohydrate contents: α-ovomucin and β-ovomucin. Insoluble ovomucin contains 84 α-ovomucin and 20 β-ovomucin molecules, while soluble ovomucin is composed of 40 α-ovomucin and 3 β-ovomucin molecules, resulting in molecular weights of 2.3×10^7 and 8.3×10^6, respectively (Sugino et al. 1997b).

Hiidenhovi et al. (1999) described a dual-column gel filtration system for the separation of ovomucin subunits that resulted in the isolation of eight peaks. Hiidenhovi et al. (2002) isolated 280-, 340-, 500-, and 520-mg ovomucin per 100 g of albumen from whole egg albumen, thick egg albumen, liquid egg albumen, and a liquid egg albumen filtration byproduct, respectively, using an isoelectric precipitation method. There was great variation between β-ovomucin contents isolated from the different albumen materials.

Whole egg albumen contained about 25 mg of β-ovomucin in 100 g of albumen, whereas thick egg albumen, liquid egg albumen, and the filtration byproduct contained about 1.5, 3, and 5 times more β-ovomucin, respectively. Toussant and Latshaw (1999) reported that the total amount of ovomucin isolated from the thick albumen of the eggs was positively related to quality as represented by higher HU (Haugh unit) scores, while a greater proportion of the more highly glycosylated β-ovomucin with its higher concentrations of sialic acid, hexoses, and hexosamines was found in eggs with inferior quality.

Ovomucin can easily be obtained by precipitation, but the precipitate itself is not soluble in ordinary buffers. Flavourzyme, a protease/peptidase complex, enabled the solubilization of ovomucin and the hydrolysis of the soluble materials (Moreau et al. 1997; Hiidenhovi et al. 2000), which were characterized as high molecular weight (>94,000 Da) glycosylated materials, a 38,100-Da glycopeptide, several peptides ranging from 34,400 to 20,800 Da, and free amino acids (Moreau et al. 1997).

In addition to structural and chemical characterization of ovomucin to understand its role in albumen viscosity and the thinning of thick albumen during storage, several studies have focused on the characteristics related to bioactive properties. For example, Tsuge et al. (1996) reported that the hemagglutination inhibition activity of ovomucin against bovine rotavirus was dependent on both α and β subunits, while activity against hen Newcastle disease virus required the β subunit only. Tsuge et al. (1997a, 1997b) found that the P1, P2, and P3 fractions from the β subunit, which were composed of *O*-glycoproteins, containing more or less clustered sialic acid moieties and NeuAc residues, had higher binding activity to Newcastle disease virus, while the peptide-rich P4 and P5 fractions derived mainly from the α-subunit had higher binding activity to the anti-ovomucin antibodies, and that the latter activity was stabilized by disulfide bonds in the structure.

Watanabe et al. (1998a) hydrolyzed gel-like ovomucin and its β-subunit with pronase at various ratios to study the abilities of the fragments to bind to antiovomucin antibodies and Newcastle disease virus. Kobayashi et al. (2004) reported that glycopeptides derived from hen egg ovomucin by pronase digestion have the ability to bind enterohemorrhagic *Escherichia coli* O157:H7.

Nagaoka et al. (2002) found that egg ovomucin attenuated hypercholesterolemia in rats and inhibited cholesterol absorption in Caco-2 cells. The suppression of cholesterol absorption by direct interaction between cholesterol mixed micelles and ovomucin in the jejunal epithelia was considered to be part of the mechanism underlying the hypocholesterolemic action of ovomucin. It was suggested that ovomucin may also inhibit the reabsorption of bile acids in the ileum, thus lowering the serum cholesterol level. Tanizaki et al. (1997) reported that sulfated glycopeptides in ovomucin, yolk membrane, and chalazae in chicken eggs could activate cultured macrophage-like cells, J774.1, and TGC-induced macrophages from the peritoneal cavity of male mice. The *O*-linked carbohydrate chains, consisting of *N*-acetylgalactosamine, galactose, and *N*-acetylneuraminic acid and sulfate, in the sulfated glycopeptide were identified as the component having macrophage-stimulating activity.

Watanabe et al. (1998b) reported antitumor effects of pronase-treated fragments, glycopeptides, from ovomucin in hen egg white in a double-grafted tumor system. Examinations of desialylated 120-kDa fragment indicated that the sialic acid residues in that fragment are not necessarily essential for direct antitumor activity but might be indispensable for regression of distant tumors. Oguro et al. (2001) also reported that a highly glycosylated 70-kDa peptide fragment (70F) in the α-subunit of pronase-treated hen egg white ovomucin was shown to have antitumor activity. These findings suggested that 70F ovomucin may have an antiangiogenic effect induced through mechanisms other than the VEGF/receptor system.

1.3.2.1.5. Ovoglobulin

The term *ovoglobulin* refers to ovoglobulin G2 and G3, each constituting about 4% of egg albumen proteins. Ovoglobulin G2 and G3 are smilar in many properties, including their molecular weight (49 kDa). In chickens, ovoglobulin G2 shows polymorphism (Asal et al. 1993). Electrophoretic patterns of albumen of stored chicken eggs show more diffuse and slightly more acidic bands for several proteins, including ovoglobulin, than do those of fresh albumen (Ogawa and Tanabe 1990).

Although the biological function of ovoglobulin is not known, it is considered to be important for the foaming property of egg white (Sugino et al. 1997b). Damodaran et al. (1998) studied the competitive adsorption of five major egg white proteins, namely, ovalbumin, ovotransferrin, ovoglobulins, ovomucoid, and lysozyme, from a bulk solution having relative protein concentration ratios similar to those in egg white to the air–water interface at low and high ionic strengths. At 0.1 ionic strength, only ovalbumin and ovoglobulins adsorbed to the interface, while ovotransferrin, ovomucoid, and lysozyme were essentially excluded from the interface.

1.3.2.1.6. Lysozyme

Egg white lysozyme (mucopeptide *N*-acetylmuramoylhydrolase, EC 3.2.1.17) is a 14.4-kDa protein consisting of 129 amino acid residues and has an isoelectric point close to 10.7. Its tertiary structure is stabilized by four disulfide bonds. Because of its basic character, lysozyme binds to ovomucin, transferrin, or ovalbumin. The chalaziferous layer and chalaza bind 2–3 times more lysozyme than do other egg white proteins.

The amino acid sequence and the complete 2-Å-resolution X-ray crystallographic three-dimensional structure of lysozyme have been known since the 1960s (Li-Chan and Nakai 1989). Yet, many researchers continue to examine further details of the structure and function of this protein. The three-dimensional structure of hen egg white lysozyme (HEWL) in a hexagonal crystal form has been determined and refined to 1.46 Å resolution (Brinkmann et al. 2006).

Numerous studies have been conducted to determine changes in lysozyme structure under various conditions, including cocrystallization in the presence of various alcohols (Deshpande et al. 2005), aqueous–organic solvent mixtures (Griebenow and Klibanov 1996), sorbitol (Petersen et al. 2004), pH 1.5 (Babu and Bhakuni 1997), various guanidine HCl concentrations (Liu et al. 2005a) with or without polyethylene glycol and dextran (Ganea et al. 2004), heat treatment after pretreatment with supercritical CO_2 (Liu et al. 2004), and refolding at low concentrations or in the presence of thiol/disulfide reagents (Raman et al. 1996). Other studies have investigated factors influencing solubility, aggregation, fibril formation, and/or activity of lysozyme, including mechanical stirring with regard to the effects of physical and molecular interfaces (Colombie et al. 2001); high-pressure treatment (Smeller et al. 2006); temperature (Sakamoto et al. 2004); exposure to alkaline pH 12.2 (Homchaudhuri et al. 2006); low pH and temperatures close to the midpoint temperature for protein unfolding (Arnaudov and de Vries 2005); succinylation (van der Veen et al. 2005), and the addition of L-arginine (Reddy et al. 2005), arginine ethylester (Shiraki et al. 2004), or other amino acid esters (Shiraki et al. 2005).

Much of the interest in lysozyme has focused on its antimicrobial activity. Lysozyme is an enzyme that catalyzes the hydrolysis of the β-1,4-glycosidic linkage between *N*-acetylmuraminic acid and *N*-acetylglucosamine in polysaccharide components of certain bacterial cell walls. The binding of *N*-acetylglucosamine oligosaccharides to lysozyme was examined by X-ray powder diffraction at room temperature (Von Dreele 2005). Marolia and D'Souza (1999) reported an increase in lysozyme activity of the hen egg white foam matrix by crosslinking in the presence of *N*-acetyl glucosamine. Lysozyme has antimicrobial activity, specifically against Gram-positive bacteria, but much less against Gram-negative bacteria because of its inability to penetrate the outer lipopolysaccharide membranes in Gram-negative bacteria.

Ibrahim et al. (2001) reported that a helix–loop–helix (HLH) peptide at the upper lip of the active-site cleft of lysozyme confers potent antimicrobial activity with membrane permeabilization action. Residues 98–112 of egg white lysozyme, part of a HLH domain, possess broad-spectrum antimicrobial activity *in vitro*, having potent microbicidal activity against Gram-positive and Gram-negative bacteria, and the fungus *Candida albicans*. The *N*-terminal helix of HLH was specifically bactericidal to Gram-positive bacteria, whereas the *C*-terminal helix was bactericidal to all strains tested.

Trziszka et al. (2004) studied the influence of factors such as age of hens, storage conditions of eggs, and thermal treatment of egg white, on lysozyme activity. The lowest activity was observed in eggs from laying hens either younger than 30 weeks or older than 60 weeks, while the highest activity was observed in laying hens aged 40–50 weeks. Thermal processing, especially pasteurization, reduced the activity of lysozyme. According to Schaefer et al. (1999), there were no significant changes in the activity and content of lysozyme during storage.

There have been many studies to increase the antimicrobial activity of lysozyme by increasing hydrophobicity or emulsifying property. These include thermal denaturation (Ibrahim et al. 1996a, 1996b), covalent attachment of saturated fatty acids (Liu et al. 2000b), glycosylation (Nakamura and Kato 2000; Nakamura et al. 2000), and reduction of disulfide bonds (Touch et al.

2004). The double-glycosylated lysozyme R21T/G49N obtained by site-directed mutagenesis to introduce two *N*-linked glycosylation sites showed better emulsifying properties than did the two single-polymannosyl lysozymes R21T and G49N (Shu et al. 1998), and lysozyme conjugates with xyloglucan or galactomannan showed increased antimicrobial activity (Nakamura et al. 2000; Nakamura and Kato 2000). It was found that lipophilization of lysozyme combined with glycosylation provided enhanced antimicrobial activity toward Gram-negative bacteria and improved yield (Liu et al. 2000a). Lipophilization broadened the bactericidal action of lysozyme to Gram-negative bacteria with little loss of enzymatic activity, and the bactericidal activity increased in proportion to the number of bound short-chain fatty acids (Liu et al. 2000b).

Studies have also suggested that lysozyme possesses antibacterial activity independent of the catalytic function, with enzymatic hyrolysis enhancing this activity through the production of antibacterial peptides including residues 98–112, 98–108, and 15–21 (Mine and Kovacs-Nolan 2006). Novel antimicrobial peptides have been isolated, purified, and characterized from lysozyme hydrolysate (Mine et al. 2004). Partial denaturation of hen egg white lysozyme at 80°C for 20 min at pH 6.0 produced strong bactericidal activity not only against Gram-positive bacteria but also against Gram-negative bacteria, probably through a membrane-disrupting mechanism of the dimeric form of lysozyme, independent of its muramidase activity (Ibrahim et al. 1996a, 1996b). It was also found that Ca^{2+} induced a conformational change allowing antimicrobial action against Gram-negative bacteria by disrupting the normal electrostatic interactions between divalent cations and components of the outer membrane (Ibrahim et al. 1997). Partial heat denaturation extended the activity spectrum of lysozyme under high-pressure conditions, suggesting enhanced uptake of partially denatured lysozyme through the outer membrane (Masschalck et al. 2001). Masschalck et al. (2002) investigated lytic and nonlytic mechanisms of inactivation of Gram-positive bacteria by lysozyme under atmospheric and high hydrostatic pressure, and showed that the nonlytic mechanism involves membrane perturbation.

Metal binding to lysozyme has also received wide interest. Lysozyme binds one to eight ions of Cu^{2+} and 1 to 6 ions of Zn^{2+} depending on the ratio of ion to lysozyme in the mixture (Moreau et al. 1995). It was reported that Ni^{2+}, Mn^{2+}, Co^{2+}, and Yb^{3+} chloride salts induce an increase in solubility of the tetragonal form in crystals of hen egg white lysozyme at high salt concentration, while Mg^{2+} and Ca^{2+} chloride salts do not (Li 2006). On the other hand, NaCl induces aggregation of lysozyme; in a 0.5 *M* NaCl solution six chloride anions and at least one sodium cation are bound to preferred sites on the surface of lysozyme (Poznanski 2006).

Hen egg white lysozyme has a sweet taste, which, like the enzymatic activity, is dependent on the tertiary structure of the molecule (Maehashi and Udaka 1998; Masuda et al. 2001). Masuda et al. (2001) found that both the sweetness and enzymatic activities were lost when the intradisulfide linkage in a lysozyme molecule was reduced and *S*-3-(trimethylated amino) propylated, or by heating at 95°C for 18h. Masuda et al. (2005) suggested that the basicity of a broad surface region formed by five positively charged residues (Lys13, Lys96, Arg14, Arg21, and Arg73) was required for lysozyme sweetness.

1.3.2.1.7. Ovomacroglobulin (Ovostatin)

Ovomacroglobulin, also known as *ovostatin*, is composed of four subunits, each having a molecular weight of 175,000, with pairs of the subunits joined by disulfide bonds. Ovomacroglobulin inhibits hemagglutination, possesses anticollagenase activity, and has inhibitory activity against diverse proteolytic enzymes representing serine, thiol, and metal proteases (Li-Chan and Nakai 1989).

Nielsen and Sottrup-Jensen (1993) investigated 53 residues of the internal sequence from the proteinase-binding hen egg white ovomacroglobulin, which correspond to residues 945–997 of human α-2-macroglobulin, and found 74% degree of conservation. Cys949, which is one constituent of the internal thiol ester site of members of the family of proteins related to α-2-macroglobulin, is an Asn residue in hen egg white ovostatin, but the other constituent, Gln952, is preserved.

Variants of the receptor binding domain of both human α-2-macroglobulin and the corresponding domain of hen egg white ovomacroglobulin have been expressed in *Escherichia coli* and refolded *in vitro* (Nielsen et al. 1996). Competition experiments with methylamine-treated α-2-macroglobulin for binding to the multifunctional α-2-macroglobulin receptor identified two Lys residues (residues 1370 and 1374 in human α-2-macroglobulin) spaced by three amino acid residues as crucial for receptor binding.

1.3.2.1.8. Ovoflavoprotein

Ovoflavoprotein, also referred to as *flavoprotein* or *riboflavin-binding protein*, is a phosphoglycoprotein that is responsible for binding most of the riboflavin (vitamin B_2) in egg white. According to Jacobs et al. (1993), flavoprotein has the highest Se content (1800ng/g) among egg white proteins. As it is abundant in low-cost egg processing byproducts, ovoflavoprotein could serve as a useful food ingredient, and various methods have been proposed for its separation from egg white (Takeuchi et al. 1992; Kodentsova et al. 1994; Rao et al. 1997).

Chicken egg white riboflavin binding protein is the prototype of a family that includes other riboflavin and folate binding proteins. An unusual characteristic of these molecules is their high degree of crosslinking by disulfide bridges and, in the case of the avian proteins, the presence of stretches of highly phosphorylated polypeptide chain. Each mole of riboflavin-binding apoprotein (apo-RBP) binds one mole of riboflavin with high-affinity constant (1.4n*M*), causing loss of the characteristic riboflavin fluorescence. The 2.5-Å-resolution crystal structure of chicken egg white riboflavin-binding protein shows a ligand-binding domain that appears to be strongly conditioned by the presence of the disulfide bridges, and a phosphorylated motif that is essential for vitamin uptake and includes two helices before and after the flexible phosphorylated region (Monaco 1997). The riboflavin molecule is bound to the protein with the isoalloxazine ring stacked in between the rings of Tyr75 and Trp156; this geometry and the proximity of other tryptophans explains the fluorescent quenching observed when riboflavin binds to the protein.

1.3.2.1.9. Ovoglycoprotein

Ovoglycoprotein is an acidic glycoprotein. The average molecular masses of ovoglycoprotein from chicken egg white, its partially and completely deglycosylated derivatives, were estimated to be about 30,000, 28,400 and 21,400, respectively, by matrix-assisted laser desorption ionization time-of-flight mass spectrometry, and their isoelectric points were in the ranges 4.37–4.51, 4.34–4.44, and 4.17–4.43, respectively, by isoelectric focusing (Haginaka et al. 2000).

The biological functions of ovoglycoprotein are still unclear. It is a stable glycoprotein, remaining soluble even after heat treatment at 100°C or by trichloroacetic acid treatment (Li-Chan et al. 1995). Chiral recognition sites existing on the protein domain of ovoglycoprotein have led to studies proposing its use for enantiomeric separations of basic drug enantiomers and other solutes (Haginaka et al. 2000; Matsunaga and Haginaka 2001). Kodama and Kimura (2004) isolated and purified glycoprotein from chicken egg albumen; the glycoprotein had inhibitory activity against *Helicobacter pylori* colonization and may be useful for the prevention or treatment of conditions such as peptic ulcers.

1.3.2.1.10. Ovoinhibitor

Ovoinhibitor belongs to the Kazal family of proteinase inhibitors (Saxena and Tayyab 1997). It is a glycoprotein with six *a*-type domains and one *b*-type domain, making it a much larger inhibitor than ovomucoid, the other Kazal inhibitor in egg albumen. The existence of multiple domains may explain the multiple inhibitory specificity of ovoinhibitor, which has been shown to inhibit serine proteinases such as trypsin and chymotrypsin as well as a variety of fungal and bacterial proteases (Li-Chan and Nakai 1989).

Nine spots with isoelectric points ranging from 6.20 to 6.40 and molecular weights from 63.6 to 69.5 kDa were assigned to ovoinhibitor by proteomic analysis using 2D electrophoresis and mass spectrometry (Guérin-Dubiard et al. 2006). The deviation of the experimentally observed molecular weight from the theoretical value of 49.4 kDa according to amino acid sequence may be attributed in part to glycosylation. Galzie et al. (1996) isolated domain I of ovoinhibitor and found it to be homogeneous by the criteria of gel chromatography, SDS-PAGE, and PAGE; the molecular weight by gel filtration (10.9 kDa) and SDS-PAGE (8.3 kDa) were slightly higher than that computed from amino acid sequence, which was attributed to the glycoprotein nature of domain I, which contains 10% neutral carbohydrate and 2% sialic acid. Domain I was found to be a potent inhibitor of bovine trypsin but had almost no activity against chymotrypsin, elastase, and proteinase K. Begum et al. (2003)

reported that recombinant ovoinhibitor with a molecular mass of 49 kDa determined by SDS-PAGE and time of flight-mass spectrometry analyses showed inhibitory activity against trypsin, chymotrypsin, and elastase, similar to native ovoinhibitor.

Sugimoto et al. (1996) purified a proteinase inhibitor from yolk of hen's ovarian follicles, tentatively termed *vitelloinhibitor*, which resembled egg white ovoinhibitor not only in inhibitory spectrum (active for bovine trypsin and bovine chymotrypsin) but also in thermal stability, pH stability, antiserum reactivity, and amino acid composition. The molecular weight of vitelloinhibitor differed from that of ovoinhibitor, but was similar to the larger of two components separated from an α-2-proteinase inhibitor preparation isolated from laying hen serum. The partial *N*-terminal amino acid sequences of vitelloinhibitor and the two components of serum inhibitor and ovoinhibitor were all similar, and it is likely that vitelloinhibitor may be an ovoinhibitor analog derived from a serum precursor such as the larger component of α-2-proteinase inhibitor.

1.3.2.1.11. Cystatin (Ficin–Papain Inhibitor)

Ovocystatin, or cystatin from chicken egg white, is a cysteine proteinase inhibitor that inhibits thiol proeinases such as ficin and papain. Chicken cystatin has a molecular weight of approximately 13 kDa and contains two disulfide bonds located near the carboxy terminus; the reactive site is well conserved compared to other cystatins (Li-Chan and Nakai 1989). Phosphorylated cystatin has a lower pI of 5.6, compared to pI 6.5 in the nonphosphorylated form (Guérin-Dubiard et al. 2006). Golab et al. (2001) reported higher concentration of cystatin from unfertilized than fertilized eggs. Immunoblot analyses showed both phosphorylated and nonphosphorylated isoforms of chicken cystatin in unfertilized egg white as well as in egg yolk and chicken serum. Significantly lower immunostaining of the phosphorylated form after day 4 of embryogenesis in egg white suggested its preferential transport into the yolk sac.

Trziszka et al. (2004) investigated the effects of age of hens, storage conditions of eggs, and thermal treatment of egg white on the activity of cystatin. The lowest activity of cystatin was observed in eggs from laying hens either younger than 30 weeks or older than 60 weeks, while the highest activity of cystatin was observed in laying hens aged 40–50 weeks. Moreover, activity decreased by 4–12% in eggs stored at 15°C for 28 days, and thermal processing such as pasteurization significantly decreased cystatin activity.

In addition to its activity as a proteinase inhibitor, several reports have revealed other bioactive properties of cystatin. Wesierska et al. (2005) demonstrated antimicrobial activity of cystatin against bacterial pathogens. Brand et al. (2004) found that cystatins reversibly inhibited bone matrix degradation in the resorption lacunae adjacent to osteoclasts, suggesting the involvement of cystatins in the modulation of osteoclastic bone degradation. Cytokine-inducing activity of family 2 cystatins was reported by Kato et al. (2000).

Family 2 cystatins, including cystatins C, SA1, SA2, S, and egg white cystatin, were reported to upregulate interleukin-6 IL-6 production by two human gingival fibroblast cell lines or murine splenocytes and also IL-8 production by gingival fibroblasts at physiological concentrations (Kato et al. 2000).

1.3.2.1.12. Avidin

Avidin is basic tetrameric glycoprotein that is best known for its biotin binding properties. Each of the four monomers binds one molecule of biotin and the avidin–biotin interaction, with dissociation constant of ~$10^{-15}\,M$, is the strongest noncovalent interaction reported between protein and ligand (Bayer and Wilchek 1994). Rosano et al. (1999) reported almost exact 222 symmetry in the crystal structures of the four avidin subunits. Each avidin chain, composed of 128 amino acid residues, is arranged in an eight-stranded antiparallel β-barrel, whose inner region defines the D-biotin binding site. A fairly rigid binding site, readily accessible in the apoprotein structure, is sterically complementary to the shape and polarity of biotin.

Swamy et al. (1996) used Fourier transform infrared (FTIR) spectroscopy to study avidin secondary structure and complexation with biotin and biotin–lipid assemblies. Analysis of the amide I stretching band of avidin showed a secondary structural content composed of approximately 66% β-sheet and extended structures, with the remainder attributed to disordered structure and β-turns. Binding of biotin or specific association with the biotinylated lipid did not result in any appreciable changes in the secondary structure content of the protein, but showed changes in hydrogen bond stability of the β-sheet or extended chain regions. Difference spectra of the bound biotin implicated a direct involvement of the ureido moiety in the ligand interaction that was consistent with hydrogen bonding to amino acid residues in the avidin protein. The complexes of avidin with both biotin and membrane-bound lipid assemblies displayed a large increase in thermal stability compared with the native protein.

Both chicken egg white avidin and its bacterial relative streptavidin are widely used as tools in a number of affinity-based separations, in diagnostic assays, and in a variety of other applications. Other applications include the potential of avidin as an insecticide or antimicrobial agent. Cooper et al. (2006) reported insecticidal activity of avidin combined with genetically engineered and traditional host plant resistance against Colorado potato beetle. Avidin was reported to have a bacteriostatic effect against *Salmonella typhimurium*, but this effect is eliminated by addition of biotin (Klasing 2002).

Avidin contains terminal *N*-acetylglucosamine and mannose residues that bind to some lectins. Since lectins are expressed at various levels on the surface of tumor cells, the conjugation of cytotoxic agents to glycoproteins such as avidin that are recognized by lectins could be useful in the treatment of tumors. Avidin has therefore been suggested as a promising vehicle for the delivery of radioisotopes, drugs, toxins, or therapeutic genes to intraperitoneal tumors (Yao et al. 1998), and in medical pretargeting cancer treatments (Hytonen et al. 2003).

1.3.2.1.13. Enzymes

In addition to lysozyme, albumen contains many enzymes, including those with phosphatase, catalase, and glycosidase activity (Sugino et al. 1997b). An aminopeptidase has also been isolated from albumen; it acts with broad specificity, hydrolyzing aliphatic, aromatic and basic aminoacyl-2-naphthylamides, and di- to hexapeptides, with a preference for methionine at the NH_2 end, and basic or bulky hydrophobic residues at the penultimate position (Skrtic and Vitale 1994). The enzyme is a hydrophilic, acidic glycoprotein with molecular weight of ~180 kDa and optimal activity at pH 7.0–7.5 and 50°C. Amastatin, bestatin, and *o*-phenanthroline were found to be strong inhibitors, and puromycin, EDTA, and iodoacetamide were less potent inhibitors, while Co^{2+} activated the enzyme (Skrtic and Vitale 1994). A dimeric glycoprotein containing one molecule of FAD per 80-kDa subunit was isolated from chicken egg white and found to have sulfhydryl oxidase activity on a range of low-molecular-weight thiols, generating hydrogen peroxide in aerobic solution (Hoober et al. 1996).

1.3.2.1.14. Other Proteins in Albumen

Albumen contains many other proteins in addition to those described above, although most of these proteins have not been identified or investigated in detail. More developments in instrumentation for proteomics analysis are enabling characterization of some of these proteins. For example, Raikos et al. (2006) has identified FLJ10305 and Fatso (fto) proteins from egg albumen. FLJ10305 is a hypothetical protein predicted from the database for *Gallus gallus* that had not been previously confirmed in *G. domesticus*, while Fatso (fto) is a protein existing in mouse but not previously reported in eggs (Raikos et al. 2006).

Clusterin is a protein that has also been identified in egg albumen (Raikos et al. 2006; Guérin-Dubiard et al. 2006). Its biological role in chicken is suggested to be the prevention of premature aggregation and precipitation of eggshell matrix components before and during assembly into the rigid protein scaffold necessary for ordered mineralization (Raikos et al. 2006). It may also have a role during incubation of the developing embryo to stabilize lysozyme, conalbumin, and other proteins (Mann et al. 2003).

The Ch21 protein, or "quiescence-specific protein," has also been confirmed to be present in hen egg white (Desert et al. 2001). This protein was initially found in chick embryo skeletal tissues, and belongs to the superfamily of lipophilic molecule carrier proteins (Larsen et al. 1999; Desert et al. 2001).

Tenp, a protein bearing strong homology with a bacterial permeability-increasing protein family (Beamer et al. 1998), and VMO-1, an outer-layer vitelline membrane protein, have been identified in egg albumen by Guérin-Dubiard et al. (2006).

Kido et al. (1995) characterized VMO-1 tightly bound to ovomucin fibrils of hen egg yolk membrane. VMO-1 is composed of 163 amino acid residues with a calculated molecular weight of 17,979 and four disulfide bonds. The sequence analyses show that VMO-1 contains three 53-residue internal repeats containing distinctive regions of turns flanked by β-sheets consistent with a new β-fold motif, the β-prism. Circular dichroism spectra in the far-UV region at room temperature resembled random-coil characteristics, while in the near-UV region, small positive peaks were observed. Unfolding at 67.5–70°C was observed by nuclear magnetic resonance (NMR) and differential scanning calorimetry (DSC).

1.3.2.2. Lipids in Albumen

Albumen contains very low (0.03% w/w) lipid content. The main fatty acids in albumen lipids are palmitic, oleic, linoleic, arachidonic, and stearic acids (Table 1.7) (Watkins et al. 2003).

1.3.2.3. Carbohydrates in Albumen

Glucose is the main "free" sugar, constituting a ~0.8–1% by weight of albumen. It is usually removed by fermentation prior to drying of egg white to prevent browning caused by Maillard reaction. Carbohydrate is also present in the form of *N*-linked or *O*-linked oligosaccharides conjugated to albumen proteins or glycoproteins (Koketsu 1997).

1.3.2.4. Minerals in Albumen

As shown in Table 1.8, the major minerals in egg white are sulfur, potassium, sodium, and chlorine; phosphorus, calcium, and magnesium are found in lower quantities, as are various other trace minerals (Sugino et al. 1997b; Watkins 1995).

Kilic et al. (2002) determined levels of some minerals in white and yolk of eggs sampled from villages or farms in Ankara, Turkey. In general, higher levels of minerals such as Cu, Zn, Mg, Ca, and Fe were found in the white from village eggs than from farm eggs. Jacobs et al. (1993) found that 56% of total Se content was associated with ovalbumins 1 and 2 (ca 500 ng/g). Of the proteins identified, flavoprotein had the highest Se content (1800 ng/g). Selenium content of other proteins ranged from 359 to 1094 ng/g. Aydin et al. (2001) reported that eggs from CLA-fed hens had greater Fe, Ca, and Zn concentrations and lower Mg, Na, and Cl concentrations in albumen relative to those from hens fed corn oil.

1.3.2.5. Vitamins in Albumen

Albumen does not contain fat-soluble vitamins, but does contain significant proportions of the water-soluble vitamins in egg, including biotin, niacin, and riboflavin (Table 1.8) (Watkins 1995). Many of the vitamins are present in a bound form with proteins in albumen, as described in Section 1.3.2.1. However, albumen and egg in general are devoid of vitamin C.

1.3.3. Egg Yolk

Egg yolk contains ~50% solids; the major constituents of the solid matter are lipids (~65–70% dry basis) and proteins (~30% dry basis) (Table 1.2). The composition of the solid matter in vitelline membrane differs from the yolk itself; it is higher in protein (87%) and carbohydrate (10%) than lipids (3%). Poser et al. (2004) found no significant effect of chicken age or storage on protein composition of the vitelline membrane. Lin and Lee (1996) studied relationships between hen age (39, 62, or 93 weeks) and egg composition, and reported that eggs laid by 93-week-old hens were ~18% heavier than those laid by 39-week-old hens, and that the yolk weight (expressed as grams per egg) also increased with hen age.

Egg yolk can be separated into plasma (supernatant) and granule fractions (precipitate) by centrifugation. The plasma makes up ~78% of total liquid yolk; it contains ~51% solids, which consist primarily of lipid (~80%), 2% ash, and 18% nonlipid material that is mostly protein (Li-Chan et al. 1995). Granules, on the other hand, contain ~34% lipid, 60% protein, and 5% ash on a moisture-free basis (Li-Chan et al. 1995). Anton and Gandemer (1997) reported that granules of egg yolk contained about half the lipids and cholesterol and about double the proteins of yolk and plasma. Yolk and granules require a minimum ionic strength of 0.3 *M* NaCl to become solubilized at pH 7.0, whereas plasma is solubilized at any ionic strength.

The composition and content of yolk may be influenced by the diet of the laying hen. Yolk weights and egg weights increased significantly after feeding 10,000- and 20,000-ppm α-tocopheryl acetate (Engelmann et al. 2001). Novak and Scheideler (2001) reported that yolk solids content is significantly increased by Ca supplementation ($p < .03$) as well as by flaxseed supplementation ($p < .06$) compared to a corn–soy control group. Kim et al. (2006) showed that addition of activated charcoal with wood vinegar in layer diet resulted in improved yolk index and yolk color, as well as higher eggshell breaking strength and HU values. Hwangbo et al. (2006) reported that cheese byproduct beneficially improved the fatty acid composition of concern to human health in the egg yolk without adverse effects on egg quality. Triglyceride, cholesterol, and low-density lipoprotein levels in egg yolk were significantly reduced by supplementing the hen's diet with red mold rice (Wang and Pan 2003), but neither plasma lipoprotein composition nor yolk cholesterol content was affected by dietary α-tocopherol and corn oil (Shafey et al. 1999).

Barron et al. (1999) reported that yolk-directed very low-density lipoprotein (VLDLy) concentration in plasma decreased in hens undergoing ovarian regression owing to food and light deprivation, prolactin treatment, or overfeeding, but declined more rapidly in food and light-deprived hens. Lien et al. (2003) reported significant effects of cod liver oil and/or chromium picolinate on the serum traits and egg yolk fatty acids and cholesterol content in laying hens. Lien et al. (2004) found that the VLDL content was markedly reduced while HDL content was significantly increased by Cu and Cr supplementation of the diet of White Leghorn layers hens.

1.3.3.1. Proteins in Yolk

Egg yolk contains ~16% proteins, consisting of proteins in solution referred to as *livetins*, and lipoprotein particles including high-density lipoproteins (HDL), low-density lipproteins (LDL), and very low-density lipoproteins (VLDL). The total amino acid composition of yolk is presented in Table 1.6, and the major protein components are shown in Table 1.10.

In addition to their nutritional and functional properties, various biological activities have more recently been attributed to the yolk proteins (Mine and Kovacs-Nolan 2006). These include antimicrobial activity, antiadhesive properties, and antioxidant properties. Furthermore, some of the bioactive properties have been ascribed to peptides generated by partial hydrolysis of yolk proteins. For example, Park et al. (2001) purified and characterized antioxidative peptides obtained by hydrolysis of lecithin-free egg yolk by the commercial enzyme Alcalase (subtilism Carlsberg, EC3.4.21.62). Two different peptides exhibiting strong antioxidative activity were composed of 10 and 15 amino acid residues, and both contained a leucine residue at their *N*-terminal positions.

1.3.3.1.1. Low-Density Lipoprotein (LDL) or Lipovitellenins

Low-density lipoprotein (LDL) represents two-thirds of the yolk solids, and is believed to be responsible for the functional properties of yolk, particularly its emulsifying ability. The low density (0.98) of LDL is attributed to its much higher lipid content compared to protein content (Burley and Vadehra 1989; Anton et al. 2003) (Table 1.11), leading to the suggestion that it should in fact be classified as VLDL.

TABLE 1.10. Proteins and Lipids in Egg Yolk

Constituent	Major Components	Relative %
Proteins[a]	Apovitellenin I–VI	37.3
	Lipovitellin apoproteins	
	α-Lipovitellin	26.7
	β-Lipovitellin	13.3
	Livetins	
	α-Livetin (serum albumin)	2.7
	β-Livetin (α-2-glycoprotein)	4.0
	γ-Livetin (γ-globulin)	2.7
	Phosvitin	13.3
	Biotin binding protein	Trace
Lipids[b]	Triglyceride	65
	Phosphatidylcholine	26
	Phosphatidylethanolamine	3.8
	Lysophosphatidylcholine	0.6
	Cholesterol	4
	Sphingomyelin	0.6

[a]Modified from Burley and Vadehra (1989).
[b]Adapted from Juneja (1997).

TABLE 1.11. Composition of Low-Density Lipoproteins from Egg Yolk

Component	g/100g Dry Matter[a]
Proteins	12.0
Total lipids	86.7
Triglycerides	62.0 (71%)
Phospholipids	21.5 (25%)
Phosphatidylcholine	18.4 (21%)
Phosphatidylethanolamine	3.0 (3%)
Cholesterol	3.2 (4%)
Fatty acid	
Palmitic acid (16:0)	(24.7%)
Oleic acid (18:1)	(41.1%)
Linoleic acid (18:2)	(16.0%)
Saturated fatty acids	(34%)
Monounsaturated fatty acids	(45%)
Polyunsaturated fatty acids	(21%)

[a]Numbers in parentheses represent percent of total lipid or fatty acid.
Source: Adapted from Anton et al. (2003).

Low-density lipoprotein is synthesized and assembled in the liver as a modified blood VLDL, whose main apoprotein is apo B (Burley et al. 1993). Apo B enters yolk by endocytosis, and furthermore yields four of the yolk–lipoprotein apoproteins (apovitellenins III–VI) on enzymatic hydrolysis. Salvante et al. (2001) reported that avian egg production is accompanied by dramatic changes in lipid metabolism, including a marked increase in hepatic production of estrogen-induced, yolk-targeted very low-density lipoprotein (VLDLy). The structure and function of plasma VLDL particles changes from the larger generic, nonlaying VLDL particles that are involved in triglyceride transport throughout the body, to the smaller VLDLy particles that supply the yolk with energy-rich lipid.

Yolk lipoproteins show polydispersity that may partially reflect varying lipid contents or composition, but in addition, there are at least two subfractions known variously as LDL1 and LDL2 (Juneja and Kim 1997), LDF1 and LDF2, or LDP1 and LDP2 (Burley and Vadehra 1989). Six apoproteins (apovitellenins I–VI) have been reported. Apovitellenin I is poorly soluble in the absence of denaturing agents such as urea, guanidine hydrochloride, or sodium dodecylsulfate, and is characterized by a lack of histidyl residues in its amino acid composition. Apovitellenin II is very soluble in salt solutions, and contains *N*-linked polysaccharide chains consisting of glucosamine, hexose, and sialic acid. At least four fractions of apovitellenin II have been reported, with varying properties and composition (Burley and Vadehra 1989). Apovitellenins III–VI have been shown to be derived by proteolysis of apo B (Burley et al. 1993), with possible additional minor apoproteins resulting from varations in the positions of proteolysis.

Mine (1997b) examined structural and functional changes of LDL resulting from enzymatic modification of its phospholipids using phospholipase A2. The ^{31}P NMR spectrum and enzyme hydrolysis profiles revealed higher susceptibility of phospholipids in LDL complex to modification compared to phospholipids in small unilamellar vesicles or emulsions, suggesting higher membrane fluidity of LDL and that the interactions of proteins with PLs are not strong. Although the modification of phospholipids in LDL did not affect secondary structure of the apoprotein or the immunologic property of LDL, the emulsions stabilized with modified LDL showed considerable heat stability, possibly through enhancement of phospholipid–protein interactions.

Mine and Bergougnoux (1998) studied adsorption properties of cholesterol-reduced egg yolk low-density lipoprotein (CR-LDL) at oil-in-water (O/W) interfaces. The protein concentration at the interface was greater for emulsions made by CR-LDL than control LDL at pH 7.0 and 3.5, which was attributed to formation of lipoprotein aggregates resulting from cholesterol removal in LDL. Removing the cholesterol from egg yolk LDL caused changes in phospholipid–protein interactions at the interface, which could explain the instability of CR-LDL emulsion. According to Mine (1998), egg yolk LDL micelles break down when the micelles come into contact with the interface, and rearrangement of lipoproteins, cholesterol, and phospholipids take place following adsorption at an O/W interface.

Anton et al. (2003) investigated the effects of hen egg yolk LDL structure and composition on emulsification properties. The extracted LDL consisted of spherical particles with mean diameter of 20–60 nm, and contained ~87% lipid and 12% protein. The solubility of the extracted LDL was >90% under all conditions tested, including a wide range of pH (3–8) and NaCl concentration (0.15–0.55 *M*). Five major apoproteins were extracted from LDL, with molecular weights of 15, 60, 65, 80, and 130 kDa. Of these, the 15-kDa apoprotein appeared have the greatest capacity to absorb at the O/W interface in emulsions, owing its high content of side chains with amphipathic α-helices.

1.3.3.1.2. High-Density Lipoprotein (HDL) or Lipovitellins

High-density lipoprotein consists of α- and β-lipovitellins, which differ in amino acid composition as well as bound phosphorus and carbohydrate. The proportion of α- and β-lipovitellins in yolk granules appears to be genetically based. It was reported that β-HDL is resistant to heat (Anton et al. 2000). The protein content of lipovitellins is about 80%, while the lipid content is about 20%, including phospholipids (60% of the lipids, primarily as lecithin), triacylglycerols (40%), and small amounts of cholesterol, sphingomyelin, and other lipids (Burley and Vadehra 1989). Both lipovitellins are glycoconjugates with mannose, galactose, glucosamine, and sialic acid, but α-lipovitellin contains much higher sialic acid content than does β-lipovitellin, explaining its relatively acidic nature (Juneja and Kim 1997). The apoprotein form of lipovitellins, sometimes referred to as *vitelline*, is present in a dimeric form thought to be linked through hydrophobic interactions, and delipidation of lipovitellin has been reported to result in loss of solubility (Juneja and Kim 1997).

Yamamoto and Omori (1994) studied antioxidative activity of egg yolk lipoproteins and apoproteins in a linoleic acid emulsion system. High-density lipoprotein was a more effective antioxidant than LDL, and apo-HDL was more effective than apo-LDL. The lipid moiety of HDL also had an antioxidative effect on linoleic acid directly or in an emulsion, and possibly enhanced the antioxidative activity of the lipoproteins.

Kassaify et al. (2005) conducted *in vitro* experiments using confluent Caco-2 cell monolayers to investigate adhesion elimination, adhesion prevention, and antimicrobial properties of various extracted granule and plasma fractions against *Salmonella enteritidis*, *S. typhimurium*, and *Escherichia coli* O157:H7. The results revealed that the granule component HDL was the yolk fraction with protective effect against the foodborne pathogens, and this protective activity was confirmed to remain intact despite peptic and tryptic enzymatic digestion and to have antiadhesive but not antimicrobial effect.

1.3.3.1.3. Phosvitin

Phosvitin is a phosphoglycoprotein that contains about 10% phosphorus, with α- and β-phosvitin containing about 2% and 9% phosphorus, respectively. It is therefore one of the most highly phosphorylated proteins occurring in nature. Castellani et al. (2003) noted the heterogeneity of phosvitin polypeptides during chromatographic isolation. Yamamura et al. (1995) reported that chicken vitellogenin, a serum lipoprotein specific for laying hens, is thought to be proteolytically cleaved into the heavy- and light-chain lipovitellins and phosvitin, the major yolk granule proteins, during or after transportation into oocyte.

About 80% of the protein-bound phosphorus in egg yolk is located in phosvitin. Serine residues are predominant in this protein, many of which are phosphorylated and occur consecutively in the primary sequence of the molecule (Goulas et al. 1996). In addition to phosphorus, the phosvitin molecule also contains 2.5% hexose, 1.0% hexosamine, and 2.0% sialic acid. However, unlike many of the other yolk proteins, it does not contain any lipid.

The unique chemical characteristics of phosvitin conferred by its high proportion of ionizable phosphorylated serine residues are accompanied by properties such as high water solubility and resistance to heat denaturation (Anton et al. 2000) and proteolytic attack (Juneja and Kim 1997). Because of the phosphate groups, phosvitin is one of the strongest naturally occurring metal binding biomolecules. Under low ionic strength and acidic conditions, phosvitin forms soluble complexes with Ca^{2+}, Mg^{2+}, Mn^{2+}, Co^{2+}, Fe^{2+}, and Fe^{3+}. Castellani et al. (2004) investigated the Fe binding capacity of phosvitin at different pH levels (3–7) and at various ionic strengths (0.1–0.6 *M*). It was found that ionic strength had different influences on the Fe binding capacity of the protein, depending on the pH value. Highest Fe binding capacity (115 μg Fe/mg phosvitin) was observed at pH 6.5 and ionic strength of 0.15 *M*. Heating to 90°C or high pressure up to 600 MPa did not lead to a loss of Fe binding capacity. Nielsen et al. (2000) reported that addition of ascorbic acid or ascorbic acid 6-palmitate gave rise to an increase in the amount of free iron Fe(II) in egg yolk dispersions, possibly owing to reaction with the phosvitin–Fe(III) complex found in egg yolk leading to release of Fe(II), which subsequently propagated lipid oxidation.

The iron chelating activity of phosvitin has been associated with protection aginst oxidative damage (Ishikawa et al. 2004, 2005). Castellani et al. (2006) also found that phosvitin showed satisfactory emulsifying capacity under conditions favoring iron fixation. It was suggested that properties of the protein moiety in addition to the electrostatic repulsive forces of phosphate groups in phosvitin affect its emulsifying properties (Khan et al. 1998). Aluko and Mine (1997) reported that granule lipoproteins are more surface-active than phosvitin and that protein mixtures containing lipoproteins and pure phosvitin would stabilize food emulsions better at pH 7.0 and 9.0 than at pH 4.0. Aggregated phosvitin better stabilized emulsions against coalescence (Castellani et al. 2005), and a Maillard-type phosvitin–galactomannan conjugate showed antioxidant activity with improved emulsifying activity, emulsion stability, and heat stability (Nakamura et al. 1998).

Novel hen egg phosvitin phosphopeptides (PPP) with molecular masses of 1–3 kDa were prepared by Jiang and Mine (2000); PPP with 35% phosphate retention was effective for enhancing calcium binding capacity and inhibiting the formation of insoluble calcium phosphate. Jiang and Mine (2001) reported that 1–3-kDa PPP fragments derived from partially dephosphorylated phosvitin by tryptic digestion showed a higher ability than did commercial casein phosphopeptides to solubilize calcium in a calcium phosphate precipitate, while Feng and Mine (2006) reported that PPP from partially dephosphorylated phosvitin increased iron uptake in a Caco-2 cell monolayer model. Choi et al. (2004) demonstrated high Ca solubilization in the presence of phosvitin or its tryptic peptides when incubated under conditions simulating those of the ileum, while Choi et al. (2005) found that phosvitin peptides improved bioavailability of Ca and thus increased incorporation of Ca into bones of rats.

Katayama et al. (2006) reported that oligophosphopeptides from hen egg yolk phosvitin has novel antioxidative activity against oxidative stress in intestinal epithelial cells and that both phosphorus and peptide structure have key roles in the activity. The protective effects of phosvitin phosphopeptides against H_2O_2-induced oxidative stress were almost the same as that of glutathione, and PPP with a high content of P exhibited higher protective activity than did those without P, although phosphoserine per se did not show any significant antioxidative stress activity.

Choi et al. (2004) reported that phosvitin and its peptides exhibited antibacterial and DNA leakage effects against *E. coli* under thermal stress at 50°C, and suggested that phosvitin peptides disrupt the bacterial cells by chelating with metals in the outer cell membranes. Khan et al. (2000) also reported antibacterial effect of phosvitin against *E. coli* under thermal stress at 50°C; the antibacterial activity was dramatically reduced by treatment with α-chymotrypsin, although the chelating effect remained.

1.3.3.1.4. Livetin
Livetin is a water-soluble protein that accounts for 30% of the plasma proteins and is composed of α-livetin (serum albumin), β-livetin (α_2-glycoprotein), and γ-livetins [γ-globulin, immunoglobulin Y (IgY)] (Sugino et al. 1997b). The mean molecular weights of α-, β-, and γ-livetins are reported to be 80,000, 45,000, and 170,000, respectively.

Chicken serum albumin (α-livetin) has been implicated as the causative allergen of the bird egg syndrome. Chicken serum albumin is a partially heat-labile inhalant. IgE reactivity to chicken serum albumin was reduced by nearly 90% by heating for 30 min at 90°C; only partial cross-reactivity between chicken serum albumin and conalbumin was observed (Quirce et al. 2001). Akita and Nakai (1995) studied allergenicity of IgY and other egg proteins, using passive cutaneous anaphylaxis. The Fab′ fragment produced by pepsin digestion was less allergenic than the whole IgY molecule. Later, Akita et al. (1999) reported that the pFc′ fragment of IgY was found to be more antigenic than the Fab′ fragment as determined by ELISA.

The γ-livetins in yolk are transported from the blood serum of hens. Of the three immunoglobulins (IgM, IgA, and IgG) found in the serum, the laying hen transfers IgG to yolk at concentrations of ~25 mg/mL, whereas IgM and IgA are transferred to the egg white at concentrations of 0.15 and 0.7 mg/mL, respectively (Hatta et al. 1997a). Morrison et al. (2002) identified several regions within the antibody molecule important for its uptake into the egg yolk. Intact Fc and hinge regions, but not the Fc-associated carbohydrate, are required for transport. It was suggested that the C_H2/C_H3 interface is recognized by the receptor responsible for Ig transport. At this interface, residues 251–254 form an exposed loop on the surface of C_H2. In chicken IgY, the sequence is LYIS. A second site important for transport is at positions 429–432 within C_H3. All transported antibodies have the sequence HEAL.

The γ-globulins or γ-livetins in yolk are referred to as *immunoglobulin Y* (IgY) to distinguish them from mammalian IgG. Although IgY is derived from hen serum IgG, it differs in many chemical and structural features from mammalian IgG (Kovacs-Nolan and Mine 2004a). Both IgG and IgY contain Asn-linked oligosaccharides, although the composition of the oligosaccharides are different. Like IgG, yolk IgY contains two heavy (H) and two light (L) chains; however, the molecular weight of the H chains of IgY are greater than those of mammalian IgG, yielding an overall molecular weight of 180 kDa compared to 150–160 kDa for mammalian IgG. Furthermore, the IgY H chain lacks a hinge region and possesses four constant domains and one variable domain, whereas the IgG H chain contains a hinge region between the first two of three constant domains, which leads to flexibility of the Fab fragments. Sun et al. (2001) reported average molecular weights of IgY, heavy-chain, light-chain, and Fab fragments of 167,250, 65,105, 18,660, and 45,359 Da, respectively, for IgY antibodies against rabies virus. Peptic digestion degrades IgY into Fab fragments, in contrast to the disulfide linked F(ab′)2 fragments generated from IgG.

The compositional differences between IgY and IgG are accompanied by differences in their functional properties. For example, IgY has a lower isoelectric point; it does not associate with mammalian complement, protein A, protein G, or rheumatoid factors; and it binds less with human and bacterial Fc receptors (Kovacs-Nolan and Mine 2004a).

Li et al. (1998) compared egg and yolk weights, percentage hen–day production, and yolk antibody (IgY) production in Single–Comb White Leghorn (SCWL) and Rhode Island Red (RIR) hens immunized with bovine serum albumin (BSA). Similar percentages of total IgY as well as BSA-specific antibodies were produced in the eggs, but SCWL hens had higher yolk weights and percentage hen–day production. Carlander et al. (2003) investigated the genetic variation in IgY from eggs of SCWL hens, RIR hens, and a cross between the two lines. Highest IgY content was found in eggs from SCWL, although there was wide variability between individual hens, and it was suggested that yolk antibody production might be increased by using a high-producing chicken line and by genetic selection within the line.

Barua et al. (2000) determined the effects of aging and estrogen treatment on the concentration of IgY in egg yolk of chicken, *Gallus domesticus*. Immunoglobulin Y concentration was significantly greater in young laying hens than in middle-aged and old laying hens, The concentration of yolk IgY and plasma IgG was significantly increased after diethylstilbestrol (DES, an estrogenic compound) treatment as compared with pretreatment ($p < .01$). Tokarzewski (2002) investigated the influence of antibiotics such as enrofloxacin and chloramphenicol on the level of IgY in serum and egg yolk after immunostimulation of hens with *Salmonella enteritidis* antigens. Both antibiotics tested decreased the level of specific IgY in laying hens immunized with living bacteria and lipopolysaccharide, suggesting that antibiotics have a suppressive effect on the immunologic system.

Wang et al. (2000) reported that dietary polyunsaturated fatty acids significantly affected laying hen lymphocyte proliferation and IgG concentration in serum and egg yolk. The linseed oil diet increased ($p < .05$) the IgG concentration in laying hen serum, while a sunflower oil diet reduced ($p < .05$) IgY content in egg yolk, indicating that the ratio of n-6 to n-3 PUFA plays a major role in modulating cell-mediated and humoral immune responses of laying hens, and that n-3 fatty acids possess different potencies of immunomodulation.

Lee (1996) examined effects of temperature and pH on stability of anti-BSA IgY antibodies. IgY was relatively heat-stable, remaining stable even after heating to 65°C for 30 min. It remained stable over the pH 5–11 range, but the antigen binding activity was rapidly lost at pH 2–3 or lower, probably because of conformational changes. Lee (1999) reported resistance of IgY to pepsin proteolysis at pH values of 3 or higher but extensive degradation at pH 2; IgY was fairly resistant to trypsin and chymotrypsin digestion. In thermal stability tests, IgY activity against *Streptococcus mutans* serotype *c* in both yolk and crude IgY decreased with increasing temperature from 70°C to 80°C; high levels of sucrose, maltose, glycerol, or 2% glycine displayed effective protection against thermal denaturation of Ig (Chang et al. 1999). Lyophilized yolk powder with 5% gum arabic showed better stability against proteases.

Jaradat and Marquardt (2000) examined stability of IgY under various conditions within a food matrix. Trehalose was the best protectant, followed by cyclodextrin and infant formula when IgY was stored for 6 or 14 weeks at various temperatures. Sucrose, lactose, and dextran were not effective as protectants under these conditions. IgY activity was completely lost after pepsin treatment in the presence of sugars or complex carbohydrates, while 34% and 40% of activity was recovered with treatment in the presence of infant formula and egg yolk, respectively. IgY was fairly stable after trypsin treatment or after heating at 50°C, 60°C, or 70°C, while complete loss of IgY activity was observed at 80°C and 90°C in the presence of all protectants except infant formula and egg yolk, which retained 5% residual activity. Cho et al. (2005) reported protective effect of microencapsulation consisting of multiple emulsification and heat gelation processes on immunoglobulin in yolk.

Akita et al. (1998) reported that the Fab′ fragment produced by peptic digestion from IgY antibodies raised against enterotoxigenic *Escherichia coli* (ETEC) strain H10407 was as effective as the whole IgY molecule in neutralizing the ETEC labile toxin. Furthermore, low pH in conjunction with high salt content resulted in complete destruction of the neutralization activity of the whole IgY antibodies, but not of the Fab′ fragments.

Isolation and purification methods for immunoglobulins from hen egg yolk are reviewed by De Meulenaer and Huyghebaert (2001). Many methods have been described for separation and purification of immunoglobulins from yolk, including filtration systems (Kim and Nakai 1996, 1998), anionic polysaccharides including pectin (Chang et al. 2000b), low-temperature ethanol precipitation followed by chromatography (Xiaohong et al. 2003), lithium sulfate precipitation (Bizhanov et al. 2004), and caprylic acid precipitation (Raj et al. 2004). Cheung et al. (2003) described purification of antibody Fab and F(ab′)2 fragments using Gradiflow® technology (Gradipore Ltd, Sydney, Australia), while various affinity chromatographic methods have been used for isolation of specific antibodies (Mine 1997a; Akita and Li-Chan 1998; Li-Chan et al. 1998; Kim and Li-Chan 1998; Kim et al. 1999; Verdoliva et al. 2000; Tu et al. 2001; Chen et al. 2002).

More recently, there have been many studies on the use of IgY for oral passive immunization against various bacteria and viruses (Kovacs-Nolan and Mine 2004a, 2004b). Oral administration of IgY has proved to be effective for the treatment of a variety of gastrointestinal pathogens such as bovine and human rotaviruses (Sarker et al. 2001), bovine coronavirus, *Yersinia ruckeri*, enterotoxigenic *Escherichia coli* (Akita et al. 1998; Amaral et al. 2002; Shin and Kim 2002; Sunwoo et al. 2002), and *Aeromonas hydrophila* (Li et al. 2005).

Helicobacter pylori is known to be a major pathogenic factor in the development of gastritis, peptic ulcer disease, and gastric cancer, and many investigations have been conducted on yolk-derived immunoglobulin Y antibodies against *H. pylori* or *H. pylori* urease (Shin et al. 2002; Shimamoto et al. 2002; Nomura et al. 2004, 2005; Horie et al. 2004; Suzuki et al. 2004).

Other studies have investigated passive immunization against dental plaque formation in humans by using a mouth rinse containing IgY-specific antibodies to *Streptococcus mutans* (Hatta et al. 1997b; Smith et al. 2001), and prevention of oral candidiasis in immunocompromised children using anti–*C. albicans* IgY (Wilhelmson et al. 2005). Potential applications in aquaculture and animal husbandry include passive protection of rainbow trout against *Yersinia ruckeri* (Lee et al. 2000) and *Vibrio anguillarum* (Arasteh et al. 2004), shrimp against white spot syndrome virus (WSSV) disease (Kim et al. 2004b), and poultry against *Salmonella enteritidis* or *S. gallinarum* (Gurtler et al. 2004; Rho et al. 2005).

Specific yolk antibodies have been applied in immunochemical assays for quantitation of IgG in milk (Li-Chan and Kummer 1997), and detection of peanut proteins (De Meulenaer et al. 2005), gamma chain of human hemoglobin (Jintaridth et al. 2006), herbicide metsulfuron methyl in water (Welzig et al. 2000), alkaline phosphatase (Chen et al. 2006), lactoperoxidase (Lee et al. 2004), bisphenol A (De Meulenaer et al. 2002), flumequine residues in raw milk (Coillie et al. 2004), *Salmonella enteritidis* and *S. typhimurium* (Lee 2003; Choi 2004), and *E. coli* (de Almeida et al. 2003).

1.3.3.1.5. Enzymes

Many enzymes have been reported in yolk, particularly in the plasma, most notably cholinesterase (EC 3.1.1.8) (Burley and Vadehra 1989). Other enzymes that have been reported in yolk include acid phosphatases, acid proteases, adenosine deaminase, alkaline phosphatase, amylase, cathepsin D, α-mannosidase, peptidase, purine N_1–C_6 hydrolase, pyruvate kinase, robinoclease, and tributylase (Burley and Vadehra 1989). Midorikawa et al. (1998) reported that aminopeptidase Ey (EC 3.4.11.20) from chicken (*Gallus gallus domesticus*) egg yolk is a homodimeric 300-kDa metalloexopeptidase containing 1.0 g atom of zinc per subunit. It has a broad specificity for *N*-terminal amino acid residues at P1 position of the substrate. Analysis of the 3196-bp (base pair) nucleotide sequence of the cDNA revealed a single open-reading frame coding for 967 amino acid residues. The predicted amino acid sequence of the enzyme showed 66%, 65%, 64%, and 63% identity to aminopeptidases N (EC 3.4.11.2) from human, pig, rabbit, and rat, respectively. Aminopeptidase Ey contained the metallobinding sequence motif, His–Glu–Xaa–Xaa–His, found in zinc metallopeptidases. Zinc-binding sites, His386, His390, and Glu409, and a catalytic site, Glu387, were conserved in the homologous aminopeptidases N.

1.3.3.1.6. Other Proteins or Peptides in Yolk

Sialylglycopeptides can be considered as one of the major components in hen egg yolk since their content in egg yolk is comparable to those of major yolk proteins such as low-density lipoprotein, lipovitellins, and phosvitin. Seko et al. (1997) investigated the sialylglycopeptide components of hen egg yolk, localized in the yolk plasma. The amino acid sequence of the peptide moiety of sialyloglycopeptide was determined to be Lys–Val–Ala–Asn–Lys–Thr, with an *N*-linked disialylbiantennary glycan at the Asn residue. Structural information on glycopeptides with neutral and sialylated *N*-glycans has been studied using positive- and negative-ion MSn spectra (Deguchi et al. 2006).

1.3.3.2. Lipids in Egg Yolk

Lipids are the main components (32–36%) of the egg yolk solids. The composition of yolk lipid is about 65% triglyceride, 28–30% phospholipid, and 4–5% cholesterol. The fatty acid composition of lipid in yolk plasma and granules is shown in Table 1.7. The composition of yolk lipids can be affected by various factors including hen age and genotype and changes in the diet of the hens.

Polyunsaturated fatty acid contents were significantly higher for eggs laid by 39-week-old hens compared with older hens, while monounsaturated fatty acid contents were significantly higher for eggs laid by 93-week-old hens (Lin and Lee 1996). The contents of long-chain (20 and 22) omega-6 and omega-3 polyunsaturated fatty acids (PUFA) were 20% and 25%, respectively, higher in egg yolks from 21-week-old hens than 57-week-old hens (Nielsen 1998). Shafey (1996) reported that egg size did not significantly affect yolk lipid or fatty acid concentration. However, lipid levels were lower while linoleic acid level was higher in yolks of eggs from hens older than 47 weeks of age, than in those produced by younger birds. The unsaturated : saturated fatty acid ratio for yolk produced at 39 and 47 weeks of age was greater than that for yolk produced at 31 weeks. However, the monounsaturated : polyunsaturated fatty acid ratio for yolk produced at 27 and 39 weeks of age was lower than that for yolk produced at 51 weeks.

Dziadek et al. (2003a, 2003b) studied the influence of the hen genotype on chemical composition of table eggs, using nine commercial lines of laying hens [Lohmann Brown, Shaver 579, AK (experimental from IZ-OBD Zakrzewo), ISA White, Messa 445, Messa 443, Astra W-1, Astra W-2, and Astra N]. Yolk lipid content ranged from 29.37% in the eggs from AK layers to 31.85% in the eggs from Astra N birds. Triglyceride content in the egg yolk varied from 199.70 mg/g in the Messa 443 group to 236.55 mg/g in the Astra N group of birds. Scheideler et al. (1998) studied the effect of strain and age on egg composition from hens fed diets rich in omega-3 fatty acids, and reported that the percentage of C18 : 0 and C18 : 1 fatty acids in the yolk was significantly affected by strain, diet, and strain–diet interaction. Latour et al. (1998) suggested that breeder age influences the utilization of yolk lipid by developing embryos, and that the type of fat (corn oil, poultry fat, or lard) provided in the diet may have an additional influence.

Sim (1998) described the development of "designer eggs" rich in omega-3 PUFA such as α-linolenic acid, eicosapentaenoic acid (EPA), and docosahexaenoic acid (DHA), which have been associated with beneficial effects for human health, including reduction of triglyceride level, blood pressure, platelet aggregation, and tumor growth. Many studies have therefore investigated the incorporation of different feed ingredients such as fish oil; vegetable oils, including flaxseed (linseed), soy oil, or canola oil; and microalgae into the diet of hens, in order to optimize the omega-3/omega-6 and PUFA/saturated fatty acid ratios of eggs for human health [e.g., see Scheideler and Froning (1996), An and Kang (1999), Han et al. (1999), Santoso et al. (1999), Yannakopoulos et al. (1999), Daenicke et al. (2000), Herstad et al. (2000), Lewis et al. (2000), Shimizu et al. (2001), Sari et al. (2002), Basmacioglu et al. (2003), Milinsk et al. (2003), Cheng et al. (2004), Zotte et al. (2005), Fredriksson et al. (2006)]. Incorporation of omega-3 PUFA was reported to occur mainly in the *sn*-2 position, particularly of phospholipids (Cossignani et al. 1994), and EPA and DHA contained in eggs were observed to be stable (Oku et al. 1996).

Products enriched with PUFAs are prone to oxidation, and enrichment with antioxidants is necessary in order to prevent the risk of oxidative damage. Grune et al. (2001) suggested supplementation of feed with least 80 IU vitamin E/kg to prevent increase in cytotoxic aldehydic lipid peroxidation during production and storage of omega-3 PUFA-enriched eggs. Gebert et al. (1998) reported that dietary vitamin E resulted in a decrease of PUFA, SFA, and total lipids in fresh yolk lipids, whereas MUFA did not change. Boruta and Niemiec (2005) reported that dietary vitamin E supplement slowed down the process of oxidation of egg yolk fatty acid during storage.

Several studies have also investigated the effect of dietary conjugated linoleic acid (CLA) on the composition of egg yolk lipids. Du et al. (1999) reported that the levels of CLA incorporated into lipid of egg yolk were proportional to levels of CLA in the diet, although more CLA was incorporated in the triglycerol than were phospholipid components, and the incorporation rates of different CLA isomers in different classes of lipids also were significantly different. Furthermore, inclusion of CLA in the diet influenced the metabolism of polyunsaturated fatty acids. Du et al. (2000) reported that the amount of arachidonic acid was decreased by CLA added to linoleic acid– and linolenic acid–rich diets, but EPA and DHA were increased in the linolenic-rich diet, indicating that synthesis or deposition of long-chain *n*-3 fatty acids was accelerated after CLA feeding. However, increases in saturated fatty acids in yolk and decreases in MUFA and PUFA by dietary CLA have also been reported (Schaefer et al. 2001; Hur et al. 2003; Watkins et al. 2003). Szymczyk and Pisulewski (2003) reported that feeding CLA-enriched diets resulted in gradually increasing deposition of CLA isomers ($p < .01$) in egg yolk lipids, while Watkins et al. (2003) suggested that feeding CLAs to hens led to accumulation of isomers in polar and neutral lipids of the egg yolk that migrated into egg albumen.

Aydin et al. (2001) reported that olive oil prevented CLA-induced increases in 16:0 and 18:0 and decrease in 18:1(ω-9) in yolk, and also prevented CLA-induced abnormal changes in the pH of albumen and yolks. Hur et al. (2003) indicated that lipid oxidation of egg yolk during cold storage could be inhibited by dietary CLA due not only to changes in fatty acid composition but also to the high concentration of CLA in egg yolk. Hwangbo et al. (2006) showed that as dietary levels of cheesemaking byproduct increased, linear increases in total CLA and *cis*-9,*trans*-11 CLA contents of egg yolk took place, together with decreases in total saturated fatty acid content. Szymczyk and Pisulewski (2005) reported that dietary vitamin E increased the rate of laying and egg production per hen and may also exert alleviating effects on fatty acid composition of CLA-enriched eggs.

Many other treatments have been reported to affect egg yolk composition. A significant linear reduction was found in plasma and yolk triglycerides (24% and 30%) as the dietary copper content was increased from 0 to 250 mg/kg (Al-Ankari et al. 1998). Reduction of egg cholesterol was obtained by inclusion of red fungus rice containing monacolin K (Wang and Pan 2003). Biswas et al. (2000) reported that levels of egg yolk cholesterol and lipid were significantly reduced by adding Japanese green tea powder in the feed. Li and Ryu (2001) reported that 0.1% wood vinegar tends to improve egg production and significantly increase PUFA (C20:4, C22:6) content in egg yolk. Kim et al. (2001b) found that supplementation of feed with 1.0% Bio-alpha® (a fermented feed containing a range of microorganisms) increased DHA content and reduced cholesterol content of egg yolks significantly compared with controls. Microwave cooking substantially reduced levels of PUFA compared to boiling or frying (Murcia et al. 1999).

Liu et al. (2005b) investigated the composition and quality of lipids in various commercial egg products, including fresh egg, Pidan (preserved egg), tea egg, simmered egg, iron egg (deeply simmered and dried egg), and salted egg. The lipid content ranged from 27% for Pidan to 46% for iron egg. Phospholipid and cholesterol contents were highest in fresh egg (~351 and 38 mg/g oil, respectively) and lowest in Pidan (~175 and 28 mg/g oil, respectively). On the other hand, fatty acid compositions of all yolk lipids were consistent, with oleic, palmitic and linoleic acids as the three most abundant fatty acids. Acid and thiobarbituric acid values were generally higher for lipids in processed eggs than those in fresh eggs.

1.3.3.2.1. Triglycerides

Yolk lipid contains about 65% triacylglycerols or triglycerides (TG). The saturated palmitic acids and stearic acids constitute 30–38% of the fatty acids in yolk lipid, while monounsaturated (primarily oleic acid) and polyunsaturated (including linoleic and arachidonic acid) fatty acids each represent another one-third of the fatty acids (Table 1.7). Position 1 of the yolk TG is occupied predominantly by saturated palmitic acid, while position 2 contains the unsaturated oleic and linoleic acids. Position 3 includes both saturated (palmitic and stearic) and unsaturated (oleic) fatty acids (Juneja 1997).

An and Kang (1999) investigated effects of dietary fat sources with omega-3 or omega-6 PUFA on lipid metabolism and fatty acid composition of egg yolk in laying hens, and reported no effect on the lipid fraction contents in the egg yolk. Du et al. (1999) indicated that the amount of CLA incorporated into total lipid, TG, phosphatidyl choline (PC), and phosphatidylethanolamine (PE) of egg yolk was proportional to the levels of CLA in the diet; more CLA was incorporated in TG than in PC and PE. The incorporation rates of different CLA isomers into different classes of lipids also were significantly different.

Lee et al. (2002b) investigated the influence of dietary tung oil, containing a high level of α-eleostearic acid (*cis*-9,*trans*-11,*trans*-13-octadecatrienoic acid, EA) on growth, egg production, and lipid and fatty acid compositions in tissues and egg yolks of laying hens in White Leghorn hens. α-EA was not deposited in the tissues and egg yolk of hens fed tung oil, but conjugated linoleic acid (CLA) was detected in all tissues and egg yolks. These results suggested that dietary tung oil affected the lipid metabolism of laying hens and could modify the lipid and fatty acid composition in tissues and eggs.

Ginzberg et al. (2000) reported that in chickens fed with biomass of the red microalga *Porphyridium* sp, linoleic acid and arachidonic acid levels increased by 29% and 24%, respectively. In addition to PUFA such as arachidonic acid and EPA, about 70% of the algal dry weight is composed of a unique combination of soluble sulfated polysaccharide. Wang and Pan (2003) indicated that inclusion of red fungus rice in chicken feed reduced egg triglyceride level and LDL concentration.

Brady et al. (2003) reported that hen egg yolk contained significant antibacterial activity, which was associated with the release of free fatty acids. Chloroform–methanol extraction on egg yolk demonstrated the activity to be lipoprotein-bound before enzymatic digestion and associated with the lipid-soluble chloroform phase afterward. Acetone extraction yielded a fraction containing 97% TG, which on treatment with pancreatin showed high antibacterial activity against *Streptococcus mutans*. Both oleic and linoleic acid were found to inhibit growth of *S. mutans*.

1.3.3.2.2. Phospholipids

The major components of egg yolk phospholipids (PL) are phosphatidylcholine (PC) and phosphatidylethanolamine (PE), which make up ~81% and 12% of egg yolk lecithin; lysophosphatidylcholine (LPC), lysophosphatidylethanolamine (LPE), and sphingomyelins are also components of yolk PL. Polyunsaturated fatty acids are especially concentrated in the *sn*-2 position, while saturated fatty acids are found in the *sn*-1 position of yolk phospholipids (Juneja 1997). The major fatty acids in egg PC are palmitic, oleic, stearic, and linoleic acids, representing 32%, 26%, 16%, and 13%, respectively; arachidonic and docosahexaenoic acids (4.8% and 4%, respectively) are also present in significant amounts (Juneja 1997).

Yolk phospholipid contents, expressed in relation to weight of egg oil or whole egg, was reported to be positively related ($p < .05$) to hen age (Lin and

Lee 1996). Eggs from hens receiving low-dose chitosan treatment contained 1.8-fold increase in yolk phospholipids level (Vrzhesinskaya et al. 2005).

Kivini et al. (2004) studied the influence of oil-supplemented feeds (containing 15% vegetable-based or fish oils) on the concentration of phospholipids and their composition in hen eggs. Although the total phospholipid contents and proportions of PC, PE, and sphingomyelin were similar for all feeding groups; supplemented feeds had a statistically significant ($p < .05$) effect on the fatty acid composition of phosphatidylcholines. Supplements decreased the proportion of saturated fatty acids in total fat, but not in the phospholipids.

Shimizu et al. (2001) investigated effects of feeding dietary fish oil to hens on the fatty acid composition of eggs. Variation in fatty acid composition in egg yolks was found in the acyl groups of PC and PE, rather than in TG. Results showed that supplementing the diets of hens with fish oil altered essential fatty acid composition, in particular by increasing DHA and decreasing arachidonic acid in egg yolk phospholipids.

Nakane et al. (2001) reported growth factor-like lipids in hen egg yolk and white, which were associated with high amounts of lysophosphatidic acid (acyl LPA) and small amounts of lysoplasmanic acid (alkyl LPA). The levels of acyl LPA in hen egg yolk (44.23 nmol/g tissue) and white (8.81 nmol/g tissue) were on the same order as or higher than the levels of acyl LPA required to elicit biological responses in various animal tissues. Egg yolk acyl LPA contained predominantly saturated fatty acids as the acyl moiety, whereas egg white acyl LPA contained primarily PUFA.

Many studies have been conducted on methods for extraction and separation of phospholipids or lecithins from egg [e.g., see Kim et al. (1995), Nielsen (2001), Yoon and Kim (2002), Palacios and Wang (2005a, 2005b)], as well as preparation of lysolecithin by the enzymatic action of phospholipase A2 (PLA2), including immobilized PLA2 (Kim et al. 2001a). In addition to providing sources of purified phospholipids for basic research, these methods have been established to meet the demand to produce purified egg lecithin for pharmaceutical, nutraceutical, and food applications. Examples of properties of yolk phospholipids with potential industrial applications as nutraceuticals and functional food ingredients include antioxidative activity (Sugino et al. 1997a) and inhibition of cholesterol absorption (Jiang et al. 2001).

1.3.3.2.3. Sphingomyelins

Sphingomyelins are present as a minor component in egg yolk lipid, constituting only ~2% of yolk phospholipid. The major component of egg yolk sphingomyelins are palmitosylsphingosines, *N*-acetyl-*O*-trimethylsilyl derivatives of long-chain base residues of natural sphingomyelin (Olsson et al. 1997).

Sphingomyelins are components of animal cell membranes, and their interactions with other membrane constituents including phospholipids and cholesterol are of considerable interest. Lindblom et al. (2006) compared the translational dynamics for bilayers with various mixtures of 1,2-

dioleoyl-*sn*-glycero-3-phosphocholine (DOPC), 1,2-dipalmitoyl-*sn*-glycero-3-phosphocholine (DPPC), and chicken egg yolk sphingomyelin, with or without cholesterol. Compared to DOPC, which has a preference for location in a disordered phase, DPPC and egg yolk sphingomyelin prefer the ordered phase. Veiga et al. (2001) studied interaction of cholesterol with sphingomyelin in mixed membranes containing PC; the results showed possible stabilization of gel-phase sphingolipid domains by cholesterol.

Sphingomyelin hydrolysis by sphingomyelinase is essential in regulating membrane levels of ceramide, a well-known metabolic signal. Since natural sphingomyelins have a gel-to-fluid transition temperature in the range of the physiological temperatures of mammals and birds, Ruiz-Arguello et al. (2002) treated pure egg sphingomyelin (gel-to-fluid crystalline transition temperature ~39°C) with sphingomyelinase in the temperature range 10–70°C. Sphingomyelinase was active on pure sphingomyelin bilayers, leading to concomitant lipid hydrolysis, vesicle aggregation, and leakage of aqueous liposomal contents.

Eckhardt et al. (2002) found that dietary sphingomyelin suppressed intestinal cholesterol absorption by decreasing thermodynamic activity of cholesterol monomers. Scarlata et al. (1996) reported that egg sphingomyelin inhibited phosphatidylinositol 4,5-bisphosphate hydrolysis catalyzed by human phospholipase C-δ-1 in model membranes and detergent phospholipid mixed micelles. Moschetta et al. (2000) reported greatly enhanced protective effects of sphingomyelin and DPPC compared to egg yolk PC against bile salt–induced cytotoxicity, which may be relevant for protection against bile salt–induced cytotoxicity *in vivo*.

1.3.3.2.4. Cholesterol

Choleterol content is about 1.6% in raw egg yolk and about 5.0% in egg yolk lipids. One yolk on the average contains about 226 mg of cholesterol, and the cholesterol content of the yolk is influenced by the genotype of the birds (Li-Chan et al. 1995). It was reported that yolk cholesterol content does not differ significantly with hen age (Lin and Lee 1996). Cholesterol content of the PC enriched fraction from nondeoiled yolk was found to be much higher than that from deoiled yolk (Palacios and Wang 2005a).

Cholesterol synthesis by the laying hen and its concentration in the egg may be influenced by diet or drugs, and considerable attempts have been made to reduce egg cholesterol level. There have been reports about the reduced level of cholesterol in egg yolk by including feed supplements such as biomass of the red microalga *Porphyridium* sp. (Ginzberg et al. 2000), dietary lipid-lowering drugs (lovastatin) (Mori et al. 1999), copper (Al-Ankari et al. 1998), copper and chromium (Lien et al. 2004), Japanese green tea powder (Biswas et al. 2000), green tea powder (Uuganbayar et al. 2005), garlic powder or copper alone and in combination (Lim et al. 2006), garlic powder (Chowdhury et al. 2002; Mottaghitalab and Taraz 2004; Yalcin et al. 2006), forage legume (Rahimi 2005), probiotics (Kurtoglu et al. 2004), cod liver oil and chromium picolinate (Lien et al. 2003), conjugated linoleic acid (Hur et al. 2003), red fungus rice

(Wang and Pan 2003), sardine oil (Santoso et al. 1999), flaxseed (Sari et al. 2002), L-carnitine and yeast chromium (Du et al. 2005), pravastatin (Kim et al. 2004a), β-cyclodextrin (Park et al. 2005a), *Portulaca oleracea* (Zotte et al. 2005), chitosan (Vrzhesinskaya et al. 2005), fish oil and flaxseed (Basmacioglu et al. 2003), and oligofructose and inulin (Chen et al. 2005). According to Szymczyk and Pisulewski (2003), the cholesterol content of egg yolks, when expressed in mg/g yolk, was not affected by the dietary conjugated linoleic acid concentrations.

Several studies have reported reduction of cholesterol through the use of microbes. Chang et al. (2000a) observed 23% reduction of cholesterol by treatment of egg yolk with *Rhodococcus equi* NCHU1 at 40°C, pH 7.5 for 3 days. Chengtao et al. (2002) described cholesterol-degrading lactic acid bacteria that were able to degrade cholesterol and had the same fermentation characteristics as lactic acid bacteria in sterilized egg yolk. Valcarce et al. (2002) investigated potential use of the nonpathogenic ciliate *Tetrahymena thermophila* for the bioconversion of cholesterol into pro–vitamin D sterols in egg suspensions. By optimizing the conditions for efficient conversion, 55% reduction in cholesterol was obtained after 24 h incubation; the cholesterol was converted into three unsaturated sterols: D_7-22-bisdehydrocholesterol, an ergosterol analog (27%); D_7-dehydrocholesterol, also known as *pro–vitamin* D_3 (4%); and δ-22-dehydrocholesterol (8%). Lv et al. (2002) described bioconversion of yolk cholesterol to cholest-4-en-3-one by extracellular cholesterol oxidase from *Brevibacterium* ODG-007 in 50 m*M* phosphate buffer.

Other studies have reported treatments for the physical removal of cholesterol, including nanofiltration (Allegre et al. 2006), β-cyclodextrin in chitosan beads (Chiu et al. 2004), and crosslinked β-cyclodextrin (Jung et al. 2005). The composition and functional properties of cholesterol-reduced (CR) egg yolk prepared by β-cyclodextrin treatment has been subject of several studies. Egg products made from CR egg yolk could have 80% lower cholesterol contents as compared to those made from regular egg yolk (Chiang and Yang 2001). Awad et al. (1997) reported lower amounts of lipid, protein, and linoleic acid and higher amounts of carbohydrate, ash, and oleic acid after cholesterol reduction. Cholesterol reduced egg yolk meal had higher emulsification capacity, foaming capacity, yellow color, and brightness, but a lower red color index than do controls (Cao and Peng 1997). Mine and Bergougnoux (1998) studied adsorption properties of cholesterol-reduced egg yolk low-density lipoprotein (CR-LDL) at oil-in-water (O/W) interfaces. Removing the cholesterol from egg yolk LDL caused changes in phospholipid–protein interactions at the interface, attributed to formation of lipoprotein aggregates; the concentration of protein at the interface was greater for emulsions made by CR-LDL than control LDL, and emulsion particles were larger.

Cholesterol oxidation in foods yields oxides that show cytotoxic, atherogenic, mutagenic, and carcinogenic properties. Tenuta-Filho et al. (2003) analyzed levels of cholesterol and their oxides 7α-hydroxycholesterol (7α-OH), 7β-hydroxycholesterol (7β-OH), 7-ketocholesterol (7-keto), and

25-hydroxycholesterol (25-OH) in various food products, and found levels varying from 22.81 ± 9.00 mg/100 g in whole-milk powder to 1843.62 ± 92.69 mg/100 g in egg yolk powder. Obara et al. (2006) investigated the effect of water activity on cholesterol oxidation in spray- and freeze-dried egg powders. Water activity was higher in egg yolk powders than in whole-egg powders. During storage, the highest oxysterol accumulation appeared in egg powders with the lowest water activity; oxysterol accumulation was higher in whole-egg powders than in egg yolk powders, and also higher in spray-dried powders than in freeze-dried ones. Five oxysterols were identified and quantified, which were found in the following order in spray-dried whole-egg powders: 5α,6α-epoxycholesterol > 7α-hydroxycholesterol > 5β,6β-epoxycholesterol > 7-ketocholesterol > 7β-hydroxycholesterol. During storage of the powders, 5,6-epoxycholesterols were produced in the highest amounts.

1.3.3.2.5. Glycolipids

Cerebrosides are classified as glycolipids. Momma et al. (1972) isolated cerebroside from egg yolk lipid by silicic acid column chromatography and examined its fatty acid composition, detecting 15 components. The main hydroxy fatty acids were hydroxylignoceric (22.0%), hydroxybehenic (19.3%), and hydroxytricosanoic (12.2%) acids; other fatty acids present in appreciable amounts were lignoceric (15.7%), palmitic (9.6%), and behenic (7.2%) acids. The bases sphingosine and dihydrosphingosine were detected. Glucose and galactose, in the ratio 1:1, formed the sugar component. Cerebroside fraction A and cerebroside fraction B represented 1.4% and 1.6% of the total polar lipid fraction, respectively (Kilikidis 1978).

Galactosylceramide is the only neural glycosphingolipid found in chicken egg yolk. Keenan and Berridge (1973) identified gangliosides as constituents of the lipid fraction from egg yolk. Approximately 0.07 μmol of ganglioside sialic acid was recovered per gram wet weight of yolk. Separation by thin-layer chromatography (TLC) revealed the presence of about eight separate ganglioside species in this fraction. Glucose, galactose, hexosamine, and sialic acid, in the molar proportions of 1.00:1.25 to 1.95:1.19, were the major carbohydrates of yolk ganglioside fractions. Of the 14 fatty acids were observed in the ganglioside fraction, the predominant ones were 16:0 (26.6%), 18:0 (9.2%), 18:1 (22.9%), and 22:0 (28.3%).

1.3.3.3. Carbohydrates in Egg Yolk

The content of carbohydrate in egg yolk is about 0.7–1.0%, including free carbohydrate estimated to be 0.3% as glucose, and the remainder in bound form as glycoproteins and glycolipids. Sialic acid (*N*-acetylneuraminic acid) is a common component of these glycoconjugates (Koketsu 1997), especially in the yolk membrane, which contains 13.7 ± 0.5 μg sialic acid/mg, compared to 5.2 ± 0.4 μg/mg in the yolk itself (Nakano et al. 1994).

Seko et al. (1997) determined the chemical structures of free sialylglycans (FSGs) and a sialylglycopeptide (SGP) in yolk plasma. The glycan moiety of SGP was found to be an *N*-linked disialylbiantennary glycan. FSGs were determined to be two free disialyl-biantennary glycans whose reducing end is either gt Man-β-1-4GlcNAc (FSG-I) or gt Man-β-4GlcNAc-β-1-4GlcNAc (FSG-II).

Nakano et al. (1996) extracted a polysaccharide fraction from dry delipidated chicken egg yolk by digestion with papain. It contained approximately 20% carbohydrates consisting of galactose, mannose, and glucosamine, and 39% protein with serine as the most abundant amino acid. No uronic acid was detected in the dimethylmethylene blue–reactive fraction.

1.3.3.4. Minerals in Egg Yolk

The content of minerals in yolk is about 1%. The major mineral in yolk is phosphorus, 61% of which is contained in phospholipid (Sugino et al. 1997b). Other minerals in yolk include Ca, Cl, K, Na, S, Mg, and Mn (Table 1.8), as well as other minerals in trace levels.

The mineral content of yolk may be altered by dietary treatment. Addition of Zn, Mn, and Cu in combination to the hen diet increased their concentrations in egg yolk and slightly decreased egg weight from older hens (Mabe et al. 2003). Selenium contents of egg increased with age of hens and also with the level of dietary Se supplements in the diets of broiler breeders (Pappas et al. 2005); at lower dietary levels, Se was preferentially deposited in yolk, whereas in high Se treatments, the Se was deposited evenly in the yolk and egg white. Aydin et al. (2001) reported that olive oil prevented CLA-induced mineral exchange between yolk and albumen, presumably by reducing the yolk saturated fatty acids, which are believed to disrupt the vitelline membrane during cold storage. The iodine concentrations in egg yolk, egg albumen, and whole egg increased with increased iodine supplementation at 3 or 6 mg/kg diet (Kahraman et al. 2004; Yalcin et al. 2004). Han et al. (1999) reported significantly higher selenium and iodine contents of mineral-enriched brand eggs (9.17 ± 1.57 and 70.52 ± 29.66 μg/g yolk, respectively), compared to regular eggs (5.73 and 6.47 μg/g yolk, respectively).

1.3.3.5. Vitamins in Egg Yolk

With the exception of niacin and riboflavin, greater amounts of vitamins are found in yolk than in albumen (Table 1.8). One egg provides ~12%, 6%, 9% and 8% of the recommended daily allowance for vitamin A, vitamin D, riboflavin, and pantothenic acid, respectively (Gutierrez et al. 1997).

Han et al. (1999) analyzed brand eggs and reported that vitamin E contents of vitamin-enriched brand eggs ranged from 102.51 to 162.39 μg/g yolk for the first 6 months and ranged from 34.27 to 151.14 μg/g yolk for the second 6 months. Vitamin A contents of vitamin-enriched brand eggs ranged from 11.98 to 23.21 μg/g yolk, which were higher than those of regular eggs (17.21 μg/g yolk).

Ternes et al. (1995) studied distribution of vitamin A, tocochromanols, vitamin D_2, and vitamin D_3 in the granule and plasma fractions of egg yolk. Vitamin A was present at higher concentration in the granules than in the plasma. High dietary tocopherol levels resulted in high α-tocopherol concentration in the yolk, with higher concentration in the plasma LDL fraction than in the granule HDL fraction. Vitamins D_2 and D_3 were present in equilibrium between plasma and granula fractions. Surai et al. (1998) found that excessive provision of vitamin A to the laying hen resulted in an adverse effect on vitamin E, carotenoids, and ascorbic acid in the embryonic and neonatal liver and can compromise the antioxidant status of the progeny. Vrzhesinskaya et al. (2005) reported that feeding of chitosan at 10 and 20 mg/kg body mass to 19-week-old laying hens over 1.5 months caused a decrease in whole-egg content of vitamins A and E but did not affect the vitamin B_2 level.

Increasing the vitamin E levels in the diet can lead to increased α-tocopherol or vitamin E concentrations in egg yolk (Barreto et al. 1999; Meluzzi et al. 2000; Galobart et al. 2001a; Mazalli et al. 2004), although transfer efficiency decreases with increasing α-tocopherol content in the diet (Galobart et al. 2001a; Flachowsky et al. 2002). Supplementation of poultry feed with 20,000 mg vitamin E/kg feed resulted in ≤254.9 mg α-tocopherol/100 g yolk, or 51.0 mg per egg (assuming 60 g/egg), compared to 7–10 mg/100 g yolk or 1.0–1.5 mg/egg when hens were fed the basal diet containing 20 mg vitamin E/kg feed (Flachowsky et al. 2002). Drotleff and Ternes (1999) reported that all four α-tocotrienol geometric isomers could be determined in egg yolk when a commercially available, synthetic α-tocotrienol was orally administered to a laying hen.

Murcia et al. (1999) reported that α-tocopherol (2.9–6.1 mg/100 g) was the predominant tocopherol isomer followed by (β + γ)-tocopherol and δ-tocopherol, while α-tocotrienol was detected in trace amounts. All of these tocopherols were reduced during cooking by up to 50% in omelettes and microwave treatment. The vitamin E content of yolk decreased after 90 days of storage at 4°C or after 28 days at 25°C (Franchini et al. 2002). Spray-drying also significantly decreased α-tocopherol content of eggs and increased lipid oxidation, but dietary supplementation with different levels of α-tocopheryl acetate significantly reduced lipid oxidation in spray-dried eggs (Galobart et al. 2001a), and a clear antioxidant effect of dietary α-tocopheryl acetate supplementation was observed in omega-3 fatty acid–enriched eggs (Galobart et al. 2001b). Koreleski et al. (2003) reported that the increases in yolk α-tocopherol content observed by dietary vitamin E supplementation were accompanied by improved fat stability during storage for 15 and 42 days. Similar stabilizing properties were observed after supplementing the feed with synthetic antioxidants (BHT, BHA, and EQ) and to a lesser degree when vitamin C was added.

Kirunda et al. (2001) investigated supplementation of hen diets with vitamin E as a means to alleviate egg quality deterioration associated with high-temperature exposure (HTE). Supplementation of HTE hen diets with 60 IU vitamin E/kg feed improved some parameters of egg production and quality, but vitamin E levels in the yolk were lower at all levels of vitamin E supplementation from HTE hens compared with controls.

Puthpongsiriporn et al. (2001) reported that vitamin E supplementation at 65 IU/kg diet may enhance production, induction of *in vitro* lymphocyte proliferation by ConA (concanavalin A) and lipopolysaccharide (LPS), and increase antioxidant properties of egg yolks and plasma of White Leghorn hens during heat stress. It was suggested that supplementation of 1,000 ppm vitamin C may further enhance *in vitro* lymphocyte proliferative responses of hens during heat stress. Meluzzi et al. (2000) reported that the different levels of dietary vitamin E slightly affected the fatty acid composition of the yolk, while Gebert et al. (1998) reported that vitamin E supplementation decreased total lipids, polyunsaturated fatty acids, and saturated fatty acids in the yolk lipids. On the other hand, Pal et al. (2002) reported that vitamin E content of the feed did not affect fatty acid composition of the egg yolk and dietary PUFA concentration did not significantly affect deposition of vitamins A and E in the yolk. Supplementation with 30 mg α-tocopheryl acetate/kg increased yolk vitamin E concentration for hens on cod liver oil and pumpkin seed oil diets. Vitamin A concentration in yolks was unaffected by feed oil type, but was increased by α-tocopheryl acetate supplementation. Oxidative stability of egg yolk lipids was enhanced by α-tocopheryl acetate in the cod liver oil feed group, but not in the pumpkin seed oil feed group.

The predominant source of vitamin D is the synthesis of cholecalciferol in the skin by the action of sunlight. Mattila et al. (1999) found a strong positive correlation between cholecalciferol content in poultry feed and cholecalciferol and 25-hydroxycholecalciferol content in egg yolk. Mattila et al. (2003) reported that the peak cholecalciferol contents in egg yolk (~30 μg/100 g) were reached 8–13 days after starting the high-cholecalciferol diet. After 112 days, the cholecalciferol content gradually decreased to ~22 μg/100 g. Mattila et al. (2004) investigated transfer of vitamin D_3 (cholecalciferol) and vitamin D_2 (ergocalciferol) from the diet to egg yolks using two different levels of both vitamins (6,000 and 15,000 IU/kg feed) relative to a control treatment (2,500 IU vitamin D_3/kg feed). Supplementing diets with vitamin D_3 increased egg yolk vitamin D content more effectively than did supplementation with vitamin D_2. Vitamin D supplements had no effect on production parameters compared with the control diet, although.vitamin D_3 improved bone strength.

Kawazoe et al. (1996) investigated the effect of supplementation of different sources of vitamin D (vitamin D_2-fortified shiitake, pure vitamin D_2, and pure vitamin D_3) on the transfer of vitamin D to egg yolk, and found no differential effect in performance among the different sources of vitamin D, eggshell quality, and concentrations of calcium and phosphorus in the plasma of laying hens. On the other hand, Chiang et al. (1996) showed that supplementary vitamin D_3 led to significantly higher increments of Ca, P, vitamin D_3, and 25-hydroxyvitamin D_3 contents in yolk. Chiang et al. (1997) suggested that dietary supplementation with 2,000–3,000 IU of vitamin D_3/kg or medium-wave ultraviolet B irradiation for 20–30 min could be recommended for maximizing the egg production and mass increments, eggshell weight, and albumen percentage in eggs. Ovesen et al. (2003) reported that eggs contain up to 1 μg/100 g yolk of the metabolite 25-hydroxyvitamin D, which is absorbed better and faster from the diet than is native vitamin D and has metabolic effects of its own in regulating cell growth and calcium metabolism.

Menaquinones (MK) are classified as vitamin K_2, and are found at low levels in the diet. The feasibility of producing eggs rich in MK-4, an MK with high biological activity, was investigated by Suzuki and Okamoto (1997). Eggs produced by hens fed vitamin K_1 diets had 104–1908 and 67–192 μg/100 g egg yolk of vitamin K_1 and MK-4, respectively. Hens on vitamin K_3 diets produced eggs rich in MK-4 (115–240 μg/100 g egg yolk). Park et al. (2005b) conducted a study to produce eggs enriched in vitamin D_3, vitamin K, and Fe contents. Peak concentrations of vitamins D_3 and K in egg yolk were 4.6- and 4.8-fold greater, respectively, than in nonsupplemented controls, while Fe supplementation led to 14% increased Fe content in the yolk.

Enrichment of eggs with folic acid through supplementation of the laying hen diet was shown to be saturable at 90% of maximal response (House et al. 2002). The folate levels were stable in both control and fortified eggs during 28 days of storage at 4°C.

Hisil and Otles (1997) studied coating variables, which may have an influence on vitamin B_1 concentrations when eggs are stored at or above refrigeration temperature. Yolk and albumen of fresh eggs contained on average 2.78 and 0.27 mg vitamin B_1/kg, respectively. Irregular decreases of vitamin B_1 were observed with elevated temperature, humidity, and extended storage period. Paraffin coating was found to be the most efficient method for preserving vitamin B_1 in eggs at various temperatures. After 3, 6, and 10 months of cold storage, the losses of vitamin B_1 in brown-shelled eggs coated with paraffin (wax) were 0.28, 0.43, and 1.21 mg/kg, respectively.

1.3.3.6. Pigments in Egg Yolk

The color of egg yolk is attributed to fat-soluble carotenoids, which are conjugated isoprene derivatives. The major components of carotenoids in yolk are xanthophylls, including lutein, zeaxanthin, and β-cryptoxanthin; minor amounts of carotenes, including β-carotene, are also found (Table 1.1).

Carotenoids cannot be synthesized by the hen, and the content and composition of pigments in the yolk are therefore influenced by the type and amount of pigments in the diet, their absorption by the laying hen, and transfer to the yolk. Furthermore, maintenance of a uniform yolk color is dependent on the quantity, coloring capacity, and stability of the dietary carotenoids (Nys 2000).

Schlatterer and Breithaupt (2006) reported that lutein and zeaxanthin were the predominant xanthophylls in egg yolks produced in accordance with ecologic husbandry (class 0), with concentrations ranging from 1274 to 2478 μg lutein/100 g yolk and from 775 to 1288 μg zeaxanthin/100 g yolk. Mean concentrations of synthetic xanthophylls in eggs of classes 1 (free-range), 2 (barn), and 3 (cage) were as follows: canthaxanthin, 707 ± 284 μg/100 g; β-apo-8′-carotenoic acid ethyl ester, 639 ± 391 μg/100 g; and citranaxanthin, 560 ± 231 μg/100 g. Experiments with boiled eggs proved that β-apo-8′-carotenoic acid ethyl ester was the xanthophyll with the highest stability, whereas lutein was degraded to the largest extent (19% loss).

Gruszecki et al. (1999) used specific antibodies raised against zeaxanthin to analyze the localization and orientation of zeaxanthin and lutein pigments in lipid membranes formed with egg yolk lecithin. It appeared that the extent of the spectral effects accompanying the interaction of the antibodies was different for the two xanthophyll pigments in spite of their similar molecular structures.

Cholesterol-reduced egg yolk was less yellow than regular egg yolk, as indicated by β-carotene concentrations and Hunter b values (Awad et al. 1997). The addition of 5.0% β-carotene increased the brightness and yellowness of low-cholesterol scrambled egg, low-cholesterol mayonnaise, and low-cholesterol spongecake made from reduced cholesterol egg yolk prepared by β-cyclodextrin treatment (Chiang and Yang 2001).

Egg yolk pigmentation was improved by supplements such as pigmented leaf extracts (Ostrowski-Meissner et al. 1995); biomass of the red microalga *Porphyridium* sp. (Ginzberg et al. 2000); shrimp head meal (Carranco et al. 2003); ascidian tunic shell (Kim 2002); red crab meal (Kim and Choi 2000); green berseem and marigold petals (Jadhao et al. 2000); yeast *Phaffia rhodozyma* (Akiba et al. 2000a, 2000b, 2000c); capsaicin from red pepper (Gonzalez et al. 1999); sunlight illumination of marigold flower meals (Delgado-Vargas, 1998); *Rhodocyclus gelatinosus* biomass as a source of xanthophyll (Ponsano et al. 2004); paprika meal (Kirkpinar and Erkek 1999); xanthophylls supplementation (Min et al. 2003); forage *Arachis glabrata, Leucaena leucocephala, Calliandra calothyrsus*, or *Desmodium* spp. leaves (Teguia 2000); dried carrot meal (Sikder et al. 1998); green tea powder (Uuganbayar et al. 2005); pigmented leaf extracts (Ostrowski-Meissner et al. 1995); and marine microalgae (Fredriksson et al. 2006).

Surai and Speake (1998) investigated the effect of carotenoid-supplemented diets on laying hens. The subsequent concentrations of carotenoids in the yolks of the newly laid fertile eggs were 13.3 and 41.1 μg carotenoids/g fresh yolk in the control diet and high-carotenoid diet, respectively. The proportion of lutein (percent by weight of total carotenoids) was far lower in the liver of the chick than in the yolk. Egg yolks from intensively housed hens showed carotenoid concentration of 11.2–14.8 μg/g (Karadas et al. 2005)

Balnave and Bird (1996) examined the relative efficiencies of deposition into egg yolk of apo-carotenoic acid ester (apo-E) and saponified marigold xanthophylls (MX) in the presence of canthaxanthin (CN), using a wheat-based diet. Apo-E was deposited with an efficiency of 50%, compared to MX, which was deposited with efficiency between 13% and 20%. The dose–response relationship for MX was curvilinear with a decreased efficiency at higher concentrations. Canthaxanthin was deposited with an efficiency of 38%

irrespective of the source of yellow xanthophyll, up to a dietary concentration of 5.5 mg/kg. At a dietary MX concentration of 8.3 mg/kg the efficiency of deposition of CN declined to 24%.

Slightly paler yolk color in eggs from chickens fed triticale variety Bogo was observed, compared to eggs from hens fed no triticale (Hermes and Johnson 2004). Yolk yellowness also decreased linearly with increasing inclusion of coconut meal in diets for laying hens (Braga et al. 2005); it was concluded that coconut meal can be included at levels of up to 15% in laying hens diets along with a source of pigment, without affecting hen performance.

Rizzi et al. (2003) studied the effects of the presence of aflatoxin B1 and the zeolite feed additive clinoptilolite in the feed of laying hens on egg quality, levels of mycotoxin residues in livers, and cytochrome P450–dependent hepatic mixed-function oxygenase activities. Eggs of treated groups were lighter in color than were those of the control group, and the tendency to yellowness in eggs was increased by clinoptilolite, probably through the affinity of red pigments for adsorbents and a consequent prevalence of yellow tonality. Color parameters were thought to be connected with interference by aflatoxin B1 on lipid metabolism and pigment deposition.

Antioxidant properties have been attributed to carotenoids; however, the chemical reactivity of a carotenoid is not the only factor that determines its ability to protect membranes against oxidation, as the position and orientation of the carotenoid in the bilayer are also of importance. Gabrielska and Gruszecki (1996) reported that zeaxanthin (dihydroxy-β-carotene) but not β-carotene rigidifies lipid membranes. Min et al. (2003) reported that dietary xanthophylls supplementation increased pigmentation and retardation of lipid oxidation in egg yolk. Sujak et al. (1999) reported that lutein and zeaxanthin are protectors of lipid membranes against oxidative damage. Woodall et al. (1997) investigated the ability of carotenoids to protect egg yolk PC against oxidation by peroxyl radicals generated from azoinitiators. In homogeneous organic solution, all the carotenoids tested ameliorated lipid peroxidation, but they were not as effective as α-tocopherol.

REFERENCES

Abdallah FB, Chahine JMEH (1998). Transferrins. Hen ovo-transferrin, interaction with bicarbonate and iron uptake. Eur J Biochem 258(3):1022–1031.

Abdallah FB, Chahine JMEH (1999). Transferrins, the mechanism of iron release by ovotransferrin. Eur J Biochem 263(3):912–920.

Ahlborn G, Sheldon BW (2005). Enzymatic and microbiological inhibitory activity in eggshell membranes as influenced by layer strains and age and storage variables. Poultry Sci 84(12):1935–1941.

Ahlborn G, Sheldon BW (2006). Identifying the components in eggshell membrane responsible for reducing the heat resistance of bacterial pathogens. J Food Protect 69(4):729–738.

Ahmed AMH, Rodriguez-Navarro AB, Vidal ML, Gautron J, Garcia-Ruiz JM, Nys Y (2005). Changes in eggshell mechanical properties, crystallographic texture and in matrix proteins induced by moult in hens. Br Poultry Sci 46(3):268–279.

Akagawa M, Wako Y, Suyama K (1999). Lysyl oxidase coupled with catalase in egg shell membrane. Biochim Biophys Acta 1434(1):151–160.

Akiba Y, Sato K, Takahashi K, Takahashi Y, Furuki A, Konashi S, Nishida H, Tsunekawa H, Hayasaka Y, Nagao H (2000a). Pigmentation of egg yolk with yeast *Phaffia rhodozyma* containing high concentration of astaxanthin in laying hens fed on a low-carotenoid diet. Jpn Poultry Sci 37(2):77–85.

Akiba Y, Sato K, Takahashi K, Toyomizu M, Takahashi Y, Konashi S, Nishida H, Tsunekawa H, Hayasaka Y, Nagao H (2000b). Improved pigmentation of egg yolk by feeding of yeast *Phaffia rhodozyma* containing high concentration of astaxanthin in laying hens. Jpn Poultry Sci 37(3):162–170.

Akiba Y, Sato K, Takahashi K, Toyomizu M, Takahashi Y, Tsunekawa H, Hayasaka Y, Nagao H (2000c). Availability of cell wall-fractured yeast, *Phaffia rhodozyma*, containing high concentration of astaxanthin for egg yolk pigmentation. Anim Sci J 71(3):255–260.

Akita EM, Nakai S (1995). Allergenicity of IgY and other egg proteins. Proc IFT Annual Mtg 1995, p 240.

Akita EM, Li-Chan ECY (1998). Isolation of bovine immunoglobulin G subclasses from milk, colostrum and whey using immobilized egg yolk antibodies. J Dairy Sci 81(1):54–63.

Akita EM, Li-Chan ECY, Nakai S (1998). Neutralization of enterotoxigenic *Escherichia coli* heat-labile toxin by chicken egg yolk immunoglobulin Y and its antigen-binding fragments. Food Agric Immunol 10(2):161–172.

Akita EM, Jang CB, Kitts DD, Nakai S (1999). Evaluation of allergenicity of egg yolk immunoglobulin Y and other egg proteins by passive cutaneous anaphylaxis. Food Agric Immunol 11(2):191–201.

Al-Ankari A, Najib H, Al-Hozab A (1998). Yolk and serum cholesterol and production traits, as affected by incorporating a supraoptimal amount of copper in the diet of the leghorn hen. Br Poultry Sci 39 (3):393–397.

Alcantara J, Schryvers AB (1996). Transferrin binding protein two interacts with both the N-lobe and C-lobe of ovotransferrin. Microb Pathogen 20(2):73–85.

Allegre C, Moulin P, Gleize B, Pieroni G, Charbit F (2006). Cholesterol removal by nanofiltration: Applications in nutraceutics and nutritional supplements. J Membr Sci 269(1–2):109–117.

Aluko RE, Mine Y (1997). Competitive adsorption of hen's egg yolk granule lipoproteins and phosvitin in oil-in-water emulsions. J Agric Food Chem 45(12):4564–4570.

Amaral JA, Tino De Franco M, Carneiro-Sampaio MMS, Carbonare SB (2002). Anti-enteropathogenic *Escherichia coli* immunoglobulin Y isolated from eggs laid by immunised Leghorn chickens. Res Vet Sci 72(3):229–234.

An BK, Kang CW (1999). Effects of dietary fat sources containing omega-3 or omega-6 polyunsaturated fatty acids on fatty acid composition of egg yolk in laying hens. Kor J Anim Sci 41(3):293–310.

Anton M, Gandemer G (1997). Composition, solubility and emulsifying properties of granules and plasma of egg yolk. J Food Sci 62(3):484–487.

Anton M, Le Denmat M, Gandemer G (2000). Thermostability of hen egg yolk granules: Contribution of native structure of granules. J Food Sci 65(4):581–584.

Anton M, Martinet V, Dalgalarrondo M, Beaumal V, David-Briand E, Rabesona H (2003). Chemical and structural characterisation of low-density lipoproteins purified from hen egg yolk. Food Chem 83(2):175–183.

Arasteh N, Aminirissehei AH, Yousif AN, Albright LJ, Durance TD (2004). Passive immunization of rainbow trout (*Oncorhynchus mykiss*) with chicken egg yolk immunoglobulins (IgY). Aquaculture 231(1–4):23–36.

Arias JL, Fink DJ, Xiao SQ, Heuer AH, Caplan AI (1993). Biomineralization and eggshells: Cell-mediated acellular compartments of mineralized extracellular matrix. Int Rev Cytol 145:217–250.

Arias JL, Cataldo M, Fernandez MS, Kessi E (1997). Effect of beta-aminopropionitrile on eggshell formation. Br Poultry Sci 38(4):349–354.

Arias JL, Fernandez MS (2001). Role of extracellular matrix molecules in shell formation and structure. World's Poultry Sci J 57(4):349–357.

Arnaudov LN, de Vries R (2005). Thermally induced fibrillar aggregation of hen egg white lysozyme. Biophys J 88(1):515–526.

Asal S, Kocabas S, Elmaci C (1993). Egg white protein polymorphism in chicken (*Gallus gallus* L.) and Japanese quail (*Coturnix coturnix japonica*). Doga Turk J Zool 17(4):259–266.

Awad AC, Bennink MR, Smith DM (1997). Composition and functional properties of cholesterol reduced egg yolk. Poultry Sci 76(4):649–653.

Aydin R, Pariza MW, Cook ME (2001). Olive oil prevents the adverse effects of dietary conjugated linoleic acid on chick hatchability and egg quality. J Nutr 131(3): 800–806.

Aydin R (2006). Effect of storage temperature on the quality of eggs from conjugated linoleic acid-fed laying hens. S Afr J Anim Sci 36(1):13–19.

Babu KR, Bhakuni V (1997). Ionic-strength-dependent transition of hen egg-white lysozyme at low pH to a compact state and its aggregation on thermal denaturation. Eur J Biochem 245(3):781–789.

Balnave D, Bird JN (1996). Relative efficiencies of yellow carotenoids for egg yolk pigmentation. Asian-Austral J Anim Sci 9(5):515–517.

Barreto SLT, Ferreira WM, Goncalves TM (1999). Protein and vitamin E levels for broiler breed hens. 2. Effects on yolk and tissue alpha-tocopherol concentration and nitrogen balance. Arq Brasil Med Vet Zootec 51(2):193–199.

Barron LG, Walzem RL, Hansen RJ (1999). Plasma lipoprotein changes in hens (*Gallus domesticus*) during an induced molt. Compar Biochem Physiol B 123(1):9–16.

Barua A, Furusawa S, Yoshimura Y (2000). Influence of aging and estrogen treatment on the IgY concentration in the egg yolk of chicken, *Gallus domesticus*. Jpn Poultry Sci 37(5):280–288.

Basmacioglu H, Cabuk M, Unal K, Ozkan K, Akkan S, Yalcin H (2003). Effects of dietary fish oil and flax seed on cholesterol and fatty acid composition of egg yolk and blood parameters of laying hens. S Afr J Anim Sci 33(4):266–273.

Bayer EA, Wilchek M (1994). Modified avidins for application in avidin-biotin technology: An improvement on nature. In: Sim JS, Nakai S, eds. Egg uses and Processing Technologies. New Developments. Wallingford, CT: CAB International, pp 158–176.

Beamer LJ, Fischer D, Eisenberg D (1998). Detecting distant relatives of mammalian LPS-binding and lipid transport proteins. Protein Sci 7:1643–1646.

Begum S, Saito A, Kato A, He J, Azakami H (2003). Expression and characterization of chicken ovoinhibitor in *Pichia pastoris*. Nahrung 47(5):359–363.

Besler M, Steinhart H, Paschke A (1997). Allergenicity of hen's egg-white proteins: IgE binding of native and deglycosylated ovomucoid. Food Agric Immunol 9(4):277–288.

Biswas MAH, Miyazaki Y, Nomura K, Wakita M (2000). Influences of long-term feeding of Japanese green tea powder on laying performance and egg quality in hens. Asian-Austral J Anim Sci 13(7):980–985.

Bizhanov G, Jonauskiene L, Hau J (2004). A novel method, based on lithium sulfate precipitation for purification of chicken egg yolk immunoglobulin Y, applied to immunospecific antibodies against Sendai virus. Scand J Lab Anim Sci 31(3): 121–130.

Borelli G, Mayer-Gostan N, Merle PL, de Pontual H, Boeuf G, Allemand D, Payan P (2003). Composition of biomineral organic matrices with special emphasis on turbot (*Psetta maxima*) otolith and endolymph. Calc Tissue Int 72(6):717–725.

Boruta A, Niemiec J (2005). The effect of diet composition and length of storing eggs on changes in the fatty acid profile of egg yolk. J Anim Feed Sci 14(Suppl 1): 427–430.

Brady D, Lowe N, Gaines S, Fenelon L, McPartlin J, O'Farrelly C (2003). Inhibition of *Streptococcus* mutans growth by hen egg-derived fatty acids. J Food Sci 68(4): 1433–1437.

Braga CV de P, Fuentes M de FF, Freitas ER, de Carvalho LE, de Sousa FM, Bastos SC (2005). Effect of inclusion of coconut meal in diets for laying hens. Rev Brasil Zootec 34(1):76–80.

Brand HS, Lerner UH, Grubb A, Beertsen W, Amerongen AVN, Everts V (2004). Family 2 cystatins inhibit osteoclast-mediated bone resorption in calvarial bone explants. Bone (NY) 35(3):689–696.

Brinkmann C, Weiss MS, Weckert E (2006). The structure of the hexagonal crystal form of hen egg-white lysozyme. Acta Crystallogr Sect D Biol Crystallogr 62(Pt 4): 349–355.

Burley RW, Vadehra DV (1989). The Avian Egg. Chemistry and Biology. New York: Wiley Interscience.

Burley RW, Evans AJ, Pearson JA (1993). Molecular aspects of the synthesis and deposition of hens' egg yolk with special reference to low density lipoprotein. Poultry Sci 72(5):850–855.

Cao JS, Peng ZY (1997). Removal of cholesterol from egg yolk by beta-cyclodextrin inclusion and its effects on the functional properties of egg yolk. Food Ferment Indust 23(3):17–21.

Carlander D, Wilhelmson M, Larsson A (2003). Immunoglobulin Y levels in egg yolk from three chicken genotypes. Food Agric Immunol 15(1):35–40.

Carranco ME, Calvo C, Arellano L, Perez-Gil F, Avila E, Fuente B (2003). Inclusion of shrimp (*Penaeus* sp.) head meal in laying hen diets. Effect on yolk red pigment concentration and egg quality. Interciencia 28(6):328–333, 363–364.

Castellani O, Martinet V, David-Brand E, Guérin-Dubiard C, Anton M (2003). Egg yolk phosvitin: Preparation of metal-free purified protein by fast protein liquid chromatography using aqueous solvents. J Chromatogr B 791(1–2):273–284.

Castellani O, Guérin-Dubiard C, David-Brand E, Anton M (2004). Influence of physicochemical conditions and technological treatments on the iron binding capacity of egg yolk phosvitin. Food Chem 85(4):569–577.

Castellani O, David-Brand E, Guérin-Dubiard C, Anton M (2005). Effect of aggregation and sodium salt on emulsifying properties of egg yolk phosvitin. Food Hydrocoll 19(4):769–776.

Castellani O, Belhomme C, David-Brand E, Guérin-Dubiard C, Anton M (2006), Oil-in-water emulsion properties and interfacial characteristics of hen egg yolk phosvitin. Food Hydrocoll 20(1):35–43.

Causeret D, Matringe E, Lorient D (1992). Mineral cations affect microstructure of egg yolk granules. J Food Sci 57(6):1323–1326.

Chang HM, Ou-Yang RF, Chen YT, Chen CC (1999). Productivity and some properties of immunoglobulin specific against *Streptococcus mutans* serotype c in chicken egg yolk (IgY). J Agric Food Chem 47(1):61–66.

Chang HS, Wang JI, Chang PP (2000a). Studies on the isolation and identification of *Rhodococcus* equi NCHU1 and its egg yolk cholesterol degradation. Taiw J Agric Chem Food Sci 38(6):555–564.

Chang HM, Lu TC, Chen CC, Tu YY, Huang JY (2000b). Isolation of immunoglobulin from egg yolk by anionic polysaccharides. J Agric Food Chem 48(4):995–999.

Chen CC, Tu YY, Chen TL, Chang HM (2002). Isolation and characterization of immunoglobulin in yolk (IgY) specific against hen egg white lysozyme by immunoaffinity chromatography. J Agric Food Chem 50(19):5424–5428.

Chen YC, Nakthong C, Chen TC (2005). Effects of chicory fructans on egg cholesterol in commercial laying hen. Int J Poultry Sci 4(2):109–114.

Chen CC, Tai YC, Shen SC, Tu YY, Wu MC, Chang HM (2006). Detection of alkaline phosphatase by competitive indirect ELISA using immunoglobulin in yolk (IgY) specific against bovine milk alkaline phosphatase. Food Chem 95(2): 213–220.

Cheng CH, Shen TF, Chen WL, Ding ST (2004). Effects of dietary algal docosahexaenoic acid oil supplementation on fatty acid deposition and gene expression in laying Leghorn hens. J Agric Sci 142(Pt 6):683–690.

Chengtao W, Tiangui N, Xiaoyu Y, Yujun M (2002). Rebuilding cholesterol-degrading lactic acid bacteria by using the protoplast fusion technique. Food Ferment Indust 28(3):1–5.

Cheung GLM, Thomas TM, Rylatt DB (2003). Purification of antibody Fab and F(ab′)2 fragments using Gradiflow technology. Protein Express Purif 32(1): 135–140.

Chiang YH, Hwang SI, Holick MF (1996). Effect of supplementing different levels of vitamin D-3 on performance, shell minerals and vitamin D3 metabolism in laying hens. Kor J Anim Nutr Feedstuffs 20(2):117–127.

Chiang YH, Hwang SI, Holick MF (1997). Effect of vitamin D-3 oral dose or ultraviolet irradiation on performance and egg parts in laying hens. Kor J Animal Nutr Feedstuffs 21(2):119–132.

Chiang YL, Yang SC (2001). Processing and quality evaluation of low cholesterol egg products. Taiw J Agric Chem Food Sci 39(2):108–116.

Chiu SH, Chung TW, Giridhar R, Wu WT (2004). Immobilization of beta-cyclodextrin in chitosan beads for separation of cholesterol from egg yolk. Food Res Int 37(3):217–223.

Cho YH, Lee JJ, Park IB, Huh CS, Baek YJ, Park J (2005). Protective effect of microencapsulation consisting of multiple emulsification and heat gelation processes on immunloglobulin in yolk. J Food Sci 70(2):E148–E151.

Choi SH (2004). Detection of *Salmonella* in milk by sandwich ELISA using anti-outer membrane protein immunoglobulins. Kor J Food Sci Anim Resources 24(2):176–181.

Choi I, Jung C, Seog H, Choi H (2004). Purification of phosvitin from egg yolk and determination of its physicochemical properties. Food Sci Biotechnol 13(4):434–437.

Choi I, Jung C, Choi H, Kim C, Ha H (2005). Effectiveness of phosvitin peptides on enhancing bioavailability of calcium and its accumulation in bones. Food Chem 93(4):577–583.

Chowdhury SR, Chowdhury SD, Smith TK (2002). Effects of dietary garlic on cholesterol metabolism in laying hens. Poultry Sci 81(12):1856–1862.

Coillie E, Block J, Reybroeck W (2004). Development of an indirect competitive ELISA for flumequine residues in raw milk using chicken egg yolk antibodies. J Agric Food Chem 52(16):4975–4978.

Colombie S, Gaunand A, Lindet B (2001). Lysozyme inactivation under mechanical stirring: Effect of physical and molecular interfaces. Enz Microb Technol 28(9/10):820–826.

Cooper SG, Douches DS, Grafius EJ (2006). Insecticidal activity of avidin combined with genetically engineered and traditional host plant resistance against Colorado potato beetle (*Coleoptera:Chrysomelidae*) larvae. J Econ Entomol 99(2):527–536.

Cossignani L, Santinelli F, Rosi M, Simonetti MS, Valfre F, Damiani P (1994). Incorporation of n-3 PUFA into hen egg yolk lipids. II. Structural analysis of triacylglycerols, phosphatidylcholines and phosphatidylethanolamines. Ital J Food Sci 6(3):293–305.

Cusack M, Fraser AC, Stachel T (2003): Magnesium and phosphorus distribution in the avian eggshell. Compar Biochem Physiol Pt B Biochem Mol Biol 134B(1):63–69.

Daengprok W, Garnjanagoonchorn W, Naivikul O, Pornsinpatip P, Issigonis K, Mine Y (2003). Chicken eggshell matrix proteins enhance calcium transport in the human intestinal epithelial cells, Caco-2. J Agric Food Chem 51(20):6056–6061.

Daenicke S, Halle I, Jeroch H, Boettcher W, Ahrens P, Zachmann R, Goetze S (2000). Effect of soy oil supplementation and protein level in laying hen diets on praecaecal nutrient digestibility, performance, reproductive performance, fatty acid composition of yolk fat, and on other egg quality parameters. Eur J Lipid Sci Technol 102(3):218–232.

Damodaran S, Anand K, Razumovsky L (1998). Competitive absorption of egg white proteins at the air-water interface: Direct evidence for electrostatic complex formation between lysozyme and other egg protein at the interface. J Agric Food Chem 46(3):872–876.

Davalos A, Miguel M, Bartolome B, Lopez-Fandino R (2004). Antioxidant activity of peptides derived from egg white proteins by enzymatic hydrolysis. J Food Protect 67(9):1939–1944.

de Almeida CMC, Quintana-Flores VM, Medina-Acosta E, Schriefer A, Barral-Netto M, Da Silva W (2003). Egg yolk anti-BfpA antibodies as a tool for recognizing and identifying enteropathogenic *Escherichia coli*. Scand J Immunol 57(6):573–582.

Deguchi K, Ito H, Takegawa Y, Shinji N, Nakagawa H, Nishimura SI (2006). Complementary structural information of positive- and negative-ion MSn spectra of glycopeptides with neutral and sialylated N-glycans. Rapid Commun Mass Spectrom 20(5):741–746.

Delgado-Vargas F, Paredes-Lopez O, Avila-Gonzalez E (1998). Effects of sunlight illumination of marigold flower meals on egg yolk pigmentation. J Agric Food Chem 46(2):698–706.

De Meulenaer B, Huyghebaert A (2001). Isolation and purification of chicken egg yolk immunoglobulins: A review. Food Agric Immunol 13(4):275–288.

De Meulenaer B, Baert K, Lanckriet H, Van Hoed V, Huyghebaert A (2002). Development of an enzyme-linked immunosorbent assay for bisphenol A using chicken immunoglobulins. J Agric Food Chem 50(19):5273–5282.

De Meulenaer B, De la Court M, Acke D, De Meyere T, Van de Keere A (2005). Development of an enzyme-linked immunosorbent assay for peanut proteins using chicken immunoglobulins. Food Agric Immunol 16(2):129–148.

Dennis JE, Xiao SQ, Agarwal M, Fink DJ, Heuer AH, Caplan AI (1996). Microstructure of matrix and mineral components of eggshells from white Leghorn chickens (*Gallus gallus*). J Morphol 228(3):287–306.

Dennis JE, Carrino DA, Yamashita K, Caplan AI (2000). Monoclonal antibodies to mineralized matrix molecules of the avian eggshell. Matrix Biol 19:683–692.

Desert C, Guérin-Dubiard C, Nau F. Jan G, Val F, Mallard J (2001). Comparison of different electrophoretic separations of hen egg white proteins. J Agric Food Chem 49(10):4553–4561.

Deshpande A, Nimsadkar S, Mande SC (2005). Effect of alcohols on protein hydration: Crystallographic analysis of hen egg-white lysozyme in the presence of alcohols. Acta Crystallogr Sect D Biol Crystallogr 61(Pt 7):1005–1008.

Dominguez-Vera JM, Gautron J, Garcia-Ruiz JM, Nys Y (2000). The effect of avian uterine fluid on the growth behavior of calcite crystals. Poultry Sci 79(6):901–907.

Drotleff AM, Ternes W (1999). Cis/trans isomers of tocotrienols: Occurrence and bioavailability. Eur Food Res Technol 210(1):1–8.

Du M, Ahn DU, Sell JL (1999). Effect of dietary conjugated linoleic acid on the composition of egg yolk lipids. Poultry Sci 78(11):1639–1645.

Du M, Ahn DU, Sell JL (2000). Effects of dietary conjugated linoleic acid and linoleic: linolenic acid ratio on polyunsaturated fatty acid status in laying hens. Poultry Sci 79(12):1749–1756.

Du R, Qin J, Wang J, Pang Q, Zhang C, Jiang J (2005). Effect of supplementary dietary L-carnitine and yeast chromium on lipid metabolism of laying hens. Asian-Austral J Anim Sci 18(2):235–240.

Dziadek K, Gornowicz E, Kielczewski K (2003a). Physico-chemical traits of table eggs from hens of different origin. Wiadom Zootech 41(3–4):39–43.

Dziadek K, Gornowicz E, Czekalski P (2003b). Chemical composition of table eggs as influenced by the origin of laying hens. Pol J Food Nutr Sci 12(1):21–24.

Eckhardt ERM, Wang DQH, Donovan JM, Carey MC (2002). Dietary sphingomyelin suppresses intestinal cholesterol absorption by decreasing thermodynamic activity of cholesterol monomers. Gastroenterology 122(4):948–956.

Engelmann D, Halle I, Rauch HW, Sallmann HP, Flachowsky G (2001). Influences of various vitamin E supplements on performance of laying hens. Archiv Geflueg 65(4):182–186.

Favier GI, Escudero ME, Mattar MA, Guzman AMS de (2000). Survival of *Yersinia enterocolitica* and mesophilic aerobic bacteria on eggshell after washing with hypochlorite and organic acid solutions. J Food Protect 63(8):1053–1057.

Feng F, Mine Y (2006). Phosvitin phosphopeptides increase iron uptake in a Caco-2 cell monolayer model. Int J Food Sci Technol 41(4):455–458.

Fernandez MS, Araya M, Arias JL (1997). Eggshells are shaped by a precise spatiotemporal arrangement of sequentially deposited macromolecules. Matrix Biol 16:13–20.

Fernandez MS, Arias JL (1999). Ultrastructural localization of molecules involved in eggshell formation in the avian oviduct. Proc 8th Eur Symp Quality of Eggs and Egg Products, Bologna, Italy; pp 81–86.

Fernandez MS, Moya A, Lopez L, Arias JL (2001). Secretion pattern, ultrastructural localization and function of extracellular matrix molecules involved in eggshell formation. Matrix Biol 19:793–803.

Fernandez MS, Escobar C, Lavelin I, Pines M, Arias JL (2003). Localization of osteopontin in oviduct tissue and eggshell during different stages of the avian egg laying cycle. J Struct Biol 143(3):171–180.

Flachowsky G, Engelman D, Suender A, Halle I, Sallmann HP (2002). Eggs and poultry meat as tocopherol sources in dependence on tocopherol supplementation of poultry diets. Food Res Int 35(2–3):239–243.

Franchini A, Sirri F, Tallarico N, Minelli G, Iaffaldano N, Meluzzi A (2002). Oxidative stability and sensory and functional properties of eggs from laying hens fed supranutritional doses of vitamins E and C. Poultry Sci 81(11):1744–1750.

Fraser AC, Bain MM, Solomon SE (1999). Transmission electron microscopy of the vertical crystal layer and cuticle of the eggshell of the domestic fowl. Br Poultry Sci 40(5):626–631.

Fredriksson S, Elwinger K, Pickova J (2006). Fatty acid and carotenoid composition of egg yolk as an effect of microalgae addition to feed formula for laying hens. Food Chem 99(3):530–537.

Gabrielska J, Gruszecki WI (1996). Zeaxanthin (dihydroxy-beta-carotene) but not beta-carotene rigidifies lipid membranes: A ^{1}H-NMR study of carotenoid-egg phosphatidylcholine liposomes. Biochim Biophys Acta 1285(2):167–174.

Galobart J, Barroeta AC, Baucells MD, Codony R, Ternes W (2001a). Effect of dietary supplementation with rosemary extract and alpha-tocopheryl acetate on lipid oxidation in eggs enriched with omega3-fatty acids. Poultry Sci 80(4):460–467.

Galobart J, Barroeta AC, Baucells MD, Cortinas L, Guardiola F (2001b). Alpha-tocopherol transfer efficiency and lipid oxidation in fresh and spray-dried eggs enriched with omega3-polyunsaturated fatty acids. Poultry Sci 80(10):1496–1505.

Galzie Z, Abul-Qasim M, Salahuddin A (1996). Isolation and characterization of domain I of ovoinhibitor. Biochim Biophys Acta 1293(1):113–121.

Ganea E, Trifan M, Laslo AC (2004). Refolding of lysozyme and leucine aminopeptidase under macromolecular crowding conditions. Rom J Biochem 41(1–2):13–23.

Gautron J, Bain M, Solomon S, Nys Y (1996). Soluble matrix of hen's eggshell extracts changes in vitro the rate of calcium carbonate precipitation and crystal morphology. Br Poultry Sci 37(4):853–866.

Gautron J, Hincke MT, Mann K, Panheleux M, Bain M, Mckee MD, Solomon SE, Nys Y (2001). Ovocalyxin-32, a novel chicken eggshell matrix protein. Isolation, amino acid sequencing, cloning, and immunocytochemical localization. J Biol Chem 276(42):39243–39252.

Gebert S, Messikommer R, Pfirter HP, Bee G, Wenk C (1998). Dietary fats and vitamin E in diets for laying hens: Effects on laying performance, storage stability and fatty acid composition of eggs. Archiv Geflueg 62(5):214–222.

Giansanti F, Massucci MT, Giardi MF, Nozza F, Pulsinelli E, Nicolini C, Botti D, Antonini G (2005). Antiviral activity of ovotransferrin derived peptides. Biochem Biophys Res Commun 331(1):69–73.

Ginzberg A, Cohen M, Sod-Moriah UA, Shany S, Rosenshtrauch A, Arad S (2000). Chickens fed with biomass of the red microalga *Porphyridium* sp. have reduced blood cholesterol level and modified fatty acid composition in egg yolk. J Appl Phycol 12(3–5):325–330.

Golab K, Gburek J, Gawel A, Warwas M (2001). Changes in chicken egg white cystatin concentration and isoforms during embryogenesis. Br Poultry Sci 42(3):394–398.

Gonzalez M, Castano E, Avila E, Gonzalez de Mejia E (1999). Effect of capsaicin from red pepper (*Capsicum* sp) on the deposition of carotenoids in egg yolk. J Sci Food Agric 79(13):1904–1908.

Goulas A, Triplett EL, Taborsky G (1996). Oligophosphopeptides of varied structural complexity derived from the egg phosphoprotein, phosvitin. J Protein Chem 15(1): 1–9.

Griebenow K, Klibanov AM (1996). On protein denaturation in aqueous-organic mixtures but not in pure organic solvents. J Am Chem Soc 118(47):11695–11700.

Grune T, Kraemer K, Hoppe PP, Siems W (2001). Enrichment of eggs with n-3 polyunsaturated fatty acids: Effects of vitamin E supplementation. Lipids 36(8): 833–838.

Gruszecki W, Sujak A, Strzalka K, Radunz A, Schmid GH (1999). Organisation of xanthophyll-lipid membranes studied by means of specific pigment antisera, spectrophotometry and monomolecular layer technique lutein versus zeaxanthin. Z Naturforsc Sect C J Biosci 54(7–8):517–525.

Guérin-Dubiard C, Pasco M, Hietanen A, del Bosque A, Nau E, Croguennec T (2005). Hen egg white fractionation by ion-exchange chromatography. J Chromatogr A 1090(1–2):58–67.

Guérin-Dubiard C, Pasco M, Molle D, Desert C, Croguennec T, Nau F (2006). Proteomic analysis of hen egg white. J Agric Food Chem 54(11):3901–3910.

Gurtler M, Methner U, Kobilke H, Fehlhaber K (2004). Effect of orally administered egg yolk antibodies on *Salmonella enteritidis* contamination of hen's eggs. J Vet Med Ser B 51(3):129–134.

Gutierrez MA, Takahashi H, Juneja LR (1997). Nutritive evaluation of hen eggs. In: Yamamoto T, Juneja LR, Hatta H, Kim M, eds. Hen Eggs: Their Basic and Applied Science. New York: CRC Press; pp 25–35.

Haginaka J, Matsunaga H, Kakehi K (2000). Separation of enantiomers on a chiral stationary phase based on ovoglycoprotein. VIII. Chiral recognition ability of partially and completely deglycosylated ovoglycoprotein. J Chromatogr B Biomed Sci Appl 745(1):149–157.

Hammershoj M, Larsen LB, Andersen AB, Qvist KB (2002). Storage of shell eggs influences the albumen gelling properties. Lebens Wissen Technol 35(1):62–69.

Han CK, Lee BH, Lee NH (1999). Analysis of biofunctional components in brand eggs. Kor J Anim Sci 41(3):343–354.

Hatta H, Ozeki M, Tsuda K (1997a). Egg yolk antibody IgY and its application. In: Yamamoto T, Juneja LR, Hatta H, Kim M, eds. Hen Eggs: Their Basic and Applied Science. New York: CRC Press; pp 151–178.

Hatta H, Tsuda K, Ozeki M, Kim M, Yamamoto T, Otake S, Hirasawa M, Katz J, Childers NK, Michalek SM (1997b). Passive immunization against dental plaque formation in humans: Effect of a mouth rinse containing egg yolk antibodies (IgY) specific to *Streptococcus* mutans. Caries Res 31(4):268–274.

Hatta H, Nomura M, Takahashi N, Hirose M (2001). Thermostabilization of ovalbumin in a developing egg by an alkalinity-regulated, two-step process. Biosci Biotechnol Biochem 65(9):2021–2027.

Hermes JC, Johnson RC (2004). Effects of feeding various levels of triticale var. Bogo in the diet of broiler and layer chickens. J Appl Poultry Res 13(4):667–672.

Herstad O, Overland M, Haug A, Skrede A, Thomassen MS, Egaas E (2000). Reproductive performance of broiler breeder hens fed n-3 fatty acid-rich fish oil. Acta Agric Scand Sect A Anim Sci 50(2):121–128.

Hiidenhovi J, Aro HS, Kankare V (1999). Separation of ovomucin subunits by gel filtration: Enhanced resolution of subunits by using a dual-column system. J Agric Food Chem 47(3):1004–1008.

Hiidenhovi J, Hietanen A, Makinen J, Huopalahti R, Ryhanen EL (2000). Enzymatic hydrolysis of ovomucin: Comparison of different proteases. Mededelingen Facult Landbouwk Toegepaste Biol Wetensch Univ Gent 65(3B):535–538.

Hiidenhovi J, Makinen J, Huopalahti R, Ryhanen EL (2002). Comparison of different egg albumen fractions as sources of ovomucin. J Agric Food Chem 50(10): 2840–2845.

Hincke MT, Gautron J, Tsang CPW, McKee MD, Nys Y (1999). Molecular cloning and ultrastructural localization of the core protein of an eggshell matrix proteoglycan, ovocleidin-116. J Biol Chem 274(46):32915–32923.

Hincke MT, Gautron J, Panheleux M, Garcia-Ruiz J, McKee MD, Nys Y (2000). Identification and localization of lysozyme as a component of eggshell membranes and eggshell matrix. Matrix Biol 19(5):443–453.

Hirose J, Kitabatake N, Kimura A, Narita H (2004). Recognition of native and/or thermally induced denatured forms of the major food allergen, ovomucoid, by human IgE and mouse monoclonal IgG antibodies. Biosci Biotechnol Biochem 68(12): 2490–2497.

Hirose J, Murakami-Yamaguchi Y, Ikeda M, Kitabatake N, Narita H (2005). Oligoclonal enzyme-linked immunosorbent assay capable of determining the major food allergen, ovomucoid, irrespective of the degree of heat denaturation. Cytotechnology 47(1–3):145–149.

Hirose J, Doi Y, Kitabatake N, Narita H (2006). Ovalbumin-related gene Y protein bears carbohydrate chains of the ovomucoid type. Biosci Biotechnol Biochem 70(1):144–151.

Hisil Y, Otles S (1997). Changes of vitamin B-1 concentrations during storage of hen eggs. Lebens Wissensch Technol 30(3):320–323.

Homchaudhuri L, Kumar S, Swaminathan R (2006). Slow aggregation of lysozyme in alkaline pH monitored in real time employing the fluorescence anisotropy of covalently labelled dansyl probe. FEBS Lett 580(8):2097–2101.

Hoober KL, Joneja B, White HB, Thorpe C (1996). A sulfhydryl oxidase from chicken egg white. J Biol Chem 271(48):30510–30516.

Horie K, Horie N, Abdou AM, Yang JO, Yun SS, Chun HN, Park CK, Kim M, Hatta H (2004). Suppressive effect of functional drinking yoghurt containing specific egg yolk immunoglobulin on *Helicobacter pylori* in humans. J Dairy Sci 87(12):4073–4079.

House JD, Braun K, Balance DM, O'Connor CP, Guenter W (2002). The enrichment of eggs with folic acid through supplementation of the laying hen diet. Poultry Sci 81(9):1332–1337.

Huntington JA, Stein PE (2001). Structure and properties of ovalbumin. J Chromatogr B 756:189–198.

Hur SJ, Kang GH, Jeong JY, Yang HS, Ha YL, Park GB, Joo ST (2003). Effect of dietary conjugated linoleic acid on lipid characteristics of egg yolk. Asian-Austral J Anim Sci 16(8):1165–1170.

Hwangbo J, Kim JH, Lee BS, Kang SW, Chang JS, Bae HD, Lee MS, Kim YJ, Choi NJ (2006). Increasing content of healthy fatty acids in egg yolk of laying hens by cheese byproduct. Asian-Austral J Anim Sci 19(3):444–449.

Hytonen VP, Laitinen OH, Grapputo A, Kettunen A, Savolainen J, Kalkkinen N, Marttila AT, Nordlund HR, Nyholm TKM, Paganelli G, Kulomaa MS (2003). Characterization of poultry egg-white avidins and their potential as a tool in pretargeting cancer treatment. Biochem J 372(1):219–225.

Ibrahim HR, Higashiguchi S, Juneja LR, Kim M, Yamamoto T (1996a). A structural phase of heat-denatured lysozyme with novel antimicrobial action. J Agric Food Chem 44(6):1416–1423.

Ibrahim HR, Higashiguchi S, Koketsu M, Juneja LR, Kim M, Yamamoto T, Sugimoto Y, Aoki T (1996b). Partially unfolded lysozyme at neutral pH agglutinates and kills Gram-negative and Gram-positive bacteria through membrane damage mechanism. J Agric Food Chem 44(12):3799–3806.

Ibrahim HR, Higashiguchi S, Sugimoto Y, Aoki T (1997). Role of divalent cations in the novel bactericidal activity of the partially unfolded lysozyme. J Agric Food Chem 45(1):89–94.

Ibrahim HR, Iwamori E, Sugimoto Y, Aoki T (1998). Identification of a distinct antibacterial domain within the N-lobe of ovotransferrin. Biochim Biophys Acta 1401(3):289–303.

Ibrahim HR, Sugimoto Y, Aoki T (2000). Ovotransferrin antimicrobial peptide (OTAP-92) kills bacteria through a membrane damage mechanism. Biochim Biophys Acta 1523(2–3):196–205.

Ibrahim HR, Thomas U, Pellegrini A (2001). A helix-loop-helix peptide at the upper lip of the active site cleft of lysozyme confers potent antimicrobial activity with membrane permeabilization action. J Biol Chem 276(47):43767–43774.

Ibrahim HR, Haraguchi T, Aoki T (2006). Ovotransferrin is a redox-dependent auto-processing protein incorporating four consensus self-cleaving motifs flanking the two kringles. Biochim Biophys Acta 1760(3):347–355.

Ino T, Hattori M, Yoshida T, Hattori S, Yoshimura K, Takahashi K (2006). Improved physical and biochemical features of a collagen membrane by conjugating with soluble egg shell membrane protein. Biosci Biotechnol Biochem 70(4):865–873.

Ishikawa SI, Suyama K, Satoh I (1999). Biosorption of actinides from dilute waste actinide solution by egg-shell membrane. Appl Biochem Biotechnol 77–79(0):521–533.

Ishikawa SI, Yano Y, Arihara K, Itoh M (2004). Egg yolk phosvitin inhibits hydroxyl radical formation from the Fenton reaction. Biosci Biotechnol Biochem 68(6):1324–1331.

Ishikawa SI, Ohtsuki S, Tomita K, Arihara K, Itoh M (2005). Protective effect of egg yolk phosvitin against ultraviolet-light-induced lipid peroxidation in the presence of iron ions. Biol Trace Elem Res 105(1–3):249–256.

Jacobs K, Shen L, Benemariya H, Deelstra H (1993). Selenium distribution in egg white proteins. Z Lebens Untersuch Forsch 196(3):236–238.

Jadhao SB, Tiwari CMC, Khan MY (2000). Effect of complete replacement of maize by broken rice in the diet of laying hens. Indian J Anim Nutr 17(3):237–242.

Jaradat ZW, Marquardt RR (2000). Studies on the stability of chicken IgY in different sugars, complex carbohydrates and food materials. Food Agric Immunol 12(4):263–272.

Jiang B, Mine Y (2000). Preparation of novel functional oligophosphopeptides from hen egg yolk phosvitin. J Agric Food Chem 48(4):990–994.

Jiang B, Mine Y (2001). Phosphopeptides derived from hen egg yolk phosvitin: Effect of molecular size on the calcium-binding properties. Biosci Biotechnol Biochem 65(5):1187–1190.

Jiang Y, Noh SK, Koo SI (2001). Egg phosphatidylcholine decreases the lymphatic absorption of cholesterol in rats. J Nutr 131(9):2358–2363.

Jintaridth P, Srisomsap C, Vichittumaros K, Kalpravidh RW, Winichagoon P, Fucharoen S, Svasti MJ, Kasinrerk W (2006). Chicken egg yolk antibodies specific for the gamma chain of human hemoglobin for diagnosis of thalassemia. Int J Hematol 83(5):408–414.

Juneja LR (1997). Egg yolk lipids. In: Yamamoto T, Juneja LR, Hatta H, Kim M, eds. Hen Eggs: Their Basic and Applied Science. New York: CRC Press; pp 73–98.

Juneja LR, Kim M (1997). Egg yolk proteins. In: Yamamoto T, Juneja LR, Hatta H, Kim M, eds. Hen Eggs: Their Basic and Applied Science. New York: CRC Press; pp 57–71.

Jung TH, Park HS, Kwak HS (2005). Optimization of cholesterol removal by cross-linked beta-cyclodextrin in egg yolk. Food Sci Biotechnol 14(6):793–797.

Kahraman Z, Yalcin S, Yalcin SS, Dedeoglu HE (2004). Effects of supplementary iodine on the performance and egg traits of laying hens. Br Poultry Sci 45(4):499–503.

Karadas F, Wood NAR, Surai PF, Sparks NHC (2005). Tissue-specific distribution of carotenoids and vitamin E in tissues of newly hatched chicks from various avian species. Compar Biochem Physiol Pt A Mol Integr Physiol 140(4):506–511.

Kassaify ZG, Li EWY, Mine Y (2005). Identification of antiadhesive fraction(s) in nonimmunized egg yolk powder: In vitro study. J Agric Food Chem 53(11):4607–4614.

Katayama S, Xu X, Fan MZ, Mine Y (2006). Antioxidative stress activity of oligophosphopeptides derived from hen egg yolk phosvitin in Caco-2 cells. J Agric Food Chem 54(3):773–778.

Kato T, Imatani T, Miura T, Minaguchi K, Saitoh E, Okuda K (2000). Cytokine-inducing activity of family 2 cystatins. Biol Chem 381(11):1143–1147.

Kato Y, Eri O, Matsuda T (2001). Decrease in antigenic and allergenic potentials of ovomucoid by heating in the presence of wheat flour: Dependence on wheat variety and intermolecular disulfide bridges. J Agric Food Chem 49(8):3661–3665.

Kawazoe T, Yuasa K, Noguchi K, Yamazaki M, Ando M (1996). Effect of different sources of vitamin D on transfer of vitamin D to egg yolk. J Jpn Soc Food Sci Technol 43(4):444–450.

Keenan TW, Berridge L (1973). Identification of gangliosides as constituents of egg yolk. J Food Sci 38(1):43–44.

Khan MAS, Babiker EE, Azakami H, Kato A (1998). Effect of protease digestion and dephosphorylation on high emulsifying properties of hen egg yolk phosvitin. J Agric Food Chem 46(12):4977–4981.

Khan MAS, Nakamura S, Ogawa M, Akita E, Azakami H, Kato A (2000). Bactericidal action of egg yolk phosvitin against *Escherichia coli* under thermal stress. J Agric Food Chem 48(5):1503–1506.

Kido S, Doi Y, Kim F, Morishita E, Narita H, Kanaya S, Ohkubo T, Nishikawa K, Yao T, Ooi T (1995). Characterization of vitelline membrane outer layer protein I, VMO-I: Amino acid sequence and structural stability. J Biochem Tokyo 117 (6): 1183–1191.

Kilic Z, Acar O, Ulasan M, Ilim M (2002). Determination of lead, copper, zinc, magnesium, calcium and iron in fresh eggs by atomic absorption spectrometry. Food Chem 76(1):107–116.

Kilikidis SD (1978). Study on the polar lipids of egg yolk. In Proc 20th World Veterinary Congress, pp 859–868.

Kim YS, Yoo IJ, Jeon KH, Kim CJ (1995). Optimal conditions for ethanol extraction of egg lecithins. Kor J Anim Sci 37(2):186–192.

Kim H, Nakai S (1996). Immunoglobulin separation from egg yolk: A serial filtration system. J Food Sci 61(3):510–512, 523.

Kim JW, Slavik MF (1996). Changes in eggshell surface microstructure after washing with cetylpyridinium chloride or trisodium phosphate. J Food Protect 59(8):859–863.

Kim H, Nakai S (1998). Simple separation of immunoglobulin from egg yolk by ultrafiltration. J Food Sci 63(3):485–490.

Kim H, Li-Chan ECY (1998). Separation of immunoglobulin G from cheddar cheese whey by avidin-biotinylated IgY chromatography. J Food Sci 63(3):429–434.

Kim H, Durance TD, Li-Chan ECY (1999). Reusability of avidin-biotinylated immunoglobulin Y columns in immunoaffinity chromatography. Anal Biochem 268:383–397.

Kim EM, Choi JH (2000). Effect of supplementation of dried red crab meal into laying hen diet on egg yolk pigmentation. J Anim Sci Technol 42(3):289–298.

Kim J, Lee CS, Oh J, Kim BG (2001a). Production of egg yolk lysolecithin with immobilized phospholipase A2. Enz Microb Technol 29(10):587–592.

Kim JH, Kim SC, Kim YM, Ha HM, Ko YD, Kim CH (2001b). Effects of dietary supplementation of fermented feed (Bio-alpha(R)) on performance of laying hens, fecal ammonia gas emission and composition of fatty acids in egg yolk. J Anim Sci Technol 43(3):337–348.

Kim EM (2002). The effects of supplementation of ascidian tunic shell into laying hen diet on egg quality. J Anim Sci Technol 44(1):45–54.

Kim MJ, Lee JW, Yook HS, Lee SY, Kim MC, Byun MW (2002). Changes in the antigenic and immunoglobulin E-binding properties of hen's egg albumin with the combination of heat and gamma irradiation treatment. J Food Protect 65(7): 1192–1195.

Kim JH, Hong ST, Lee HS, Kim HJ (2004a). Oral administration of pravastatin reduces egg cholesterol but not plasma cholesterol in laying hens. Poultry Sci 83(9): 1539–1543.

Kim DK, Jang IK, Seo HC, Shin SO, Yang SY, Kim JW (2004b). Shrimp protected from WSSV disease by treatment with egg yolk antibodies (IgY) against a truncated fusion protein derived from WSSV. Aquaculture 237(1–4):21–30.

Kim KE, You SJ, Ahn BK, Jo TS, Ahn BJ, Choi DH, Kang CW (2006). Effects of dietary activated charcoal mixed with wood vinegar on quality and chemical composition of egg in laying hens. J Anim Sci Technol 48(1):59–68.

Kimura M, Obi M (2005). Ovalbumin-induced IL-4, IL-5 and IFN-gamma production in infants with atopic dermatitis. Int Arch Allergy Immunol 137(2):134–140.

Kirkpinar F, Erkek R (1999). The effects of some natural and synthetic pigment materials on egg yolk pigmentation and production in yellow corn diets. Turk J Vet Anim Sci 23(1):15–21.

Kirunda DFK, Scheideler SE, McKee SR (2001). The efficacy of vitamin E (DL-alpha-tocopheryl acetate) supplementation in hen diets to alleviate egg quality deterioration associated with high temperature exposure. Poultry Sci 80(9):1378–1383.

Kivini H, Jarvenpaa EP, Aro H, Huopalahti R, Ryhanen EL (2004). Qualitative and quantitative liquid chromatographic analysis methods for the determination of the effects of feed supplements on hen egg yolk phospholipids. J Agric Food Chem 52(13):4289–4295.

Klasing KC (2002). Role for chicken avidin in nutritional immunity. Fed Am Soc Exp Biol J 16(4):A621.

Kobayashi K, Hattori M, Hara-Kudo Y, Okubo T, Yamamoto S, Takita T, Sugita-Konishi Y (2004). Glycopeptide derived from hen egg ovomucin has the ability to bind enterohemorrhagic *Escherichia coli* O157:H7. J Agric Food Chem 52(18):5740–5746.

Kodama Y, Kimura N (2004). Glycoprotein having inhibitory activity against *Helicobacter pylori* colonization. Official Gazette of the United States Patent Trademark Office Patents; 1289(1).

Kodentsova VM, Vrzhesinskaya OA, Risnik VV, Sokol'nikov AA, Spirichev VB (1994). Isolation of a riboflavin-binding apoprotein from egg white and its use for riboflavin determination in biological samples. Appl Biochem Microbiol 30(4/5):489–493.

Koketsu M (1997). Glycochemistry of hen eggs. In: Yamamoto T, Juneja LR, Hatta H, Kim M, eds. Hen Eggs: Their Basic and Applied Science. New Yolk: CRC Press; pp 99–115.

Koreleski J, Swiatkiewicz S, Iwanowska A (2003). Lipid fatty acid composition and oxidative susceptibility in eggs of hens fed a fish fat diet supplemented with vitamin E, C, or synthetic antioxidant. J Anim Feed Sci 12(3):561–572.

Kovacs-Nolan J, Zhang JW, Hayakawa S, Mine Y (2000). Immunochemical and structural analysis of pepsin-digested egg white ovomucoid. J Agric Food Chem 48(12): 6261–6266.

Kovacs-Nolan J, Mine Y (2004a). Avian egg antibodies: Basic and potential applications. Avian Poultry Biol Rev 15(1):25–46.

Kovacs-Nolan J, Mine Y (2004b). Passive immunization through avian egg antibodies. Food Biotechnol 18(1):39–62.

Kurokawa H, Dewan JC, Mikami B, Sacchettini JC, Hirose M (1999). Crystal structure of hen apo-ovotransferrin: Both lobes adopt an open conformation upon loss of iron. J Biol Chem 274(40):28445–28452.

Kurtoglu V, Kurtoglu F, Seker E, Coskun B, Balevi T, Polat ES (2004). Effect of probiotic supplementation on laying hen diets on yield performance and serum and egg yolk cholesterol. Food Add Contam 21(9):817–823.

Lakshminarayanan R, Joseph JS, Kini RM, Valiyaveettil S (2005). Structure-function relationship of avian eggshell matrix proteins: A comparative study of two major eggshell matrix proteins, ansocalcin and OC-17. Biomacromolecules 6(2):741–751.

Lambert LA, Perri H, Halbrooks PJ, Mason AB (2005). Evolution of the transferrin family: Conservation of residues associated with iron and anion binding. Compar Biochem Physiol Pt B Biochem Mole Biol 142(2):129–141.

Larsen LB, Hammershoj M, Rasmussen JT (1999). Identification of Ch21, a developmentally regulated chicken embryo protein in egg albumen. Proc 8th Eur Symp Quality of Eggs and Egg Products, vol II, Bologna Italy; pp 61–67.

Latour MA, Peebles ED, Doyle SM, Pansky T, Smith TW, Boyle CR (1998). Broiler breeder age and dietary fat influence the yolk fatty acid profiles of fresh eggs and newly hatched chicks. Poultry Sci 77(1):47–53.

Lavelin I, Meiri N, Pines M (2000). New insight in eggshell formation. Poultry Sci 79(7):1014–1017.

Lee KA (1996). Studies on the stability of hen's egg yolk immunoglobulins. J Kor Soc Food Sci 12(1):54–59.

Lee KA (1999). Resistance of hen's egg yolk immunoglobulins in livetin to digestive enzymes. J Kor Soc Food Sci Nutr 28(2):438–443.

Lee SB, Mine Y, Stevenson RMW (2000). Effects of hen egg yolk immunoglobulin in passive protect of rainbow trout against *Yersinia ruckeri*. J Agric Food Chem 48(1):110–115.

Lee JW, Yook HS, Cho KH, Lee SY, Byun MW (2001). The changes of allergenic and antigenic properties of egg white albumin (Gal d 1) by gamma irradiation. J Kor Soc Food Sci Nutr 30(3):500–504.

Lee JW, Lee KY, Yook HS, Lee SY, Kim HY, Jo C, Byun MW (2002a). Allergenicity of hen's egg ovomucoid gamma irradiated and heated under different pH conditions. J Food Protect 65(7):1196–1199.

Lee JS, Takai J, Takahasi K, Endo Y, Fujimoto K, Koike S, Matsumoto W (2002b). Effect of dietary tung oil on the growth and lipid metabolism of laying hens. J Nutr Sci Vitaminol 48(2):142–148.

Lee SB (2003). Detection of IgY specific to *Salmonella enteritidis* and *Salmonella typhimurium* in the yolk of commercial brand eggs using ELISA. Kor J Food Sci Anim Resources 23(2):161–167.

Lee SB, Choi SH, Choi JW (2004). Indirect ELISA method for measurement of lactoperoxidase using IgY antibody. Kor J Food Sci Anim Resources 24(2):182–188.

Lee NY, Cheng JT, Enomoto T, Nakano Y (2006). One peptide derived from hen ovotransferrin as pro-drug to inhibit angiotensin converting enzyme. J Food Drug Anal 14(1):31–35.

Lewis NM, Seburg S, Flanagan NL (2000). Enriched eggs as a source of n-3 polyunsaturated fatty acids for humans. Poultry Sci 79(7): 971–974.

Li X, Nakano T, Sunwoo HH, Paek BH, Chae HS, Sim JS (1998). Effects of egg and yolk weights on yolk antibody (IgY) production in laying chickens. Poultry Sci 77(2):266–270.

Li HL, Ryu KS (2001). Effect of feeding various wood vinegar on performance and egg quality of laying hens. J Anim Sci Technol 43(5):655–662.

Li XL, Yin ZZ, Qian Y, Zhu ZG, Fang WH (2005). Crude extraction and purification of IgY against *Aeromonas hydophila* and its antibacterial activity in vitro. J Zhejiang Univ (Agric Life Sci) 31(4):503–506.

Li SJ (2006). Structural details at active site of hen egg white lysozyme with di- and trivalent metal ions. Biopolymers 81(2):74–80.

Li-Chan E, Nakai S (1989). Biochemical basis for the properties of egg white. Crit Rev Poultry Biol 2(1):21–58.

Li-Chan, ECY, Powrie WD, Nakai S (1995). The chemistry of eggs and egg products. In: Stadelman WJ, Cotterill OJ, eds. Egg Science and Technology, 4th ed. New York: Food Products Press; pp 105–175.

Li-Chan ECY, Kummer A (1997). Influence of standards and antibodies in immunochemical assays for quantitation of immunoglobulin G in bovine milk. J Dairy Sci 80(6):1038–1046.

Li-Chan ECY, Ler SS, Kummer A, Akita EM (1998). Isolation of lactoferrin by immunoaffinity chromatography using yolk antibodies. J Food Biochem 22(3): 179–195.

Lien TF, Wu CP, Lu JJ (2003). Effects of cod liver oil and chromium picolinate supplements on the serum traits, egg yolk fatty acids and cholesterol content in laying hens. Asian Austral J Anim Sci 16(8):1177–1181.

Lien TF, Chen KL, Wu CP, Lu JJ (2004). Effects of supplemental copper and chromium on the serum and egg traits of laying hens. Bri Poultry Sci 45(4):535–539.

Lim KS, You SJ, An BK, Kang CW (2006). Effects of dietary garlic powder and copper on cholesterol content and quality characteristics of chicken eggs. Asian-Austral J Anim Sci 19(4):582–586.

Lin LY, Lee MH (1996). Effect of hen's age on the composition of yolk lipid. Food Sci Taiw 23(2):168–173.

Lin H, Mertens K, Kemps B, Govaerts T, De Ketelaere B, De Baerdemaeker J, Decuypere E, Buyse J (2004). New approach of testing the effect of heat stress on eggshell quality: Mechanical and material properties of eggshell and membrane. Br Poultry Sci 45(4):476–482.

Lindblom G, Oradd G, Filippov A (2006). Lipid lateral diffusion in bilayers with phosphatidylcholine, sphingomyelin and cholesterol—an NMR study of dynamics and lateral phase separation. Chem Phys Lipids 141(1–2):179–184.

Liong JWW, Frank JF, Bailey S (1997). Visualization of eggshell membranes and their interaction with *Salmonella enteritidis* using confocal scanning laser microscopy. J Food Protect 60(9):1022–1028.

Liu S, Azakami H, Kato A (2000a). Improvement in the yield of lipophilized lysozyme by the combination with Maillard-type glycosylation. Nahrung 44(6):407–410.

Liu ST, Sugimoto T, Azakami H, Kato A (2000b). Lipophilization of lysozyme by short and middle chain fatty acids. J Agric Food Chem 48(2):265–269.

Liu HL, Hsieh WC, Liu HS (2004). Molecular dynamics simulations to determine the effect of supercritical carbon dioxide on the structural integrity of hen egg white lysozyme. BioTechnol Progress 20(3):930–938.

Liu W, Cellmer T, Keerl D, Prausnitz JM, Blanch HW (2005a). Interactions of lysozyme in guanidinium chloride solutions from static and dynamic light-scattering measurements. Biotechnol Bioeng 90(4):482–490.

Liu LY, Yang MH, Lin JH, Lee MH (2005b). Lipid profile and oxidative stability of commercial egg products. J Food Drug Anal 13(1):78–83.

Lu C, Tang Y, Wang L, Ji W, Chen Y, Yang S, Wang W (2002). Bioconversion of yolk cholesterol by extracellular cholesterol oxidase from *Brevibacterium* sp. Food Chem 77(4):457–463.

Mabe I, Rapp C, Bain MM, Nys Y (2003). Supplementation of a corn-soybean meal diet with manganese, copper, and zinc from organic or inorganic sources improves eggshell quality in aged laying hens. Poultry Sci 82(12):1903–1913.

Maehashi K, Udaka S (1998). Sweetness of lysozymes. Biosci Biotechnol Biochem 62(3):605–606.

Mann K (1999). Isolation of a glycosylated form of the chicken eggshell protein ovocleidin and determination of the glycosylation site. Alternative glycosylation/phosphorylation at an N-glycosylation sequon. FEBS Lett 463(1–2):12–14.

Mann K, Siedler F (1999). The amino acid sequence of ovocleidin 17, a major protein of the avian eggshell calcified layer. Biochem Mol Biol Int 47(6):997–1007.

Mann K, Hincke MT, Nys Y (2002). Isolation of ovocleidin-116 from chicken eggshells, correction of its amino acid sequence and identification of disulfide bonds and glycosylated Asn. Matrix Biol 21(5):383–387.

Mann K, Gautron J, Nys Y, McKee MD, Bajari T, Schneider WJ, Hincke MT (2003). Disulfide-linked heterodimeric clusterin is a component of the chicken eggshell matrix and egg white. Matrix Biol 22(5):397–407.

Marolia KZ, D'Souza SF (1999). Enhancement in the lysozyme activity of the hen egg white foam matrix by cross-linking in the presence of N-acetyl glucosamine. J Biochem Biophys Meth 39(1/2):115–117.

Masschalck B, Van Houdt R, Van Haver EGR, Michiels CW (2001). Inactivation of gram-negative bacteria by lysozyme, denatured lysozyme, and lysozyme-derived peptides under high hydrostatic pressure. Appl Env Microbiol 67(1):339–344.

Masschalck B, Deckers D, Michiels CW (2002). Lytic and nonlytic mechanism of inactivation of Gram-positive bacteria by lysozyme under atmospheric and high hydrostatic pressure. J Food Protect 65(12):1916–1923.

Masuda T, Ueno Y, Kitabatake N (2001). Sweetness and enzymatic activity of lysozyme. J Agric Food Chem 49(10):4937–4941.

Masuda T, Ide N, Kitabatake N (2005). Structure-sweetness relationship in egg white lysozyme: Role of lysine and arginine residues on the elicitation of lysozyme sweetness. Chem Senses 30(8):667–681.

Matoba N, Yamada Y, Yoshikawa M (2003). Design of a genetically modified soybean protein preventing hypertension based on an anti-hypertensive peptide derived from ovalbumin. Curr Med Chem Cardiovasc Hematol Agents 1(2):197–202.

Matsunaga H, Haginaka J (2001). Separation of basic drug enantiomers by capillary electrophoresis using ovoglycoprotein as a chiral selector: Comparison of chiral resolution ability of ovoglycoprotein and completely deglycosylated ovoglycoprotein. Electrophoresis 22(15):3251–3256.

Mattila P, Lehikoinen K, Kiiskinen T, Piironen V (1999). Cholecalciferol and 25-hydroxycholecalciferol content of chicken egg yolk as affected by the cholecalciferol content of feed. J Agric Food Chem 47(10):4089–4092.

Mattila P, Rokka T, Konko K, Valaja J, Rossow L, Ryhanen EL (2003). Effect of cholecalciferol-enriched hen feed on egg quality. J Agric Food Chem 51(1):283–287.

Mattila P, Valaja J, Rossow L, Venalainen E, Tupasela T (2004). Effect of vitamin D2- and D3-enriched diets on egg vitamin D content, production, and bird condition during an entire production period. Poultry Sci 83(3):433–440.

Mazalli MR, Faria DE, Salvador D, Ito DT (2004). A comparison of the feeding value of different sources of fat for laying hens: 2. Lipid, cholesterol and vitamin E profiles of egg yolk. J Appl Poultry Res 13(2):280–290.

McKenzie HA, Frier RD (2003a). The behavior of R-ovalbumin and its individual components A1, A2, and A3 in urea solution: Kinetics and equilibria. J Protein Chem 22(3):207–214.

McKenzie HA, Frier RD (2003b). Behavior of S1- and S2-ovalbumin and S-ovalbumin A1 in urea solution: Kinetics and equilibria. J Protein Chem 22(3):215–220.

Meluzzi A, Sirri F, Manfreda G, Sirri F, Manfreda G, Tallarico N, Franchini A (2000). Effects of dietary vitamin E on the quality of table eggs enriched with n-3 long-chain fatty acids. Poultry Sci 79(4):539–545.

Messens W, Grijspeerdt K, Herman L (2005). Eggshell characteristics and penetration by *Salmonella enterica* serovar enteritidis through the production period of a layer flock. Br Poultry Sci 46(6):694–700.

Midorikawa T, Abe R, Yamagata Y, Nakajima T, Ichishima E (1998). Isolation and characterization of cDNA encoding chicken egg yolk aminopeptidase Ey. Comp Biochem Physiol. Pt B Biochem Mol Biol 119B(3):513–520.

Miguel M, Aleixandre A (2006). Antihypertensive peptides derived from egg proteins. J Nutr 136(6):1457–1460.

Miguel M, Alexandre MA, Ramos M, Lopez-Fandino R (2006). Effect of simulated gastrointestinal digestion on the antihypertensive properties of ACE-inhibitory peptides derived from ovalbumin. J Agric Food Chem 54(3):726–731.

Miksik I, Charvatova J, Eckhardt A, Deyl Z (2003). Insoluble eggshell matrix proteins: Their peptide mapping and partial characterization by capillary electrophoresis and high-performance liquid chromatography. Electrophoresis 24(5):843–852.

Milinsk MC, Murakami AE, Gomes STM, Matsushita M, de Souza NE (2003). Fatty acid profile of egg yolk lipids from hens fed diets rich in n-3 fatty acids. Food Chem 83(2):287–292.

Min BJ, Lee KH, Lee SK (2003). Effect of dietary xanthophylls supplementation on pigmentation and antioxidant properties in the egg yolks. J Anim Sci Technol 45(5):847–856.

Mine Y (1997a). Separation of *Salmonella enteritidis* from experimentally contaminated liquid eggs using a hen IgY immobilized immunomagnetic separation system. J Agric Food Chem 45(10):3723–3727.

Mine Y (1997b). Structural and functional changes of hen's egg yolk low-density lipoproteins with phospholipase A2. J Agric Food Chem 45(12):4558–4563.

Mine Y (1998). Adsorption behavior of egg yolk low-density lipoproteins in oil-in-water emulsions. J Agric Food Chem 46(1):36–41.

Mine Y, Bergougnoux M (1998). Adsorption properties of cholesterol-reduced egg yolk low-density lipoprotein at oil-in-water interfaces. J Agric Food Chem 46(6):2153–2158.

Mine Y, Zhang JW (2001). The allergenicity of ovomucoid and the effect of its elimination from hen's egg white. J Sci Food Agric 81(15):1540–1546.

Mine Y, Zhang JW (2002a). Comparative studies on antigenicity and allergenicity of native and denatured egg white proteins. J Agric Food Chem 50(9):2679–2683.

Mine Y, Zhang JW (2002b). Identification and fine mapping of IgG and IgE epitopes in ovomucoid. Biochem Biophys Res Commun 292(4):1070–1074.

Mine Y, Rupa P (2003). Fine mapping and structural analysis of immunodominant IgE allergenic epitopes in chicken egg ovalbumin. Protein Eng 16(10):747–752.

Mine Y, Sasaki E, Zhang JW (2003). Reduction of antigenicity and allergenicity of genetically modified egg white allergen, ovomucoid third domain. Biochem Biophys Res Commun 302(1):133–137.

Mine Y, Rupa P (2004a). Immunological and biochemical properties of egg allergens. World's Poultry Sci J 60:321–330.

Mine Y, Rupa P (2004b). Genetic attachment of undecane peptides to ovomucoid third domain can suppress the production of specific IgG and IgE antibodies. J Allergy Clin Immunol 113(2 Suppl):S234.

Mine Y, Ma F, Lauriau S (2004). Antimicrobial peptides released by enzymatic hydrolysis of hen egg white lysozyme. J Agric Food Chem 52(5):1088–1094.

Mine Y, Kovacs-Nolan J (2006). New insights in biologically active proteins and peptides derived from hen egg. World's Poultry Sci J 62(1):87–95.

Mineki M, Kobayashi M (1997). Microstructure of yolk from fresh eggs by improved method. J Food Sci 62(4):757–761.

Mineki M, Kobayashi M (1998). Microstructural changes in stored hen egg yolk. Jpn Poultry Sci 35(5):285–294.

Mizutani K, Yamashita H, Oe H, Hirose M (1997). Structural characteristics of the disulfide-reduced ovotransferrin N-lobe analyzed by protein fragmentation. Biosci Biotechnol Biochem 61(4):641–646.

Mizutani K, Yamashita H, Kurokawa H, Mikami B, Hirose M (1999). Alternative structural state of transferrin: The crystallographic analysis of iron-loaded but domain-opened ovotransferrin N-lobe. J Biol Chem 274(15):10190–10194.

Mizutani K, Yamashita H, Mikami B, Hirose M (2000). Crystal structure at 1.9 ANG resolution of the apoovotransferrin N-lobe bound by sulfate anions: Implications for the domain opening and iron release mechanism. Biochemistry 39(12):3258–3265.

Momma H, Nakano M, Fujino Y (1972). Studies on the lipids of egg yolk. IV. Cerebroside in egg yolk. Jpn J Zootech Sci 43(4):198–202.

Monaco HL (1997). Crystal structure of chicken riboflavin-binding protein. Eur Mol Biol Org J 16(7):1475–1483.

Moreau S, Awade AC, Molle D, Le-Graet Y, Brule G (1995). Hen egg white lysozyme-metal ion interactions: Investigation by electrospray ionization mass spectrometry. J Agric Food Chem 43(4):883–889.

Moreau S, Nau F, Piot M, Guerin C, Brule G (1997). Hydrolysis of hen egg white ovomucin. Z Lebens Unter Forsch A/Food Res Technol 205(5):329–334.

Mori AV, Mendonca CX Jr, Santos COF (1999). Effect of dietary lipid-lowering drugs upon plasma lipids and egg yolk cholesterol levels of laying hens. J Agric Food Chem 47(11):4731–4735.

Morrison SL, Mohammed MS, Wims LA, Trinh R, Etches R (2002). Sequences in antibody molecules important for receptor-mediated transport into the chicken egg yolk. Mol Immunol 38(8):619–625.

Moschetta A, van Berge-Henegouwen GP, Portincasa P, Palasciano G, Groen AK, van Erpecum KJ (2000). Sphingomyelin exhibits greatly enhanced protection compared with egg yolk phosphatidylcholine against detergent bile salts. J Lipid Res 41(6): 916–924.

Mottaghitalab M, Taraz Z (2004). Garlic powder as blood serum and egg yolk cholesterol lowering agent. J Poultry Sci 41(1):50–57.

Muralidhara BK, Hirose M (2000). Structural and functional consequences of removal of the interdomain disulfide bridge from the isolated C-lobe of ovotransferrin. Protein Sci 9(8):1567–1575.

Murcia MA, Martinez-Tome M, del Cerro I, Sotillo F, Ramirez A (1999). Proximate composition and vitamin E levels in egg yolk: Losses by cooking in a microwave oven. J Sci Food Agric 79(12):1550–1556.

Nadeau OW, Falick AM, Woodworth RC (1996). Structural evidence for an anion-directing track in the hen ovotransferrin N-lobe: Implications for transferrin synergistic anion binding. Biochemistry 35(45):14294–14303.

Nagaoka S, Masaoka M, Zhang Q, Hasegawa M, Watanabe K (2002). Egg ovomucin attenuates hypercholesterolemia in rats and inhibits cholesterol absorption in Caco-2 cells. Lipids 37(3):267–272.

Nakamura S, Ogawa M, Nakai S, Kato A, Kitts DD (1998). Antioxidant activity of a Maillard-type phosvitin-galactomannan conjugate with emulsifying properties and heat stability. J Agric Food Chem 46(10):3958–3963.

Nakamura S, Kato A (2000). Multi-functional biopolymer prepared by covalent attachment of galactomannan to egg-white proteins through naturally occurring Maillard reaction. Nahrung 44(3):201–206.

Nakamura S, Saito M, Goto T, Saeki H, Ogawa M, Gotoh M, Gohya Y, Hwang JK (2000). Rapid formation of biologically active neoglycoprotein from lysozyme and xyloglucan hydrolysates through naturally occurring Maillard reaction. J Food Sci Nutr 5(2):65–69.

Nakane S, Tokumura A, Waku K, Sugiura T (2001). Hen egg yolk and white contain high amounts of lysophosphatidic acids, growth factor-like lipids: Distinct molecular species compositions. Lipids 36(4):413–419.

Nakano K, Nakano T, Ahn DU, Sim JS (1994). Sialic acid contents in chicken eggs and tissues. Can J Anim Sci 74(4):601–606.

Nakano T, Lien KA, Fenton M, Sim JS (1996). Investigation of chicken egg yolk glycosaminoglycans. Biomed Res (Tokyo) 17(6):499–503.

Nakano T, Ikawa N, Ozimek L (2001). Extraction of glycosaminoglycans from chicken eggshell. Poultry Sci 80(5):681–684.

Nakano T, Ikawa NI, Ozimek L (2003). Chem composition of chicken eggshell and shell membranes. Poultry Sci 82(3):510–514.

Nau F, Pasco M, Desert C, Molle D, Croguennec T, Guerin-Dubiard C (2005). Identification and characterization of ovalbumin gene Y in hen egg white. J Agric Food Chem 53(6):2158–2163.

Naulia U, Singh KS (1998). Effect of dietary fishmeal and phosphorus levels on the performance, egg quality and mineral balances in layers. Indian J Poultry Sci 33(2):153–157.

Ngoka DA, Froning GW, Babji AS (1983). Effect of temperature on egg yolk characteristics of eggs from young and old laying hens. Poultry Sci 62(4):718–720.

Nielsen KL, Sottrup-Jensen L (1993). Evidence from sequence analysis that hen egg-white ovomacroglobulin (ovostatin) is devoid of an internal beta-Cys-gamma-Glu thiol ester. Biochim Biophys Acta 1162(1/2):230–232.

Nielsen KL, Holtet TL, Etzerodt M, Moestrup SK, Gliemann J, Sottrup-Jensen L, Thogersen HC (1996). Identification of residues in alpha-macroglobulins important for binding to the alpha-2-macroglobulin receptor/low density lipoprotein receptor-related protein. J Biol Chem 271(22):12909–12912.

Nielsen H (1998). Hen age and fatty acid composition of egg yolk lipid. Br Poultry Sci 39(1):53–56.

Nielsen JH, Kristiansen GH, Andersen HJ (2000). Ascorbic acid and ascorbic acid 6-palmitate induced oxidation in egg yolk dispersions. J Agric Food Chem 48(5): 1564–1568.

Nielsen H (2001). In situ solid phase extraction of phospholipids from heat-coagulated egg yolk by organic solvents. Lebens Wissensch Technol 34(8):526–532.

Nimtz M, Conradt HS, Mann K (2004). LacdiNAc (GalNAcβ1–4GlcNAc) is a major motif in N-glycan structures of the chicken eggshell protein ovocleidin-116. Biochim Biophys Acta 1675(1–3):71–80.

Nomura S, Suzuki H, Masaoka T, Kurabayashi K, Minegishi Y, Goshima H, Kodama Y, Kitajima M, Nomoto K, Ishii H (2004). Effect of dietary anti-urease immumoglobulin Y on *Helicobacter pylori* infection. Gastroenterology 126(4 Suppl 2): A75.

Nomura S, Suzuki H, Masaoka T, Kurabayashi K, Ishii H, Kitajima M, Nomoto K, Hibi T (2005). Effect of dietary anti-urease immunoglobulin Y on *Helicobacter pylori* infection in Mongolian gerbils. Helicobacter 10(1):43–52.

Novak C, Scheideler SE (2001). Long-term effects of feeding flaxseed-based diets. 1. Egg production parameters, components, and eggshell quality in two strains of laying hens. Poultry Sci 80(10):1480–1489.

Nys Y, Hincke MT, Arias JL, Garcia-Ruiz JM, Solomon SE (1999). Avian eggshell mineralization. Poultry Avian Biol Rev 10:142–166.

Nys Y (2000). Dietary carotenoids and egg yolk coloration: A review. Archiv Gefluegelk 64(2):45–54.

Nys Y, Gautron J, Mckee MD, Garcia-Ruiz JM, Hincke MT (2001). Biochemical and functional characterization of eggshell matrix proteins in hens. World's Poultry Sci J 57(4):401–413.

Obara A, Obiedzinski M, Kolczak T (2006). The effect of water activity on cholesterol oxidation in spray- and freeze-dried egg powders. Food Chem 95(2):173–179.

Ogawa N, Tanabe H (1990). Effects of washing and oiling on electrophoretic patterns of albumen of the stored chicken eggs. Jpn Poultry Sci (Nihon Kakin Gakkai shi) 27(1):16–20.

Oguro T, Ohaki Y, Asano G, Ebina T, Watanabe K (2001). Ultrastructural and immunohistochemical characterization on the effect of ovomucin in tumor angiogenesis. Jpn J Clin Electron Microsc 33(2):89–99.

Oku T, Kato H, Kunishige-Taguchi T, Hattori M, Wada K, Hayashi M (1996). Stability of fat soluble components such as n-3 polyunsaturated fatty acids and physicochemical properties in EPA- and DHA-enriched eggs. Jpn J Nutr 54(2):39–49.

Okubo T, Akachi S, Hatta H (1997). Structure of hen eggs and physiology of egg laying. In: Yamamoto T, Juneja LR, Hatta H, Kim M, eds. Hen Eggs: Their Basic and Applied Science. New York: CRC Press; pp 1–12.

Olsson NU, Kaufmann P, Dzeletovic S (1997). Preparation and gas chromatographic-mass spectrometric analysis of N-acetyl-O-trimethylsilyl derivatives of long-chain base residues of natural sphingomyelin. J Chromatogr B 698(1–2):1–8.

Ostrowski-Meissner H, Ohshima M, Yokota H (1995). Hypocholesterolemic activity of a commercial high-protein leaf extract used as a natural source of pigments for laying hens and growing chickens. Jpn Poultry Sci 32(3):184–193.

Ovesen L, Brot C, Jakobsen J (2003). Food contents and biological activity of 25-hydroxyvitamin D: A vitamin D metabolite to be reckoned with. Ann Nutr Metab 47(3/4):107–113.

Pal L, Dublecz K, Husveth F, Wagner L, Bartos A, Kovacs G (2002). Effect of dietary fats and vitamin E on fatty acid composition, vitamin A and E content and oxidative stability of egg yolk. Archiv Gefluegelk 66(6):251–257.

Palacios LE, Wang T (2005a). Extraction of egg-yolk lecithin. J Am Oil Chem Soc 82(8):565–569.

Palacios LE, Wang T (2005b). Egg-yolk lipid fractionation and lecithin characterization. J Am Oil Chem Soc 82(8):571–578.

Panheleux M, Bain M, Fernandez MS, Morales I, Gautron J, Arias JL, Solomon SE, Hincke M, Nys Y (1999). Organic matrix composition and ultrastructure of eggshell: A comparative study. Br Poultry Sci 40(2):240–252.

Panheleux M, Nys Y, Williams J, Gautron J, Boldicke T, Hincke MT (2000). Extraction and quantification by ELISA of eggshell organic matrix proteins (ovocleidin-17, ovalbumin, ovotransferrin) in shell from young and old hens. Poultry Sci 79(4): 580–588.

Pappas AC, Acamovic T, Sparks NHC, Surai PF, McDevitt RM (2005). Effects of supplementing broiler breeder diets with organic selenium and polyunsaturated fatty acids on egg quality during storage. Poultry Sci 84(6):865–874.

Park PJ, Jung WK, Nam KS, Shahidi F, Kim SK (2001). Purification and characterization of antioxidative peptides from protein hydrolysate of lecithin-free egg yolk. J Am Oil Chem Soc 78(6):651–656.

Park BS, Kang HK, Jang A (2005a). Influence of feeding beta-cyclodextrin to laying hens on the egg production and cholesterol content of egg yolk. Asian-Austral J Anim Sci 18(6):835–840.

Park SW, Namkung H, Ahn HJ, Paik IK (2005b). Enrichment of vitamins D3, K and iron in eggs of laying hens. Asian-Austral J Anim Sci 18(2):226–229.

Pellegrini A, Hulsmeier A, Hunziker P, Thomas U (2004). Proteolytic fragments of ovalbumin display antimicrobial activity. Biochim Biophys Acta 1672(2):76–85.

Petersen SB, Jonson V, Fojan P, Wimmer R, Pedersen S (2004). Sorbitol prevents the self-aggregation of unfolded lysozyme leading to an up to 13°C stabilisation of the folded form. J Biotechnol 114(3):269–278.

Ponsano EHG, Pinto MF, Garcia-Neto M, Lacava PM (2004). *Rhodocyclus gelatinosus* biomass for egg yolk pigmentation. J Appl Poultry Res 13(3):421–425.

Poser R, Stuebinger M, Kroeckel L, Schwaegele F (2003). Influence of hen age on chemical and physical changes of egg components during storage. Mitteil Bund Fleischforsch Kulmbach 42(161):233–240.

Poser R, Stuebinger M, Kroeckel L, Schwaegele F (2004). Egg components during storage. Influence of hen age on chemical and physical changes. Fleischwirtschaft 84(7):113–116.

Poznanski J (2006). NMR-based localization of ions involved in salting out of hen egg white lysozyme. Acta Biochim Polonica 53(2):421–424.

Puthpongsiriporn U, Scheideler SE, Sell JL, Beck MM (2001). Effects of vitamin E and C supplementation on performance, in vitro lymphocyte proliferation, and antioxidant status of laying hens during heat stress. Poultry Sci 80(8):1190–1200.

Quirce S, Maranon F, Umpierrez A, Heras MDL, Fernandez-Caldas E, Sastre J (2001). Chicken serum albumin (Gal d 5*) is a partially heat-labile inhalant and food allergen implicated in the bird-egg syndrome. Allergy 56(8):754–762.

Rahimi G (2005). Dietary forage legume (*Onobrychis altissima* grossh.) supplementation on serum/yolk cholesterol, triglycerides and egg shell characteristics in laying hens. Int J Poultry Sci 4(10):772–776.

Raikos V, Hansen R, Campbell L, Euston SR (2006). Separation and identification of hen egg protein isoforms using SDS-PAGE and 2D gel electrophoresis with MALDI-TOF mass spectrometry. Food Chem 99(4):702–710.

Raj GD, Latha B, Chandrasekhar MS, Thiagarajan V (2004). Production, characterization and application of monoclonal antibodies against chicken IgY. Veterinarski Arhiv 74(3):189–199.

Raman B, Ramakrishna T, Mohan-Rao C (1996). Refolding of denatured and denatured/reduced lysozyme at high concentrations. J Biol Chem 271(29):17067–17072.

Rao PF, Liu ST, Wei Z, Li JC, Chen RM, Chen GR, Zheng YQ (1997). Isolation of flavoprotein from chicken egg white by a single-step DEAE ion-exchange chromatographic procedure. J Food Biochem 20(6):473–479.

Reddy RC, Lilie H, Rudolph R, Lange C (2005). L-Arginine increases the solubility of unfolded species of hen egg white lysozyme. Protein Sci 14(4):929–935.

Reyes-Grajeda JP, Jauregui-Zuniga D, Rodriguez-Romero A, Hernandez-Santoyo A, Bolanos-Garcia VM, Moreno A (2002). Crystallization and preliminary X-ray analysis of ovocleidin-17 a major protein of the *Gallus gallus* eggshell calcified layer. Protein Peptide Lett 9(3):253–257.

Reyes-Grajeda JP, Moreno A, Romero A (2004). Crystal structure of ovocleidin-17, a major protein of the calcified *Gallus gallus* eggshell—implications in the calcite mineral growth pattern. J Biol Chem 279(39):40876–40881.

Rho JH, Kim MH, Kim YB, Sung KS, Lee NH (2005). Formation and processing properties of anti-*Salmonella gallinarum* specific IgY from yolk. J Anim Sci Technol 47(4):637–646.

Rizzi L, Simioli M, Roncada P, Zaghini A (2003). Aflatoxin B1 and clinoptilolite in feed for laying hens: Effects on egg quality, mycotoxin residues in livers, and hepatic mixed-function oxygenase activities. J Food Protect 66(5):860–865.

Roberts JR, Choct M (2006). Effects of commercial enzyme preparations on egg and eggshell quality in laying hens. Br Poultry Sci 47(4):501–510.

Romanoff AL, Romanoff A (1949). The Avian Egg. New York: John Wiley and Sons.

Rosano C, Arosio P, Bolognesi M (1999). The X-ray three-dimensional structure of avidin. Biomol Eng 16(1–4):5–12.

Rubilar OE, Healy MG, Healy A (2006). Bioprocessing of avian eggshells and eggshell membranes using lactic acid bacteria. J Chem Technol Biotechnol 81(6):900–911.

Ruiz J, Lunam CA (2000). Ultrastructural analysis of the eggshell: Contribution of the individual calcified layers and the cuticle to hatchability and egg viability in broiler breeders. Br Poultry Sci 41(5):584–592.

Ruiz-Arguello MB, Veiga MP, Arrondo JLR, Goni FM, Alonso A (2002). Sphingomyelinase cleavage of sphingomyelin in pure and mixed lipid membranes. Influence of the physical state of the sphingolipid. Chem Phys Lipids 114(1):11–20.

Rupa P, Mine Y (2006). Engineered recombinant ovomucoid third domain can modulate allergenic response in Balb/c mice model. Biochem Biophys Res Commun 342(3):710–717.

Sakamoto R, Nishikori S, Shiraki K (2004). High temperature increases the refolding yield of reduced lysozyme: Implication for the productive process for folding. Biotechnol Progress 20(4):1128–1133.

Salvante KG, Wallowitz M, Walzem RL, Williams TD (2001). Characterization of very-low density lipoprotein particle size in egg-laying birds. Am Zool 41(6):1575.

Samata T, Kubota Y (1997). Two-dimensional gel electrophoresis of the organic matrix in chicken eggshell. Jpn J Electroph 41(6):357–360.

Santoso U, Cahyanto MN, Sulistiawati D, Zuprizal Irianto HE (1999). High omega-3 fatty acid eggs produced by laying hens fed with sardine oil. Indon Food Nutr Progress 6(2):39–43.

Sari M, Aksit M, Ozdogan M, Basmacioglu H (2002). Effects of addition of flaxseed to diets of laying hens on some production characteristics, levels of yolk and serum cholesterol, and fatty acid composition of yolk. Archiv Gefluegelk 66(2): 75–79.

Sarker SA, Casswall TH, Juneja LR, Hoq E, Hossain I, Fuchs GJ, Hammarstrom L (2001). Randomized, placebo-controlled, clinical trial of hyperimmunized chicken egg yolk immunoglobulin in children with rotavirus diarrhea. J Pediatr Gastroenterol Nutr 32(1):19–25.

Saxena I, Tayyab S (1997). Protein proteinase inhibitors from avian egg whites. CMLS Cell Mol Life Sci 53(1):13–23.

Scarlata S, Gupta R, Garcia P, Keach H, Shah S, Kasireddy CR, Bittman R, Rebecchi MJ (1996). Inhibition of phospholipase C-delta-1 catalytic activity by sphingomyelin. Biochemistry 35(47):14882–14888.

Schaefer A, Drewes W, Schwaegele F (1999). Effect of storage temperature and time on egg white protein. Nahrung 43(2):86–89.

Schaefer K, Maenner K, Sagredos A, Eder K, Simon O (2001). Incorporation of dietary linoleic and conjugated linoleic acids and related effects on eggs of laying hens. Lipids 36(11):1217–1222.

Scheideler SE, Froning GW (1996). The combined influence of dietary flaxseed variety, level, form, and storage conditions on egg production and composition among vitamin E-supplemented hens. Poultry Sci 75(10):1221–1226.

Scheideler SE, Jaroni D, Froning G (1998). Strain and age effects on egg composition from hens fed diets rich in n-3 fatty acids. Poultry Sci 77(2):192–196.

Schlatterer J, Breithaupt DE (2006). Xanthophylls in commercial egg yolks: Quantification and identification by HPLC and LC-(APCI)MS using a C30 phase. J Agric Food Chem 54(6):2267–2273.

Scruggs P, Filipeanu CM, Yang J, Chang JK, Dun NJ (2004). Interaction of ovokinin (2-7) with vascular bradykinin 2 receptors. Regul Peptides 120(1–3):85–91.

Seko A, Koketsu M, Nishizono M, Enoki Y, Ibrahim HR, Juneja LR, Kim M, Yamamoto T (1997). Occurrence of a sialylglycopeptide and free sialylglycans in hen's egg yolk. Biochim Biophys Acta 1335(1–2):23–32.

Seo JH, Lee JW, Lee YS, Lee SY, Kim MR, Yook HS, Byun MW (2004). Change of an egg allergen in a white layer cake containing gamma-irradiated egg white. J Food Protect 67(8):1725–1730.

Shafey TM (1996). The relationship between age and egg production, egg components and lipoprotein, lipids and fatty acids of the plasma and eggs of laying hens. J Appl Anim Res 10(2):155–162.

Shafey TM, Dingle JG, Kostner K (1999). Effect of dietary alpha-tocopherol and corn oil on the performance and on the lipoproteins, lipids, cholesterol and alpha-tocopherol concentrations of the plasma and eggs of laying hens. J Appl Anim Res 16(2):185–194.

Shimamoto C, Tokioka S, Hirata I, Tani H, Ohishi H, Katsu K (2002). Inhibition of *Helicobacter pylori* infection by orally administered yolk-derived anti-*Helicobacter pylori* antibody. Hepato-Gastroenterology 49(45):709–714.

Shimizu Y, Arai K, Ise S, Shimasaki H (2001). Dietary fish oil for hens affects the fatty acid composition of egg yolk phospholipids and gives a valuable food with an ideal balance of n-6 and n-3 essential fatty acids for human nutrition. J Oleo Sci 50(10): 797–803.

Shin SO, Kim JW (2002). Production of a specific yolk antibody against enterotoxigenic *E. coli* F41 fimbrial antigen. J Anim Sci Technol 44(5):633–642.

Shin JH, Yang M, Nam SW, Kim JT, Myung NH, Bang WG, Roe IH (2002). Use of egg yolk-derived immunoglobulin as an alternative to antibiotic treatment for control of *Helicobacter pylori* infection. Clin Diagn Lab Immunol 9(5):1061–1066.

Shiraki K, Kudou M, Nishikori S, Kitagawa H, Imanaka T, Takagi M (2004). Arginine ethylester prevents thermal inactivation and aggregation of lysozyme. Eur J Biochem 271(15):3242–3247.

Shiraki K, Kudou M, Sakamoto R, Yanagihara I, Takagi M (2005). Amino acid esters prevent thermal inactivation and aggregation of lysozyme. Biotechnol Progress 21(2):640–643.

Shu Y, Maki S, Nakamura S, Kato A (1998). Double-glycosylated lysozyme at positions 19 and 49 constructed by genetic modification and its surface functional properties. J Agric Food Chem 46(6):2433–2438.

Sikder AC, Chowdhury SD, Rashid MH, Sarker AK, Das SC (1998). Use of dried carrot meal (DCM) in laying hen diet for egg yolk pigmentation. Asian-Austral J Anim Sci 11(3):239–244.

Sim JS (1998). Designer eggs and their nutritional and functional significance. World Rev Nutr Dietet 83:89–101.

Skrtic I, Vitale L (1994). Methionine-preferring broad specificity aminopeptidase from chicken egg-white. Compar Biochem Physiol. Pt B Biochem Mol Biol 107(3): 471–478.

Smeller L, Meersman F, Heremans K (2006). Refolding studies using pressure: The folding landscape of lysozyme in the pressure-temperature plane. Biochim Biophys Acta 1764(3):497–505.

Smith DJ, King WF, Godiska R (2001). Passive transfer of immunoglobulin Y antibody to *Streptococcus mutans* glucan binding protein B can confer protection against experimental dental caries. Infect Immun 69(5):3135–3142.

Souza P, Souza HBA de, Barbosa JC, Gardini CHC, Neves M das (1997). Effect of laying hens age on the egg quality maintained at room temperature. Ciencia Tecnol Aliment 17(1):49–52.

Stadelman WJ, Cotterill OJ, eds (1995). Egg Science and Technology, 4th ed. Binghamton, NY: Food Products Press, Haworth Press.

Sugimoto Y, Kusakabe T, Nagaoka S, Nirasawa T, Tatsuguchi K, Fujii M, Aoki T, Koga K (1996). A proteinase inhibitor from egg yolk of hen is an ovoinhibitor analog. Biochim Biophys Acta 1295(1):96–102.

Sugimoto Y, Sanuki S, Ohsako S, Higashimoto Y, Kondo M, Kurawaki J, Ibrahim HR, Aoki T, Kusakabe T, Koga K (1999). Ovalbumin in developing chicken eggs migrates from egg white to embryonic organs while changing its conformation and thermal stability. J Biol Chem 274(16):11030–11037.

Sugino H, Ishikawa M, Nitoda T, Koketsu M, Raj-Juneja L, Kim M, Yamamoto T (1997a). Antioxidative activity of egg yolk phospholipids. J Agric Food Chem 45(3):551–554.

Sugino H, Nitoda T, Juneja LR (1997b). General chemical composition of hen eggs. In: Yamamoto T, Juneja LR, Hatta H, Kim M, eds. Hen Eggs: Their Basic and Applied Science. New York: CRC Press; pp 13–24.

Sujak A, Gabrielska J, Grudzinski W, Borc R, Mazurek P, Gruszecki W (1999). Lutein and zeaxanthin as protectors of lipid membranes against oxidative damage: The structural aspects. Arch Biochem Biophys 371(2):301–307.

Sun S, Mo W, Ji Y, Liu S (2001). Preparation and mass spectrometric study of egg yolk antibody (IgY) against rabies virus. Rapid Commun Mass Spectrom 15(9):708–712.

Sunwoo HH, Lee EN, Menninen K, Suresh MR, Sim JS (2002). Growth inhibitory effect of chicken egg yolk antibody (IgY) on *Escherichia coli* O157:H7. J Food Sci 67(4): 1486–1494.

Surai PF, Speake BK (1998). Distribution of carotenoids from the yolk to the tissues of the chick embryo. J Nutr Biochem 9(11):645–651.

Surai PF, Ionov IA, Kuklenko TV, Kostjuk IA, MacPherson A, Speake BK, Noble RC, Sparks NHC (1998). Effect of supplementing the hen's diet with vitamin A on the accumulation of vitamins A and E, ascorbic acid and carotenoids in the egg yolk and in the embryonic liver. Br Poultry Sci 39(2):257–263.

Surai PF (2002). Selenium on poultry nutrition 2. Reproduction, egg and meat quality and practical applications. World's Poultry Sci J 58:431–450.

Suzuki Y, Okamoto M (1997). Production of hen's eggs rich in vitamin K. Nutr Res 17(10):1607–1615.

Suzuki H, Nomura S, Masaoka T, Goshima H, Kamata N, Kodama Y, Ishii H, Kitajima M, Nomoto K, Hibi T (2004). Effect of dietary anti-*Helicobacter pylori*-urease immunoglobulin Y on *Helicobacter pylori* infection. Aliment Pharmacol Ther 20(Suppl 1):185–192.

Swamy MJ, Heimburg T, Marsh D (1996). Fourier-transform infrared spectroscopic studies on avidin secondary structure and complexation with biotin and biotin-lipid assemblies. Biophys J 71(2):840–847.

Szymczyk B, Pisulewski PM (2003). Effects of dietary conjugated linoleic acid on fatty acid composition and cholesterol content of hen egg yolks. Br J Nutr 90(1):93–99.

Szymczyk B, Pisulewski PM (2005). Effects of dietary conjugated linoleic acid isomers and vitamin E on fatty acid composition and cholesterol content of hen egg yolks. J Anim Feed Sci 14(1):109–123.

Takagi K, Teshima R, Okunuki H, Itoh S, Kawasaki N, Kawanishi T, Hayakawa T, Kohno Y (2005). Kinetic analysis of pepsin digestion of chicken egg white ovomucoid and allergenic potential of pepsin fragments. Int Arch Allergy Immunol 136(1):23–32.

Takahashi K, Shirai K, Kitamura M, Hattori M (1996). Soluble egg shell membrane protein as a regulating material for collagen matrix reconstruction. Biosci Biotechnol Biochem 60(8):1299–1302.

Takeuchi S, Saito T, Itoh T (1992). Rapid analysis of chicken egg white proteins via high-performance liquid chromatography. Anim Sci Technol 63(6):598–600.

Tanizaki H, Tanaka H, Iwata H, Kato A (1997). Activation of macrophages by sulfated glycopeptides in ovomucin, yolk membrane, and chalazae in chicken eggs. Biosci Biotechnol Biochem 61(11):1883–1889.

Teguia A (2000). A note on the effect of feeding local forages to commercial layers on egg production and yolk colour. J Anim Feed Sci 9(2):391–396.

Tenuta-Filho A, Morales-Aizpurua IC, Procopio de Moura AF, Kitahara SE (2003). Cholesterol oxides in foods. Rev Brasil Ciencias Farmaceut 39(3):319–325.

Ternes W, Kraemer P, Menzel R, Zeilfelder K (1995). Distribution of fat-soluble vitamins (A, E, D2, D3) and carotenes (lutein, zeaxanthin) in the lipids from granule and plasma of yolk. Archiv Gefluegelk 59(5):261–268.

Tokarzewski S (2002). Influence of enrofloxacin and chloramphenicol on the level of IgY in serum and egg yolk after immunostimulation of hens with *Salmonella enteritidis* antigens. Pol J Vet Sci 5(3):151–158.

Touch V, Hayakawa S, Saitoh K (2004). Relationships between conformational changes and antimicrobial activity of lysozyme upon reduction of its disulfide bonds. Food Chem 84(3):421–428.

Toussant MJ, Latshaw JD (1999). Ovomucin content and composition in chicken eggs with different interior quality. J Sci Food Agric 79(12):1666–1670.

Trziszka T, Saleh Y, Kopec W, Wojciechowska-Smardz I, Oziemblowski M (2004). Changes in the activity of lysozyme and cystatin depending on the age of layers and egg treatment during processing. Archiv Gefluegelk 68(6):275–279.

Tsai WT, Yang JM, Lai CW, Cheng YH, Lin C, Yeh CW (2006). Characterization and adsorption properties of eggshells and eggshell membrane. Bioresource Technol 97(3):488–493.

Tsuge Y, Shimoyamada M, Watanabe K (1996). Differences in hemagglutination inhibition activity against bovine rotavirus and hen Newcastle disease virus based on the subunits in hen egg white ovomucin. Biosci Biotechnol Biochem 60(9):1505–1506.

Tsuge Y, Shimoyamada M, Watanabe K (1997a). Structural features of Newcastle disease virus- and anti-ovomucin antibody-binding glycopeptides from pronase-treated ovomucin. J Agric Food Chem 45(7):2393–2398.

Tsuge Y, Shimoyamada M, Watanabe K (1997b). Bindings of ovomucin to Newcastle disease virus and anti-ovomucin antibodies and its heat stability based on binding abilities. J Agric Food Chem 45(12):4629–4634.

Tu YY, Chen CC, Chang HM (2001). Isolation of immunoglobulin in yolk (IgY) and rabbit serum immunoglobulin G (IgG) specific against bovine lactoferrin by immunoaffinity chromatography. Food Res Int 34(9):783–789.

USDA (U.S. Department of Agriculture), Agricultural Research Service. (2006). USDA National Nutrient Database for Standard Reference, release 19. Nutrient Data Lab Home Page, `http://www.ars.usda.gov/ba/bhnrc/ndl` (accessed Dec 5, 2006).

USDA (2000). Egg Grading Manual. Agricultural Handbook 75. `www.ams.usda.gov/poultry/pdfs/EggGrading%20manual.pdf` (accessed Dec 5, 2006).

Uuganbayar D, Bae IH, Choi KS, Shin IS, Firman JD, Yang CJ (2005). Effects of green tea powder on laying performance and egg quality in laying hens. Asian-Austral J Anim Sci 18(12):1769–1774.

Valcarce G, Nusblat A, Florin-Christensen J, Nudel BC (2002). Bioconversion of egg cholesterol to pro-vitamin D sterols with *Tetrahymena thermophila*. J Food Sci 67(6):2405–2409.

Van der Plancken I, Van Remoortere M, Van Loey A, Hendrickx ME (2004). Trypsin inhibition activity of heat-denatured ovomucoid: A kinetic study. Biotechnol Progress 20(1):82–86.

Van der Veen M, Norde W, Cohen Stuart M (2005). Effects of succinylation on the structure and thermostability of lysozyme. J Agric Food Chem 53(14):5702–5707.

Veiga MP, Arrondo JLR, Goni FM, Alonso A, Marsh D (2001). Interaction of cholesterol with sphingomyelin in mixed membranes containing phosphatidylcholine, studied by spin-label ESR and IR spectroscopies. A possible stabilization of gel-phase sphingolipid domains by cholesterol. Biochemistry 40(8):2614–2622.

Verdoliva A, Basile G, Fassina G (2000). Affinity purification of immunoglobulins from chicken egg yolk using a new synthetic ligand. J Chromatogr B 749(2): 233–242.

Von Dreele RB (2005). Binding of N-acetylglucosamine oligosaccharides to hen egg-white lysozyme: A powder diffraction study. Acta Crystallogr Sect D Biol Crystallogr 61(Pt 1):22–32.

Vrzhesinskaya OA, Philimonova IV, Kodentsova OV, Beketova NA, Kodentsova VM (2005). Influenze of chitosan feeding of laying hens on egg vitamin and cholesterol content. Voprosy Pitaniya 74(3):28–31.

Wang JJ, Pan TM (2003). Effect of red mold rice supplements on serum and egg yolk cholesterol levels of laying hens. J Agric Food Chem 51(16):4824–4829.

Wang YW, Cherian G, Sunwoo HH, Sim JS (2000). Dietary polyunsaturated fatty acids significantly affect laying hen lymphocyte proliferation and immunoglobulin G concentration in serum and egg yolk. Can J Anim Sci 80(4):597–604.

Watanabe K, Tsuge Y, Shimoyamada M (1998a). Binding activities of pronase-treated fragments from egg white ovomucin with anti-ovomucin antibodies and Newcastle disease virus. J Agric Food Chem 46(11):4501–4506.

Watanabe K, Tsuge Y, Shimoyamada M, Ogama N, Ebina T (1998b). Antitumor effects of pronase-treated fragments, glycopeptides, from ovomucin in hen egg white in a double grafted tumor system. J Agric Food Chem 46(8):3033–3038.

Watkins BA (1995). The nutritive value of the egg. In: Stadelman WJ, Cotterill OJ, eds. Egg Science and Technology, 4th ed. New York: Food Products Press; pp 177–194.

Watkins BA, Feng S, Strom AK, DeVitt AA, Yu L, Li Y (2003). Conjugated linoleic acids alter the fatty acid composition and physical properties of egg yolk and albumen. J Agric Food Chem 51(23):6870–6876.

Weiner KXB, Hayes GR, Lucas JJ (1997). Binding specificity of avian heat shock protein 108. Biochem Biophys Res Commun 240(3):673–676.

Welzig E, Pichler H, Krsaka R, Knopp D, Niessner R (2000). Development of an enzyme immunoassay for the determination of the herbicide metsulfuron-methyl based on chicken egg yolk antibodies. Int J Env Anal Chem 78(3/4):279–288.

Wesierska E, Saleh Y, Trziszka T, Kopec W, Siewinski M, Korzekwa K (2005). Antimicrobial activity of chicken egg white cystatin. World J Microbiol Biotechnol 21(1):59–64.

Wilhelmson M, Carlander D, Kreuger A, Kollberg H, Larsson A (2005). Oral treatment with yolk antibodies for the prevention of *C-albicans* infections in chemotherapy treated children. A feasibility study. Food Agric Immunol 16(1):41–45.

Wong YC, Herald TJ, Hachmeister KA (1996). Evaluation of mechanical and barrier properties of protein coatings on shell eggs. Poultry Sci 75(3):417–422.

Wong PYY, Kitts DD (2003). Physicochemical and functional properties of shell eggs following electron beam irradiation. J Sci Food Agric 83(1):44–52.

Woodall AA, Britton G, Jackson MJ (1997). Carotenoids and protection of phospholipids in solution or in liposomes against oxidation by peroxyl radicals: Relationship between carotenoid structure and protective ability. Biochim Biophys Acta 1336(3):575–586.

Xiaohong C, Fengcai W, Xiaoli B, Yan Z (2003). Purification of IgY antibody from hen egg yolks by low-temperature ethanol. Food Ferment Indust 29(3):66–70.

Xiaoting Z, Zirong X, Yulong M (1998). Effect of diludine on laying performance and approach to mechanism of the effects in hens. J Zhejiang Agric Univ 24(3):297–302.

Yajima H, Yamamoto H, Nagaoka M, Nakazato K, Ishii T, Niimura N (1998). Small-angle neutron scattering and dynamic light scattering studies of N- and C-terminal fragments of ovotransferrin. Biochim Biophys Acta 1381(1):68–76.

Yalcin S, Kahraman K, Yalcin S, Yalcin SS, Dedeoglu HE (2004). Effects of supplementary iodine on the performance and egg traits of laying hens. Br Poultry Sci 45(4):499–503.

Yalcin S, Ebru-Onbasilar E, Reisli Z, Yalcin S (2006). Effect of garlic powder on the performance, egg traits and blood parameters of laying hens. J Sci Food Agric 86(9):1336–1339.

Yamamoto Y, Omori M (1994). Antioxidative activity of egg yolk lipoproteins. Biosci Biotechnol Biochem 58(9):1711–1713.

Yamamoto Y, Juneja LR, Hatta H, Kim M, eds (1997). Hen Eggs. Their Basic and Applied Science. Boca Raton: CRC Press.

Yamamoto K, Tanaka T, Fujimori K, Kang CS, Ebihara H, Kanamori J, Kadowaki S, Tochikura T, Kumagai H (1998). Characterization of *Bacillus* sp. endo-beta-N-acetylglucosaminidase and its application to deglycosylation of hen ovomucoid. Biotechnol Appl Biochem 28(3):235–242.

Yamamoto H, Takahashi N, Yamasaki M, Arii Y, Hirose M (2003). Thermostabilization of ovalbumin by an alkaline treatment: Examination for the possible implications of an altered serpin loop structure. Biosci Biotechnol Biochem 67(4):830–837.

Yamamura JI, Adachi T, Aoki N, Nakajima H, Nakamura R, Matsuda T (1995). Precursor-product relationship between chicken vitellogenin and the yolk proteins: The 40 kDa yolk plasma glycoprotein is derived from the C-terminal cysteine-rich domain of vitellogenin II. Biochim Biophys Acta 1244(2/3):384–394.

Yannakopoulos AL, Tserveni-Gousi AS, Yannakakis S (1999). Effect of feeding flaxseed to laying hens on the performance and egg quality and fatty acid composition of egg yolk. Archiv Gefluegelk 63(6):260–263.

Yao Z, Zhang M, Sakahara H, Saga T, Arano Y, Konishi J (1998). Avidin targeting of intraperitoneal tumor xenografts. J Natl Cancer Instit (Bethesda, MD) 90(1):25–29.

Yoon TH, Kim IH (2002). Phosphatidylcholine isolation from egg yolk phospholipids by high-performance liquid chromatography. J Chromatogr A 949(1/2):209–216.

Yoshino K, Sakai K, Mizuha Y, Shimizuike A, Yamamoto S (2004). Peptic digestibility of raw and heat-coagulated hen's egg white proteins at acidic pH range. Int J Food Sci Nutr 55(8):635–640.

Zhao R, Xu GY, Liu ZZ, Li JY, Yang N (2006). A study on eggshell pigmentation: Biliverdin in blue-shelled chickens. Poultry Sci 85(3):546–549.

Zotte AD, Tomasello F, Andrighetto I (2005). The dietary inclusion of *Portulaca oleracea* to the diet of laying hens increases the n-3 fatty acids content and reduces the cholesterol content in the egg yolk. Ital J Anim Sci 4(Suppl 3):157–159.

2

BIOSYNTHESIS AND STRUCTURAL ASSEMBLY OF EGGSHELL COMPONENTS

Maxwell T. Hincke and Olivier Wellman-Labadie
Department of Cellular and Molecular Medicine, University of Ottawa, Ottawa, Ontano, Canada

Marc D. McKee
Faculty of Dentistry and Department of Anatomy and Cell Biology, McGill University, Montreal, Quebec, Canada

Joël Gautron and Yves Nys
INRA, UR83 Recherches Avicoles, Nouzilly, France

Karlheinz Mann
Max-Planck Institut für Biochemie, Martinsried, Germany

2.1. INTRODUCTION

The avian egg is a reproductive structure that has been shaped through evolution to resist physical, microbial, and thermal attack from an external and possibly aggressive environment, while satisfying the needs of the developing embryo. Many studies have been conducted on avian eggs and most have centered on the egg of the domestic chicken. This considerable body of work has provided insight into the function and structure of the eggshell. In conjunction with the eggshell membranes, the eggshell provides protection against physical damage, microorganisms, and small predators. This complex structure also regulates the exchange of metabolic gases and water, and provides calcium to the developing embryo. In this chapter we review the results of recent proteomic and genomic analyses of the eggshell and draw attention to the impact of these data on the current understanding of eggshell function. The egg's

Egg Bioscience and Biotechnology Edited by Yoshinori Mine

natural defenses have two components: (1) the shell, which acts as a physical barrier; and (2) a chemical system composed of endogenous antibacterial proteins that have been identified mainly in egg white and eggshell membranes. Changes in eggshell properties are directly related to increasing risk of egg contamination and risk of foodborne outbreaks for the consumer.

2.2. OVERVIEW OF EGGSHELL BIOSYNTHESIS

The egg is assembled as it passes through specialized regions of the oviduct. It is composed of a central yolk surrounded by the albumen, eggshell membranes, calcified eggshell and cuticle. The combination of albumen and yolk provide a perfect balance of the nutrients needed for embryonic growth; the egg therefore has a very high nutritive value for humans because of its diversity of components and ready availability. The yolk is the main source of energy and liposoluble vitamins for the developing embryo because it contains all lipidic components of the egg (Burley and Vadehra 1989). The composition of the yolk is approximately 33% lipids, 17% protein, and 48% water. The remaining components include free carbohydrates and inorganic elements. During the 2–3-h period when the yolk–ovum complex travels down the largest portion of the oviduct, the magnum, it progressively acquires the albumen (Nys et al. 1999, 2004).

The albumen is composed of 88% water and 10% protein, 54% of which is ovalbumin, 13% ovotransferrin, 11% ovomucoid, 8% ovoglobulins, 3.5% lysozyme, 3% β-ovomucin and 1.5% α-ovomucin (Burley and Vadehra 1989). The albumen, while providing water, salts, and proteins to the developing embryo, prevents the growth of microorganisms. The alkaline pH of the albumen and the presence of proteins such as ovotransferrin and lysozyme significantly reduce the growth of microorganisms (Deeming 2002). The albumen also stabilizes the developing embryo within the egg. Two fiber-like structures within the albumen, the chalazae, anchor the yolk in the thick egg white between the blunt and rounded poles. This keeps the yolk in the middle of the egg and permits limited rotation and some lateral displacement (Whittow 2000). This mechanism ensures that the blastoderm remains above the yolk even when the egg is rotated (Burley and Vadehra 1989).

As the embryo–yolk–albumen complex travels through the proximal (white) isthmus, the inner and outer shell membranes are acquired in a 1–2-h period. These are considered to be the innermost layers of the eggshell. The inner membranes remain uncalcified, while the fibers of the outer shell membrane penetrate the mammillary cones of the calcified shell (Arias et al. 1993; Nys et al. 2004). The membranes are constructed of a network of fibers, composed of 10% collagens (types I, V, and X) and 70–75% of other proteins and glycoproteins containing lysine-derived crosslinks (Wong et al. 1984; Arias et al. 1991; Fernandez et al. 1997). The membranes envelop the albumen, but are semipermeable and allow the exchange of gases and water while retaining the albumen proteins. In addition, they act as a physical and chemical barrier to bacteria. In the distal portion of the isthmus, or red isthmus, organic aggregates are deposited on the surface of the outer eggshell membranes in a quasiperiodic array, where calcium carbonate begins to aggregate and the mammillary knobs originate (Nys et al. 2004).

The incomplete egg then enters the uterus, or shell gland, where fluid is pumped into the albumen, causing it to swell to its final size at oviposition. During the 16–17-h period when the ovum rotates in the shell gland, calcification of the egg occurs, followed by the deposition of shell pigments and eggshell cuticle on the immobilized egg during the last 1.5 h before oviposition. This process occurs in the uterine fluid, which bathes the forming egg, and these components are secreted by the cells that line the uterus. The ionic and organic constituents of the uterine fluid change progressively during eggshell formation and can be subdivided into the stages of initiation (5 h), growth (12 h), and termination (1.5 h) of eggshell mineralization.

The calcified portion of the eggshell consists primarily of calcite, the most stable polymorph of calcium carbonate, and is progressively composed of the inner mammillary cone layer, central palisades, and the outer vertical crystal layers (Nys et al. 1999, 2004) (Figs. 2.1 and 2.2). The mammillary layer is composed of a regular array of cones or knobs, with highly organic cores, into which are embedded the individual fibers of the outer eggshell membrane. Within the mammillary cone layer, microcrystals of calcite are arranged with spherulitic texture, which facilitates the propagation of cracks during piping as well as the mobilization of calcium to nourish the embryo by dissolution of highly reactive calcite microcrystals (Nys et al. 2004). The palisade layer is made up of groups of columns that are perpendicular to the eggshell surface and extend outward from the mammillary cones. This layer ends at the vertical single crystal layer, which has a crystalline structure of a density higher than that of the palisade region. The outer region of the palisade layer is a tough structure made of large crystals where the external impacts are absorbed by thin intercrystalline organic layers that make intracrystalline crack propagation difficult (Nys et al. 2004). Pores throughout the eggshell permit the diffusion of metabolic gases and water vapor.

The outermost layer is the eggshell cuticle, a noncalcified organic layer that is deposited on the mineral surface. The cuticle is also known as "bloom" since it gives the freshly laid egg a glossy appearance (Burley and Vadehra 1989). At oviposition, the cuticle on the chicken's egg has a moist luster that disappears within 2–3 min and cannot be restored through subsequent wetting (Sparks 1994). The cuticle is a relatively thin layer of variable thickness (0.5–12.8 μm), is composed of glycoprotein, polysaccharides, lipids, and inorganic phosphorus, including hydroxyapatite crystals (Dennis et al. 1996; Whittow 2000; Fernandez et al. 2001). Eggs of some bird species, such as pigeons, lack a cuticle, while in other species, such as chickens, the cuticle may be absent in a small proportion of eggs (Burley and Vadehra 1989).

The cuticle is thought to play a role in controlling water exchange by repelling water or preventing its loss, and may function in limiting microbial colonization of the eggshell surface (Whittow 2000). This layer, as well as the outer

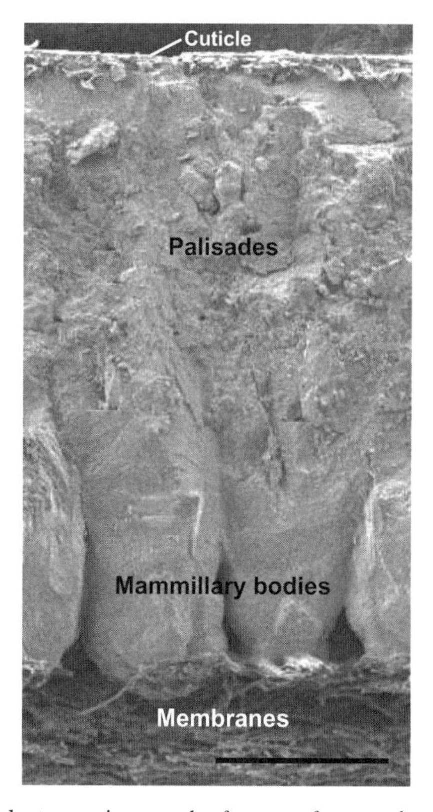

Figure 2.1. Scanning electron micrograph of a cross-fractured eggshell revealing internal structure. Major structural compartments include the outermost cuticle residing at the surface of the eggshell, the palisades region constituting more than half of the eggshell thickness, and the rounded mammillary bodies that abut directly against the eggshell membrane. Magnification bar equals 100 μm.

portion of the calcified shell, contains the eggshell pigments, which serve as camouflage, temperature control, and possibly as factors in parental recognition (Sparks 1994). Sparks and Board (1984) noted that chicken eggs lacking a cuticle readily absorbed water and that newly laid fowl eggs, where the cuticle was incomplete, were less able to resist bacterial penetration of the shell (Deeming 2002).

2.3. PROTEOMICS OF THE EGGSHELL

2.3.1. Biochemistry of Eggshell Matrix Proteins

Proteins are the functional units of biological processes, and consequently their study is of great interest. The term *proteome* refers to the complete set of proteins produced by a given cell or organism under defined conditions.

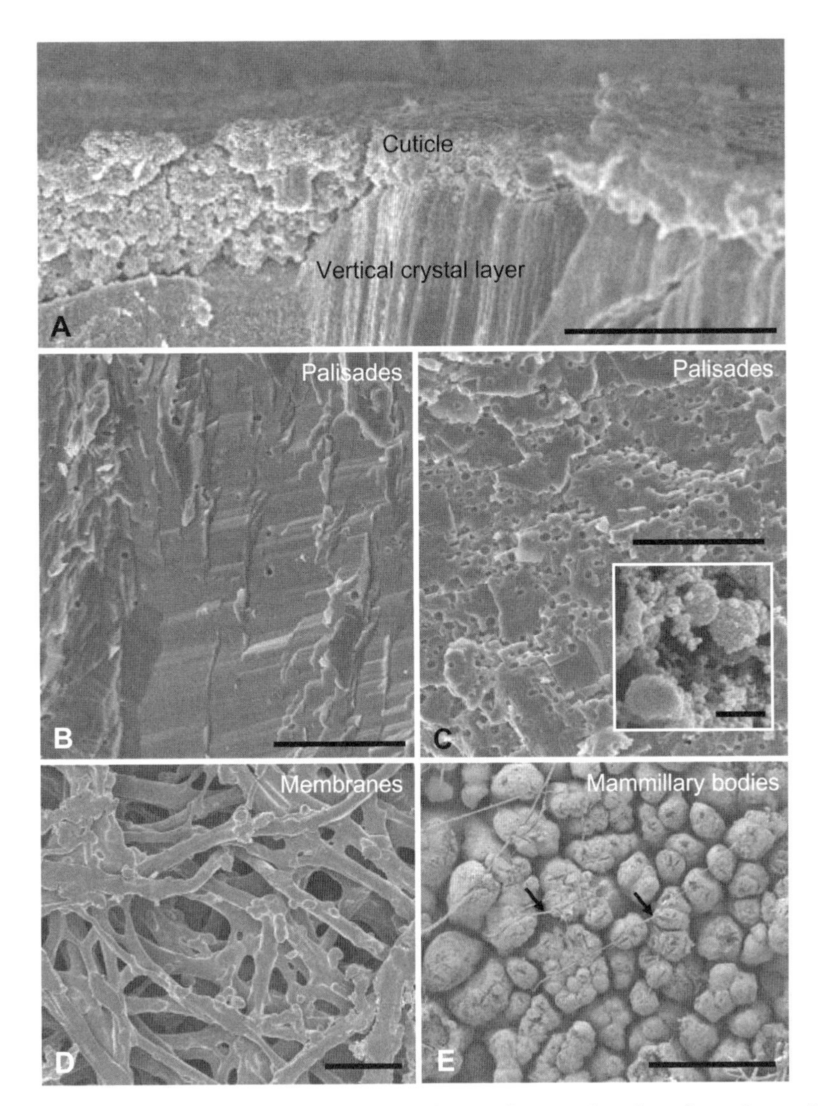

Figure 2.2. Scanning electron micrographs of cross-fractured and en face views of laid avian eggshell and membranes. (a) A mineralized, or partly mineralized, cuticle covers the outermost surface of the eggshell, residing directly apposed to the vertical crystal layer. This cuticle is heterogeneous in form and thickness, with considerable variations across the surface of the eggshell. The vertical crystal layer is characterized by vertical striations resulting from aligned crystallographic and proteinaceous orientations near the surface of the eggshell. (b, c) The bulk of the eggshell resides within the palisades layer, where large calcitic columns show multiple, terraced mineral cleavage planes when cross-fractured. In undecalcified eggshell, within the interiors of calcitic columns exposed by the fracture planes, vesicular voids can be observed ranging in abundance from occasional (b) to numerous (c). In fully decalcified eggshell palisades, organic, protein-rich vesicular structures create (or line) the voids [inset in (c)]. (d) En face view of the interior of the eggshell membranes after mechanical removal (peeling) from eggshell. This network of interconnecting fibers provides the substratum on which mineralization occurs and the eggshell is formed. (e) En face view of the tips of the mammillary bodies forming the interior lining of the eggshell after mechanical removal (peeling) of the membranes. A few residual fibers are apparent interlacing with the tips of the mamillary bodies (arrows). These mineralized mammillary bodies nucleate from the shell membranes shown in panel (d). Magnification bars equal (a) 100 μm; (b, c) 5 μm (inset 200 nm); (d) 10 μm; (e) 100 μm.

Proteomics is the study of the proteome, with the goal of identifying all proteins and their posttranslational modifications. The proteome is not static, but can change with time or under the influence of environmental conditions.

The eggshell mineral is associated with an organic matrix composed of proteins, glycoproteins, and proteoglycans, which are thought to influence the fabric of this biomaterial or to participate in its antimicrobial defenses. These eggshell matrix proteins can be released for study by demineralization of the eggshell. In the "classical" approach, the polypeptides contained in extra- and intramineral extracts, and in the precursor milieu (uterine fluid) where eggshell calcification takes place, have been analyzed by one-dimensional (1D)-electrophoresis (SDS-PAGE). A complex array of distinct protein bands was demonstrated in the soluble intra- and extramineral compartments (Hincke et al. 1992; Gautron et al. 1996) and in the uterine fluid, showing different patterns between the three stages of the eggshell calcification process [initial, growth, and terminal (Gautron et al. 1997)]. *N*-Terminal sequencing of the electrophoretic bands allowed the identification of egg white proteins such as ovalbumin, lysozyme, and ovotransferrin (Hincke 1995; Hincke et al. 2000a; Gautron et al. 2001b). However, *N*-Terminal and internal amino acid sequencing of protein bands revealed that a number of them did not correspond to previously identified sequences, and these were subject to more intensive investigation. Efficient purification schemes using ion exchange [diethylaminoethyl (DEAE)–sepharose and carboxymethyl (CM)–sepharose] and hydroxyapatite columns were developed to purify to homogeneity ovocleidin-17 (Hincke et al. 1995) and ovocalyxin-32 (Hincke et al. 2003) from eggshell extracts.

Other matrix proteins were characterized by a combination of molecular cloning, immunochemistry, and bioinformatics. A cDNA library from pooled RNA extracted from chicken uteruses that were harvested during the midphase of shell calcification was successfully used for the cloning of novel eggshell matrix proteins (Hincke et al. 1999; Gautron et al. 2007). Expression screening of this library, using polyclonal antisera raised to partially purified eggshell matrix proteins, allowed clones with the corresponding cDNA sequences to be identified. After purification, the plasmid inserts containing the cDNA coding for the corresponding protein were sequenced. The conceptual amino acid sequence was compared to partial amino acid sequencing data for proteins present in uterine fluid and eggshell extracts. This method allowed the identification of two novel eggshell matrix proteins: ovocleidin-116 (Hincke et al. 1999) and ovocalyxin-36 (Gautron et al. 2007). Another associated approach was to compare the available expression sequence tag (EST) sequences to partial protein or nucleotide sequences from egg components. This method was successfully used to identify a 32-kDa band abundant in uterine fluid at the terminal phase of shell calcification (Gautron et al. 2001a). The 32-kDa protein band was microsequenced to yield *N*-terminal and internal peptide sequences. Database searching using tBlastN with these sequences allowed the identification of several corresponding ESTs that were assembled to obtain a full-length cDNA sequence containing all peptides previously sequenced. This protein is novel and was named *ovocalyxin-32* (OCX32) (Gautron et al. 2001a).

Results of studies with these "classical" methods led to the concept that eggshell matrix protein components can be divided into three characteristic groups: (1) "egg white" proteins, which are also present in the eggshell—these include ovalbumin (Hincke 1995), lysozyme (Hincke et al. 2000a), and ovotransferrin (Gautron et al. 2001b); (2) ubiquitous proteins that are found in many tissues—this group includes osteopontin, a phosphorylated glycoprotein present in bone and other hard tissues (Pines et al. 1994; Hincke and St. Maurice 2000; Fernandez et al. 2003), and clusterin, a widely distributed secretory glycoprotein that is also found in chicken egg white (Mann et al. 2003); and (3) matrix proteins unique to the shell calcification process that are secreted by cells in specific regions of the oviduct where eggshell mineralization is initiated and continues to completion (red isthmus and uterus). More recently, a high-throughput tandem mass spectrometry approach (MS/MS) allowed the identification of >500 eggshell matrix proteins (Mann et al. 2006), including the most abundant proteins that were already known (see Section 2.3.2). Functional interpretation of these results will certainly lead to greater insight into eggshell formation and properties.

A more detailed description of the eggshell-specific matrix proteins is provided in the following section. These matrix components are termed *ovocleidins* (*ovo*, Latin—egg; *kleidoun*, Greek—to lock in) or *ovocalyxins* (*ovo*, Latin—egg; *calyx*, Latin—shell).

Ovocleidin-17 (OC17) was the first eggshell-specific matrix protein to be isolated and characterized (Hincke et al. 1995). Its X-ray structure has also been determined (Reyes-Grajeda et al. 2004). It is a *c*-type lectin-like phosphoprotein of 17 kDa (Mann and Siedler 1999) that occurs in glycosylated (23 kDa) and nonglycosylated forms in the shell matrix (Mann 1999). It displays sequence homology to mammalian lithostathine (De Reggi and Gharib 2001) and to fish type II antifreeze proteins (Ewart and Fletcher 1993). The properties of purified OC17 and its goose homolog (ansocalcin) have been investigated (Lakshminarayanan et al. 2002, 2003; Reyes-Grajeda et al. 2004). Moreover, a family of homologous eggshell matrix proteins has been identified in a large number of avian species. These studies revealed that ansocalcin (goose), struthiocalcin-1,2 (SCA1, SCA2; ostrich), dromaiocalcin-1,2 (DCA1, DCA2; emu), and rheacalcin-1,2 (RCA1, RCA2; rhea) form two groups on the basis of sequence identity, serine phosphorylation, and conservation of cysteine residues (Mann and Siedler 2004). Goose ansocalcin aligns reasonably well with proteins of group 1 (63–70% identity with SCA1, DCA1, RCA1), but OC17 has much less sequence identity with group 2 (37–39% with SCA2, DCA2, RCA2). It remains unclear why ratites differ from goose and chicken, in that they possess two forms of the *-calcin* matrix protein as their predominant eggshell matrix proteins. It is suggested that these differences are due to loss of one gene in modern birds (goose, chicken) (Mann and Siedler 2006). Functionally, OC17 and ansocalcin do not appear to be completely equivalent in their effect on calcite crystal growth *in vitro* (Lakshminarayanan et al. 2005; Reyes-Grajeda et al. 2004). Moreover, the fact that ovocleidin-17, but not ansocalcin, is largely destroyed by treating eggshell powder with bleach suggests a different location, or a different kind of interaction with mineral, of these proteins (Lakshminarayanan et al. 2005).

Ovocleidin-116 (OC116) is the first eggshell matrix protein that was cloned (Hincke et al. 1999). It is the core protein of ovoglycan, a dermatan sulfate proteoglycan localized in the shell of a number of galliform species (Panheleux et al. 1999), and is likely to play a fundamental role in eggshell formation since it potently modifies calcite crystal growth *in vitro* [(reviewed in Nys et al. (1999, 2004)]. The OC116 protein corresponds to the 120/200-kDa eggshell dermatan sulfate proteoglycan, which is recognized by a monoclonal antibody that is specific for an epitope on the core protein of avian versican (Fernandez et al. 2001). However, its sequence is quite unlike those of other calcified tissue proteoglycans (Hincke et al. 1999). Low-stringency Blast searching with the OC116 protein sequence generated restricted and poorly significant alignments to mammalian and chicken collagens (types I, II, VII, and IX), human perlecan (heparan sulfate proteoglycan), chicken aggrecan (chondroitin sulfate proteoglycan), chicken bone sialoprotein, and lustrin A (component of molluscan shell extracellular matrix). No homology with avian versican was apparent by Blast searching. A number of the common structural characteristics [i. e., domains that are EGF-like, *C*-type lectin-like and complement regulatory protein (CRP)-like] that are found found in aggrecan, PG-M/versican, neurocan, and brevican (Watanabe et al. 1998) are absent in OC116. However, significant homology between the *N*-termini of OC116 and a mammalian matrix extracellular phosphoglycoprotein (MEPE) was detected (see text below).

Investigations into the evolutionary genetics of vertebrate tissue mineralization suggest that OC116 is a member of the secretory calcium-binding phosphoprotein (SCPP) family (Kawasaki and Weiss 2006). A strong genetic linkage was detected between single-nucleotide polymorphisms (SNPs) in the osteopontin and OC116 genes during studies investigating the possible correlation between eggshell biomechanical properties and eggshell matrix protein SNPs (Hincke et al., personal communication). Osteopontin is an eggshell matrix protein (Pines et al. 1994) associated with many biominerals such as bone and teeth (McKee and Nanci 1995). In the eggshell, it may function during mineralization by inhibiting calcium carbonate precipitation in a phosphorylation-dependent manner (Hincke and St. Maurice 2000). The release of the chicken genome sequence in 2004 revealed that the chromosomal localization of OC116 is adjacent to that of osteopontin, within the SIBLING mineralization gene locus on chromosome 4 (Fig. 2.3). These genes are contiguous with other mineralization-specific genes (bone sialoprotein, dentin matrix protein 1), suggesting that the SIBLING mineralization locus is preserved in the avian genome, as a subset of the SCPP family of proteins that function in tetrapod mineralization. Therefore, OC116 is predicted to be the ortholog of mammalian MEPE (matrix extracellular phosphoglycoprotein). Blast searching reveals that the *N*-terminus of OC116 possesses about 35%

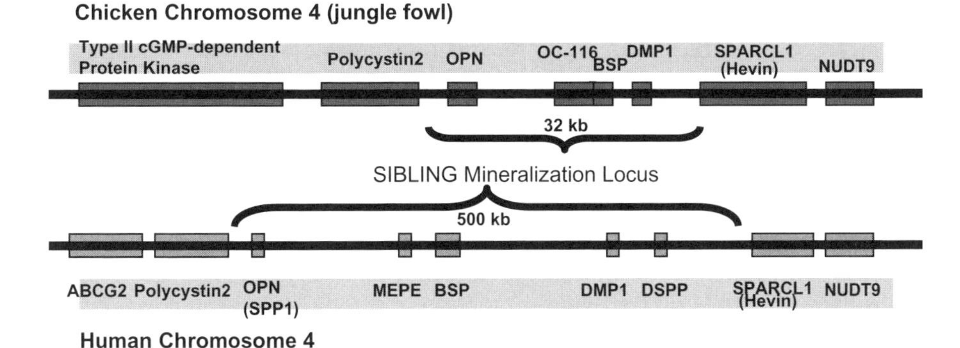

Figure 2.3. Comparison of the SIBLING (*s*mall *i*ntegrin-*b*inding *li*gand, *n*-linked glycoprotein) mineralization loci in chicken and human. The osteopontin and ovocleidin-116 genes are immediately adjacent to each other on chicken chromosome 4. These genes are contiguous with other mineralization-specific genes (BSP, DMP1), suggesting that the mammalian SIBLING mineralization locus is preserved in the avian genome. No chicken sequences with homology to the human DSPP1 gene could be identified, although the genes immediately adjacent to the mineralization locus (hevin/SPARCL1 and NUDT9) are preserved in their relative positions in this region of synteny between the chicken and human chromosomes. [osteopontin (OPN), bone sialoprotein (BSP), dentin matrix protein 1 (DMP1), dentin sialophosphoprotein (DSPP), matrix extracellular phosphoglycoprotein (MEPE)]. (See insert for color representation.)

homology with the *Macaca fascicularis* MEPE protein sequence (ACCESSION: AB046056), in agreement with this suggestion. Among the SCPP family, OC116 is uniquely specialized as an *eggshell-specific* matrix protein and is therefore predicted to play a key role in mineralization of the avian shell.

Ultrastructural immunocytochemistry indicates that OC116 (ovoglycan) is synthesized and secreted from the granular cells of the uterine epithelium, and is incorporated into, and widely distributed throughout, the palisade region of the calcified eggshell. Proteoglycans have the potential to function in biomineralization since their glycosaminoglycan units consist of repeating disaccharides with carboxylate and/or sulfate moieties. It remains to be determined whether the underlying mechanisms by which ovoglycan acts are similar to those by which proteoglycans promote cartilage calcification and collagen mineralization *in vitro* (Hunter 1991; Hunter and Szigety 1992).

Ovocalyxin-32 (OCX32) was purified from eggshell and sequenced to identify a novel 32-kDa protein that is present at high levels in the uterine fluid during the terminal phase of eggshell formation and is localized predominantly in the outer eggshell (Gautron et al. 2001a; Hincke et al. 2003). Database searches identified ESTs whose alignment yielded the complete cDNA. The OCX32 protein possesses limited identity (32%) to two distinct mammalian proteins, latexin, an inhibitor of carboxypeptidase A activity that is expressed in rat cerebral cortex and mast cells, and to a skin protein that is encoded by a retinoic acid receptor -responsive gene, TIG1 (Liu et al. 2000; Uratani et al. 2000). The timing of OCX32 secretion into the uterine fluid has suggested that it may play a role in the termination of mineral deposition (Gautron et al. 1997). To obtain sufficient material for further studies of its function, recombinant OCX32 protein was expressed in *Escherichia coli* (Xing et al. 2007). The protein was extracted from inclusion bodies and purified by sequential DEAE sepharose and Ni^{2+} metal ion affinity chromatographies as a 58-kDa GST fusion protein. The refolded GST-OCX32 significantly inhibited bovine carboxypeptidase activity and also inhibited the growth of *Bacillus subtilis*. The results suggest that OCX32 may show activity similar to that of the fusion protein and reinforce the notion that eggshell matrix proteins possess antimicrobial properties that provide protection to the developing avian embryo. An interesting more recent observation is that the OCX32 gene is expressed at higher levels in a low-egg-production strain (compared to a high-production strain) of Taiwanese country chickens (Yang et al. 2007). The biochemical/physiological basis for this correlation remains to be determined.

Ovocalyxin-36 (OCX36) protein is detected only in the regions of the oviduct, where eggshell formation takes place (isthmus and uterus). Moreover, uterine OCX36 message is strongly upregulated during eggshell calcification (Gautron et al. 2007). The OCX36 protein is observed to localize to the calcified eggshell predominantly in the inner part of the shell, and also to the shell membranes. BlastN database searching indicates that there is no mammalian version of OCX36. However, the protein sequence is 20–25% identical to proteins associated with the innate immune response, such as lipopolysaccharide binding proteins (LBP); bactericidal permeability increasing proteins (BPI); and *p*alate, *lu*ng, and *n*asal epithelium *c*lone (PLUNC) family proteins (Gautron et al. 2007). These proteins are well known in mammals for their involvement in defense against bacteria. They belong to the superfamily of proteins known to be key components of the innate immune system, which act as the first line of host defense (Bingle and Craven 2004). Moreover, the genomic organization of these proteins and OCX36 appear to be highly conserved. These observations suggest that OCX36 is a novel and specific chicken eggshell protein related to the superfamily of LBP/BPI and PLUNC proteins. Therefore OCX36 may participate in natural defense mechanisms that keep the egg free of pathogens.

Osteopontin is found in all vertebrate calcium biominerals, and its role in eggshell mineralization is believed to be very significant. The oviduct expression of osteopontin is entirely uterine-specific and is temporally associated with eggshell calcification (Pines et al. 1994). Osteopontin plays a fundamental role in bone formation, where it likely acts as an inhibitor of calcium phosphate accretion by regulating crystal growth. Osteopontin localization within the mineralized eggshell has been investigated (Hincke et al. 2000b; Fernandez et al. 2003). In eggshell it is not localized with the hydroxyapatite crystallites of the inner cuticle layer, but is enriched in the palisade layer (Hincke et al. 1999).

Transmission electron microscopy (TEM) of the organic matrix of the avian eggshell reveals two structural features within the palisade layer; vesicular structures with electron-lucent cores intermingle between flocculent sheets of organic material. Osteopontin immunolabeling is almost exclusively associated with the diffuse organic sheets (Hincke et al. 2000b), which may reflect its association with the surfaces of growing calcite crystals during eggshell mineralization. This likely corresponds to the extramineral intersection between adjacent calcitic columns in the palisade layer. In contrast, OC116 is associated predominately with the periphery of the vesicular structures that probably correspond to the walls of microvesicular holes (voids) in the calcitic eggshell (Hincke et al. 1999) (Fig. 2.2). The osteopontin localization is structurally homologous to that seen in other calcified tissues, where it collocalizes with early calcification foci in bone, cartilage, and teeth, and subsequently accumulates at high levels at sites of mineral confluence. There is a striking accumulation of osteopontin at matrix–matrix/mineral interfaces in bones, teeth, and renal calculi. These distinct planar accumulations of organic matrix at external and internal mineralized tissue surfaces and interfaces are termed "cement lines" (teeth, bone) or *lamellae* and *striations* (renal calculi), and immunolabel intensely for osteopontin (McKee et al. 1995; McKee and Nanci 1996). The proposed roles for osteopontin at these surfaces include the notions that it functions as an inhibitor of mineralization and/or as a mediator of cell–matrix and matrix–matrix/mineralization at tissue interfaces (McKee and Nanci 1995). Eggshell formation occurs in an acellular milieu, so the possibility that eggshell osteopontin participates in cellular events seems unlikely.

2.3.2. Proteomic Analysis of Chicken Calcified Eggshell Organic Matrix

Until relatively recently, the known eggshell matrix proteins had been identified and characterized by classical techniques such as sequence analysis by Edman degradation, sequence analysis by cDNA sequencing and immunochemical methods. Edman sequence analysis is still a valuable tool for the detection and complete sequence analysis of moderately sized proteins with sequences not contained in databases (Mann and Siedler 1999, 2004, 2006). Furthermore, this technique can identify many sites of posttranslational modification or even identify the modifications themselves. Blocked proteins, however, have to be deblocked or may be identified by internal sequences after cleavage and peptide separation. In combination with protein separation by electrophoresis and blotting on sequencing membranes, this method is useful for the detection of major sequences in a protein mixture of limited complexity as has been shown, for instance, for the organic matrix of several ratite eggshells (Mann 2004). However, Edman sequence analysis is slow and relatively insensitive and thus not suited for high-throughput analysis, especially with complex mixtures. In a 2006 survey, 16 chicken eggshell proteins were detected by Edman sequence analysis after separation of chicken eggshell matrix proteins by ion exchange chromatography combined with reverse-phase HPLC, while 520 proteins were identified by mass spectrometric high-throughput methods (Mann et al. 2006). Edman sequence analysis is relatively straightforward and does not need sophisticated computer programs.

Mass spectrometry relies on entirely different principles to identify proteins and to sequence peptides (Mann et al. 2001; Burgess 2003; Steen and Mann 2004).

The mass spectra obtained are interpreted by dedicated computer programs that try to match the experimental data against theoretical data obtained by *in silico* processing of all of the protein sequences stored in databases. This approach is highly dependent on the quality of databases. Since the chicken genome sequence has become available (International Chicken Genome Sequencing Consortium 2004), many proteomic studies of chicken tissues or body fluids have been published, as, for instance, the analysis of the blood plasma proteome (Corzo et al. 2004; Huang et al. 2006), the embryonic cerebrospinal fluid proteome (Parada et al. 2006), the skeletal muscle proteome (Doherty et al. 2004), or the whole stage 29 embryo proteome (Garcillán et al. 2005). However, it has to be kept in mind that the chicken genome sequence is not yet complete (Burt 2005) and that the sequenced part of it (90–95%) is not always of the high quality needed for proteomic studies (Mann et al. 2006).

With this approach, a large number of proteins associated with the eggshell have been identified (Table 2.1). The 520 identified chicken eggshell matrix proteins (Mann et al. 2006) can be ranked according to abundance in the shell matrix calculated using the exponentially modified *protein abundance index* (emPAI), a method that takes advantage of the relationship between protein concentration and the ratio of observed peptides, or rather observed unique ions, to theoretically observable peptides (Ishihama et al. 2005). The emPAI correlated reasonably well with the approximate concentrations of some of the eggshell proteins estimated from Edman sequencing yields (Mann et al. 2006) except for osteopontin, which was a major protein by protein concentration estimates, but was classified as of low abundance by emPAI calculation. This discrepancy was most probably due to the multiple phosphorylation of this protein.

Phosphopeptides are notoriously difficult to find in complex peptide mixtures without prior enrichment and dedicated mass spectrometric techniques (Mann et al. 2002) and thus did not contribute to the number of observed ions in the general proteomic survey published. Guided by the comparison of protein sequencing yields and emPAI values, the eggshell proteins could be distributed into three abundance groups. Thirty-two proteins (6% of total) were classified as of high abundance (emPAI $\geq$ 9; Table 2.1). This group contained all of the eggshell proteins known before, except osteopontin. Other proteins found in this abundance group were several proteins implicated in lipid metabolism, cytoplasmic proteins, more egg white proteins, and some previously uncharacterized proteins such as proteins with sequence similarity to certain mammalian integrin binding proteins, an annexin-related protein, a protein containing bactericidal permeability-increasing domains and a cathepsin B-like protease. Several components, namely, serum albumin, vitamin D–binding protein, exFABP, hemopexin, actin, and several egg white proteins, were also previously identified in other body fluids (Corzo et al. 2004; Parada et al. 2006). The group of proteins of intermediate abundance (emPAI 8.9–2.1;

TABLE 2.1. The Most Abundant Proteins in the Chicken Eggshell Calcified Layer Organic Matrix

Protein[a]	emPAI	Previously Known Locations and Activities	
Ovocalyxin-21 (similar to RIKEN cDNA 1810036H07; IPI00574331.1)	176.8	Specific eggshell protein	
Lysozyme C (IPI00600859.1; P00698)	128.2	Egg white, eggshell, body fluids; bacteriolytic	
Serum albumin (IPI00574195.1; P19121)	113.5	Blood plasma, egg yolk, eggshell	
Ovalbumin (IPI00583974.1; P01012)	85.6	Egg white, eggshell, body fluids; storage protein	P
51-kDa protein (similar to Edil (EGF-like repeats and discoidin I-like domains) 3/Del (developmental endothelial locus)-1; IPI00588727.1)	83.8	Edil3/Del-1: extracellular matrix, $\alpha v\beta 3$ integrin binding	
Ovocalyxin-32 (IPI00578622.1; Q90YI1)	71.0	Specific eggshell protein	P
Ovocleidin-116 (IPI00581368.1; Q9PUT1)	65.3	Specific eggshell protein	P
Ovocleidin-17 (IPI00572756.1; Q9PRS8)	64.8	Specific eggshell protein	P
Cystatin (IPI00576782.1; P01038)	45.4	Egg white, body fluids; protease inhibitor	P
Clusterin (IPI00604279.1; Q9YGP0)	42.7	Body fluids, eggshell, egg white; extracellular chaperone	P
Similar to Ig μ heavy-chain disease protein BOT, partial (~25%; IPI00595925.1)	30.6	Extracellular	
Hypothetical protein XP_429301 (similar to cathepsin B; precursor); (IPI00577962.1 and 00573387.1)	24.8	Cathepsin B: cytosol, lysosome, nucleus and extracellular; protease	
Extracellular fatty acid binding protein (IPI00600069.1; P21760)	26.8	Extracellular, body fluids; lipid binding protein, lipocalin family	
Ovocalyxin-36 (IPI00573506.2; Q53HW8)	23.2	Specific eggshell protein	
Ovotransferrin (IPI00578012.1; P02789)	22.9	Egg white, eggshell, body fluids; iron transport, bacteriostatic	
Similar to fibronectin (IPI00590535.1)	18.9	Fibronectin: extracellular matrix, body fluids	P
Apolipoprotein D (IPI00600265.1; Q5G8Y9)	18.3	Extracellular; lipid transport, lipocalin family	
Similar to hypothetical protein HSPC117 (IPI00571823.1; emPAI: aa 1–450; domains: bactericidal permeability increasing 1,2)	16.8		
50-kDa protein; similar to MGC75581 protein (IPI00575013.1 and IPI00588573.1; similar to Edil 3/Del-1)	15.9	See 51-kDa protein above	

TABLE 2.1. *(Continued)*

Protein[a]	emPAI	Previously Known Locations and Activities	
Dickkopf-related protein 3 (IPI00578016.1; Q90839)	14.0	Extracellular; inhibitor of Wnt signaling pathway	P
Peptidyl-prolyl *cis–trans* isomerase B (cyclophilin B; IPI00583337.1; P24367)	13.7	Endoplasmic reticulum lumen; assists protein folding	
Similar to hypothetical protein A430083B19 (IPI00578305.1; domains: S-100/intestinal calcium binding protein-like)	12.9		
GM2 activator protein (IPI00599943.1 and IPI00587080.1, partial GM2 activator domains combined);	12.9	Lysosomal and extracellular, cofactor required for degradation of ganglioside GM2	
Ovoinhibitor (IPI00587313.2; P10184)	12.1	Egg white; protease inhibitor	
Hemopexin (IPI00596012.1; IPI00573473.1; P20057)	11.6	Blood plasma; heme scavenger	
Similar to nexin 1 (IPI00594564.1)	11.1	Nexin 1: extracellular; protease inhibitor	
Similar to fibronectin 1 isoform 3 (IPI00601771.1)	9.0	Fibronectin: extracellular matrix, body fluids	
Prosaposin (proactivator polypeptide precursor; IPI00584371.1; O13035)	9.0	Extracellular; cleaved to saposins, essential cofactors of lysosomal sphingolipid metabolism	
Vitamin D binding protein (IPI00573327.1; Q9W6F5)	9.0	Body fluids; binding and transport of Vit D metabolites, actin scavenger, macrophage activation	
Actin, cytoplasmic 1 (IPI00594778.1; P60706)	9.0	Cytoplasm; cytoskeletal protein	
Ubiquitin/ribosomal protein S27a/ polyubiquitin (IPI00581002.1; 00584240.1; 00585698.1; 00589085.1; P62973)	9.0	Cytoplasm, nucleus	
Calcyclin (S100A6; IPI00572547.1; Q98953)	9.0	Intracellular, EF-hand Ca^{2+} binding	
Osteopontin; emPAI probably too low because of many modifications (IPI00585901.1; P23498)	(1.8)	Extracellular matrix, eggshell, body fluids; biomineralization, cell adhesion, cytokine activity	P

[a]The proteins listed here (with IPI number and Swiss/Prot/TrEMBL primary number) are ordered by decreasing exponentially modified *protein abundance index* (emPAI) (Ishihama et al. 2005). For the complete list of eggshell matrix proteins, see material supplementary to Mann et al. (2006); P—phosphoprotein (Mann et al. 2007).

Source: Mann et al. (2006).

14% of total) comprised many signal transduction components, as, for instance, the growth factor pleiotrophin, chemokine ah221, IGF-II, and the bone morphogenetic protein antagonists chordin and tsukushi. Other proteins in this group were immune system–related, such as immunoglobulin chains, the extracellular domain of polymeric immunoglobulin receptor or Ig linker protein J. An interesting observation was the presence of the antimicrobial defensins β-defensin-11 and gallinacin-8, which, together with lysozyme and other antimicrobial proteins, may account for the *in vitro* antimicrobial activity of the eggshell matrix. Also in this group were several extracellular matrix proteins or fragments thereof. The bulk of the newly detected eggshell matrix components was, however, of low abundance (emPAI < 2.1) (Mann et al. 2006).

It is likely that only a minority of the proteins that are secreted by uterine cells, for instance, the eggshell-specific ovocleidins and ovocalyxins, can be assumed to have specific functions in the eggshell or during eggshell assembly. Since proteins produced in other tissues, such as clusterin, lysozyme, or osteopontin, are also expressed and secreted by cells of the uterine wall, eggshell localization should not be taken as a necessary condition for predicting a unique function in the eggshell or during eggshell formation. Many of the proteins may also be assimilated into the eggshell merely because they were in the uterine fluid at the time of mineralization. These proteins may be remnants of processes that took place in other parts of the oviduct, such as egg white or eggshell membrane deposition, and may have been released as components of secretion processes along the length of oviduct, or may be derived from turnover of the cells and basement membrane that line the oviduct. Finally it should be noted that relative abundance does not necessarily reflect degree of importance. Many of the new proteins identified as eggshell matrix components have enzymatic activity or are involved in signal transduction, and may therefore exert their possible function at low concentrations. Given the high number of potentially interesting proteins identified, it will take some time to study their origin, functions, and interactions.

A proteomic inventory of eggshell components can be considered as only a first and necessary step to learn more about the eggshell because it tells us which proteins are present and have to be considered in future research. Further inquiry is necessary to locate the site of production of these proteins along the oviduct, to determine possible interactions and catalytic or signaling properties and posttranslational modifications. However, in many cases it will also be necessary to isolate the protein in question using conventional biochemical methods and to study its properties *in vitro* by using, for instance, binding assays, crystallization assays, and similar techniques.

2.4. EGGSHELL BIOSYNTHESIS: MINERALIZATION

Calcified matrices in vertebrate and invertebrate biology are biphasic, robust composites usually containing collagenous and noncollagenous elements in intimate contact with mineral (Robey 1996). While the avian eggshell is a complex and highly structured calcitic bioceramic with extensive intermingling of both its organic and inorganic phases, it also demonstrates a spatial separation between its organic framework and mineralized components, where the eggshell membrane interfaces with the calcified eggshell (Arias et al. 1992; Dennis et al. 1996; Nys et al. 1999). During avian egg and eggshell development, the egg sequentially acquires all of its components as it descends through specialized regions of the oviduct. The innermost structure associated with the eggshell is a meshwork of interwoven fibers known as the *shell membranes*. This structure, organized into morphologically distinct inner and outer layers, is composed of collagen-like material including types I, V, and X collagens, which are deposited after the egg enters the proximal (white) isthmus and are stabilized by crosslinking (Wong et al. 1984; Arias et al. 1991; Fernandez et al. 1997). Eggshell mineralization is subsequently initiated in the distal (red) isthmus by calcification at distinct nucleation sites on the surface of the outer eggshell membrane, which is the origin of the mammillary knobs. The mechanisms that prevent calcification toward the inner membranes and albumen are not well understood; one proposal is that collagen type X prevents a generalized calcification of the shell membrane (Arias et al. 1997). Any modification of the eggshell membranes due to inhibition of fiber formation or crosslinking alters eggshell formation and its mechanical properties. For example, inhibition of the lysine-derived crosslinking of eggshell membrane by aminopropionitrile or by a copper deficiency (Chowdhury 1990) affects the pattern and mechanical properties of the eggshell structure.

The egg then enters the uterus (shell gland), where rapid calcium carbonate deposition continues outward to give rise to the inner mammillary body (cone) layer and outer palisade (calcitic) layer during approximately 20 h of shell formation (Parsons 1982; Hamilton 1986; Burley and Vadehra 1989; Nys et al. 1999, 2004). This mineralization occurs in the uterine fluid, an acellular milieu containing ionized calcium and bicarbonate greatly in excess of the solubility product for calcite (Nys et al. 1991). Interestingly, shell formation concludes with the deposition of a hydroxyapatite-containing cuticle at the surface of the shell (Dennis et al. 1996). Eggshell formation occurs in an acellular milieu, in contrast to other mineralized tissues, and its particular mineral structure results from a self-organization of mineral and organic precursors that are secreted into the milieu bathing the eggshell during its formation.

The mineralized shell is about 96% calcium carbonate; the remaining components include the organic matrix (3.5%), approximately half of which can be readily solubilized after decalcification. The native and soluble precursors of the eggshell matrix are present in the uterine fluid, from which they become incorporated into the calcifying shell. The ultrastructure and crystallography of the compact mineral layer can be partially explained by a single model of competition for crystal growth; growth of crystals from the nucleation site occurs initially in all directions but, owing to competition for space between adjacent sites of growth, only crystals growing perpendicular to the egg surface have space to grow. This model explains the appearance of preferred crystal orientation in the outer part of the eggshell, but is based on the hypothesis that the crystal growth is anisotropic. This anisotropy results from inhibition of crystal growth on the faces parallel to the *c* axis, resulting in an elongation of the calcite crystal. This inhibition is likely to result from some organic components that are present in the uterine fluid, and then integrated into the eggshell. These are termed *matrix proteins*, and are released by demineralization of the shell. These components are suspected to influence the texture of the eggshell by controlling the size, shape, and orientation of this polycrystalline structure and therefore the mechanical properties of this material.

One of the first observations in favor of a role of the matrix proteins in control of egg calcification was that the protein composition of uterine fluid varies during the initial, calcification, and terminal phases of eggshell deposition (Gautron et al. 1997). Subsequently, a large collaborative effort was carried out to identify the organic components of the eggshell matrix and establish their involvement in eggshell calcification. The eggshell matrix proteins that were identified by classic approaches were subdivided into at least three groups. *Eggshell-specific proteins* such as the ovocleidins and ovocalyxins have been identified by *N*-terminal amino acid sequencing, immunochemistry, and EST database mining (Hincke et al. 1995, 1999; Gautron et al. 2001a, 2007). The *egg white proteins* ovalbumin, lysozyme, and ovotransferrin are also present in the uterine fluid, and are localized primarily in the shell membranes and mammillary cone layer of the eggshell (Hincke 1995; Gautron et al. 2001b). Finally, ubiquitous proteins, such as osteopontin and clusterin (Mann et al. 2003; Pines et al. 1994; Hincke et al. 2000b; Fernandez et al. 2003), are also found in a number of tissues. Sequential incorporation of matrix proteins into the calcifying eggshell results in their differential localization between the inner (mammillary) and outer (palisade) layers of the mineralized shell (Hincke et al. 1992). We hypothesize that their specific localization pattern provides functional insight into the role of such proteins during eggshell formation (Fig. 2.4).

In general, the soluble matrix proteins of calcitic biomaterials modify crystal growth, and therefore regulate the macroscopic properties of the resulting bioceramic. For example, in the mollusk shell, specific proteins control phase switching between the calcite and aragonite forms of calcium carbonate (Belcher et al. 1996). A number of experimental observations support the role of the eggshell matrix proteins in determining the fabric of the eggshell and therefore influencing its resulting mechanical properties. Egg calcification takes place in uterus in three distinct phases (initiation, active calcification, and termination of shell calcification). In the uterus, the egg bathes an acellular milieu (uterine fluid) that contains the organic precursors of the matrix. The organic composition of the uterine fluid changes during fabrication of the eggshell. Each phase of shell mineralization is associated with a specific protein electrophoretic profile in the uterine fluid, suggesting specific roles for the organic contents during the calcification process (Gautron et al. 1997). The

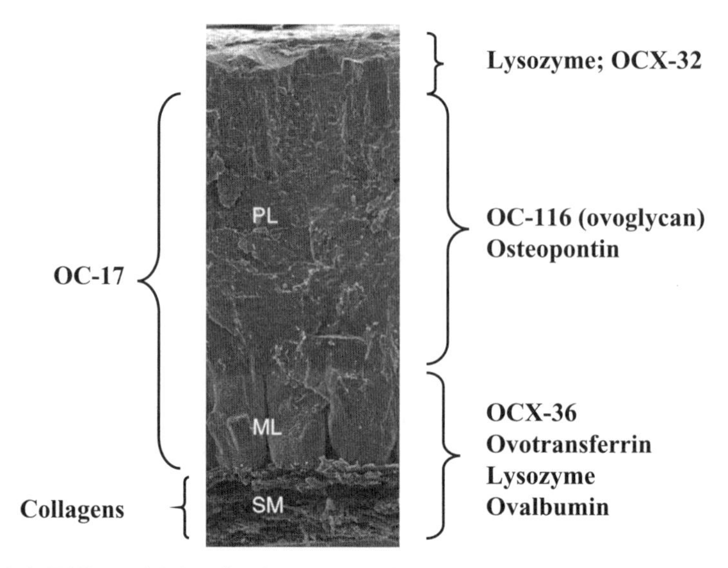

Figure 2.4. Differential localization pattern for eggshell matrix proteins within the eggshell. Data from a number of studies are summarized to demonstrate the localization of some eggshell-specific proteins (OC17, OC116, OCX32, OCX36) and egg white proteins (lysozyme, ovotransferrin).

nature of the interactions between the mineral phase and the eggshell matrix proteins has been extensively investigated. The presence of calcium binding proteins has been reported in eggshell extracts and in uterine fluid (Abatangelo et al. 1978; Cortivo et al. 1982; Hincke et al. 1992; Gautron et al. 1997). The effect of the whole uterine fluid on the precipitation kinetics, as well as on the size and the morphology of crystals grown *in vitro*, was also investigated using a micromethod favoring a good reproducibility of crystal size and morphology (Dominguez-Vera et al. 2000; Jimenez-Lopez et al. 2003). Proteins in the uterine fluid modify the kinetics of calcium carbonate precipitation *in vitro* (Dominguez-Vera et al. 2000). The lag time for calcium carbonate precipitation is reduced by the uterine fluid harvested during the initial and growth stages of eggshell mineralization, suggesting that these matrix precursors promote crystal nucleation. To a lesser extent, the uterine fluid collected during the growth phase also enhances precipitation kinetics. In contrast, the total uterine fluid harvested at the terminal stage of calcification inhibits calcite precipitation (Gautron et al. 1997). In the presence of uterine fluid, all calcite crystals formed *in vitro* were found to be calcite (Fig. 2.5), demonstrating that this milieu promotes the calcium carbonate polymorph of the shell [see Rodriguez-Navarro et al., cited in Gautron et al. (2005)].

In agreement with these observations, partially purified eggshell matrix proteins inhibit calcium carbonate precipitation and alter patterns of calcite crystal growth, leading to morphological modifications of rhombohedric calcite

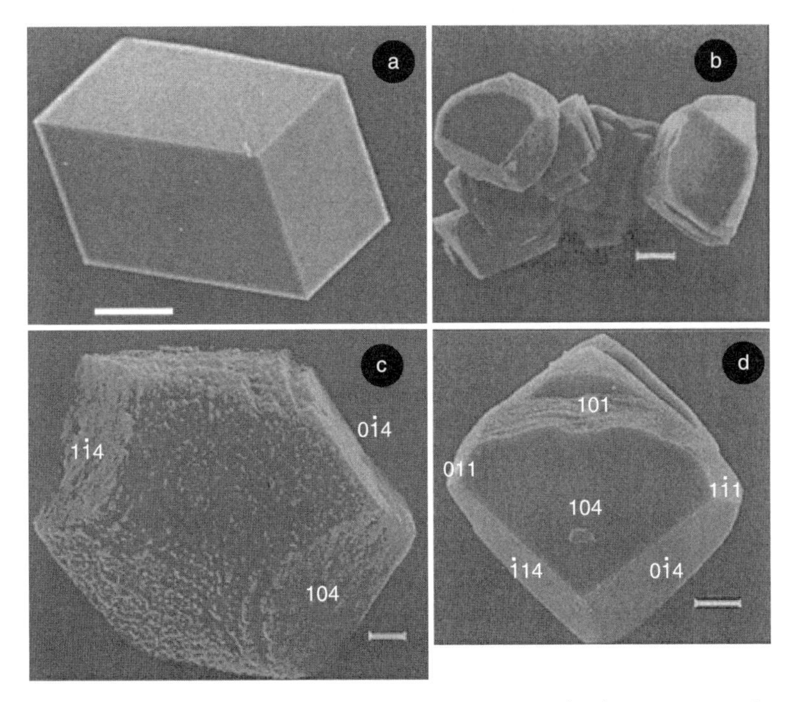

Figure 2.5. Morphology of calcite crystals grown *in vitro* in the presence of uterine fluid proteins harvested from different phases of eggshell formation (all 50 μg/mL): (a) control experiments showing exclusively the (104) rhombohedra faces; (b) uterine fluid from the initial phase of eggshell formation; (c) uterine fluid from the growth phase of eggshell formation; (d) uterine fluid from the terminal phase of eggshell formation. Scale bar: (a) 20 μm; (b) 10 μm; (c) 2 μm; (d) 10 μm. [Modified from Dominguez-Vera et al. (2000), with permission.]

crystals grown *in vitro* [reviewed in Nys et al. (1999, 2004)]. Low concentrations of ansocalcin (up to 10 μg/mL) induce calcite crystals with screw dislocations, while at higher concentrations (>50 μg/mL), polycrystalline calcite aggregates are nucleated. These results differ from those obtained with calcite crystal growth in the presence of purified ovocleidin-17, where at low levels (10–100 μg/mL) calcite crystals were twinned and less aggregated (Lakshminarayanan et al. 2005). On the other hand, other researchers found a different concentration-dependent aggregation of calcite crystals grown in the presence of ovocleidin-17 (50–200 μg/mL) (Reyes-Grajeda et al. 2004). These differing results for ovocleidin-17 may reflect differences in the experimental conditions of crystal growth. Ovotransferrin leads to smaller crystals and at 0.5 mg/mL promotes the development of elongated crystals (Gautron et al. 2001b). Lysozyme at high concentration (>10 mg/mL) affects mainly the calcite faces parallel to the *c* axis, by inhibition of growth on {110} faces (Hincke et al. 2000a; Rodriguez-Navarro et al. 2000).

Finally, pure glycoaminoglycans also affect calcite morphology, leading to crystal elongation (Arias et al. 2002). Therefore, highly sulfated proteoglycans such as ovoglycan are likely to influence mineralization by electrostatic interactions. Protein phosphorylation is another posttranslational modification that may be crucial, since partially purified eggshell osteopontin strongly inhibits calcium carbonate precipitation in a phosphorylation-dependent manner, suggesting that it could be a potent regulator of eggshell calcification (Hincke and St. Maurice 2000). Moreover, Mann et al. (2007) have demonstrated that the major phosphorylated eggshell matrix proteins include osteopontin, ovocleidin-17, ovocleidin-116, and ovocalyxin-32.

The results of *in vitro* experiments are supported by *in vivo* observations. If eggshell matrix participates in establishing the morphology of calcite crystals, it would affect the texture (crystal size and orientation) of the eggshell and influence its mechanical properties. This hypothesis was confirmed by quantifying components of matrix proteins in parallel with variations in eggshell mechanical properties (Panheleux et al. 2000; Ahmed et al. 2005). The well-known improvement in shell quality (breaking strength) after molting is correlated with reduced levels of ovocleidin-116 and ovotransferrin in the eggshell matrix, and a decrease in calcite grain size, which could be responsible for the improved mechanical properties (Ahmed et al. 2003, 2005; Panheleux et al. 2000).

A complementary approach to establish the role of matrix proteins in the variability of the eggshells physical and mechanical properties has been taken using genetic and genomic approaches (Dunn et al. 2007). This study reveals a number of significant associations between alleles of some eggshell matrix proteins (ovocleidin-116 and ovocalyxins, osteopontin, ovalbumin) and measurements of eggshell solidity. In summary, matrix component play an active role in the control of calcite growth kinetics and crystal morphology during eggshell mineralization. In this manner, the matrix proteins regulate the textural properties and resulting biomechanical strength of the eggshell.

2.5. EGGSHELL ANTIMICROBIAL DEFENSES

The avian eggshell is a complex, multifunctional biomineral composed of calcium carbonate and proteins. Multiple eggshell matrix proteins are suspected to play a role in eggshell formation, but the role of individual eggshell proteins in the eggshell remains to be clearly understood. The eggshell is in direct contact with the environment, and this strategic location hints at a role in antimicrobial defense.

Mine et al. (2003) noted that protein extracted from the eggshell of the domestic chicken demonstrated activity against *Pseudomonas aeruginosa, Bacillus cereus, Staphylococcus aureus, Escherichia coli*, and *Salmonella enteritidis*. Ovary and oviduct tissue extracts of the domestic hen were found to show antimicrobial activity against both Gram-positive and Gram-negative bacteria (Silphaduang et al. 2006). Moreover, eggshell cuticle protein extract, from chicken, duck, goose, Canada goose, mute swan, wood duck, and hooded merganser, demonstrated antimicrobial activity against *P. aeruginosa, B. subtilis* and *S. aureus* (Wellman-Labadie et al. 2007a, 2007b).

Although antimicrobial activity has been identified in protein extracts from avian eggshell and in particular outer eggshell, the identity of specific proteins responsible for the activity remains to be elucidated. Multiple proteins possessing antimicrobial activity, including egg white proteins, eggshell matrix proteins, β-defensins/gallinacins, and others have been identified in the avian egg at the shell or membrane level and may be involved in antimicrobial defense. These are discussed further in this section.

2.5.1. Role of Albumen Proteins in Antimicrobial Activity of the Avian Eggshell

Egg white proteins, such as lysozyme and ovotransferrin, have been extensively studied and deemed responsible for the characteristic antimicrobial activity of albumen. Lysozyme, or *N*-acetylmuramideglycanohydrolase, is an enzyme that splits the bond between the glycosidic β-1,4-linked residues of *N*-acetylneuramic acid (NAM) and *N*-acetylglucosamine (NAG) in the peptidoglycan structure of most Gram-positive bacteria and is present in high levels in avian egg white (Bera et al. 2005; Burley and Vadehra 1989). Ovotransferrin, another major albumen component, is an 80-kDa glycoprotein composed of two domains that each reversibly bind one molecule of iron (Valenti et al. 1981) and thereby impede the growth of bacteria by limiting levels of this essential element. Because of their potent activity in albumen, the ovotransferrin and lysozyme detected within the avian eggshell and membranes also represent likely candidates for eggshell antimicrobial activity.

c-Type lysozyme is a component of the chicken eggshell matrix, and is localized within the surface cuticle and also associated with the shell membranes (Hincke et al. 2000a). Vadehra et al. (1972) previously reported that *Gallus gallus* eggshell contained active *c*-type lysozyme and proposed that it played a role in antimicrobial defence. This observation was extended to eggs from a number of galliform species (Panheleux et al. 1999). Moreover, Wellman-Labadie et al. (2007a, 2007b) has reported *c*-type lysozyme as a component of the outer eggshell of chicken, duck, goose, Canada goose, mute swan, wood duck, and hooded merganser and detected lysozyme enzymatic activity in these extracts. The identification of active *c*-type lysozyme across a wide range of avian species indicates a fundamental avian need for this protein at this site and suggests a role in antimicrobial defense. *c*-type lysozyme of the avian eggshell may therefore prevent the establishment of lysozyme-sensitive microorganisms on this surface. Considering that 92% of chicken eggshell contaminants are Gram-positive bacteria (Deeming 2002), it is likely that the avian egg has evolved to effectively defend against the attack of such lysozyme-sensitive microorganisms.

Ovotransferrin, the iron-binding glycoprotein responsible for the bacteriostatic activity of avian albumen, has also been identified as a component of the eggshell in chicken and other avian species (Panheleux et al. 1999; Gautron et al. 2001b; Nys et al. 2001). Von Hunolstein et al. (1992) demonstrated inhibition of bacteria by transferrins, such as ovotransferrin, at concentrations of ≥1 mg/mL. Valenti et al. (1981) demonstrated that in the presence of salts, the binding of iron and ovotransferrin would be favored and lead to an inhibition of bacterial growth. Although ovotransferrin is found in low levels in the outer avian eggshell, sufficient salts may be present in this environment for ovotransferrin to contribute somewhat to antimicrobial defense of the egg.

2.5.2. Novel Eggshell Proteins as Antimicrobial Candidates

Ovocalyxin-32 (OCX32) is a chicken eggshell matrix protein sharing properties with latexin, a carboxypeptidase inhibitor of the rat, and TIG1, a skin protein (Hincke et al. 2003; Gautron et al. 2001a). Recombinant *Gallus gallus* OCX32 has been shown to possess carboxypeptidase inhibitory activity and to inhibit the growth of *B. subtilis* (Xing et al. 2007). Moreover, a 32-kDa band was detected in eggshell surface extracts of several avian species that reacted with antiserum raised against *G. gallus* OCX32 (Wellman-Labadie et al., manuscript in preparation). The association of OCX32 with the eggshell cuticle of multiple avian species provides additional support for its role in antimicrobial defense. Proteinase inhibitors such as SLPI and elafin have been reported to possess antimicrobial activity through inhibition of microbial proteases (Hagiwara et al. 2003). Ghim et al. (1998) reported that the product of the *yodJ* gene of *B. subtilis* shows homology to a carboxypeptidase of *Enterococcus faecilis* suggesting the presence of a bacterial carboxypepidase in *B. subtilis*. Carboxypeptidase activity has been found to contribute to the virulence and antibiotic resistance of *E. faecilis* (Arthur et al. 1994). It appears that the recombinant OCX32 protein, through inhibition of carboxypeptidase activity, may inhibit the growth of *B. subtilis*. This proposed role of carboxypeptidase inhibitors in innate immunity would provide an explanation for the wide distribution of these proteins.

Ovocalyxin-36 (OCX36) is a novel eggshell matrix protein that is localized in the shell and membranes of the chicken eggshell (Gautron et al. 2007). It shows homology to lipopolysaccharide binding proteins (LBP), bactericidal permeability-increasing proteins (BPI), and to members of the *p*alate, *lu*ng, and *n*asal epithelium *c*lone family of proteins (PLUNC). These proteins are well known in mammals for their involvement in defense against bacteria. They belong to the superfamily of proteins known to be key components of the innate immune system that act as the first line of host defense (Bingle and Craven 2004). Homology between OCX36 and the LBP/BPI family of mammalian proteins is further reinforced by comparison of their gene structure. The exon–intron organization of the OCX36 gene is very similar to that of the highly conserved *LBP* and *BPI* genes (Hubacek et al. 1997), with most corresponding exons possessing identical sizes.

The BPI protein is composed of *N*-terminal and *C*-terminal domains that display homology to the *N*- and *C*-termini of OCX36 (Gautron et al. 2007). Therefore, OCX36 is predicted to possess overall structure and protein folding similar to those seen in the human BPI crystal structure (Beamer et al. 1997). Another chicken BPI analog, Tenp, possesses 18% identity with human BPI and was shown by proteomic analysis to be a hen egg white component (Guerin-Dubiard et al. 2006). Surprisingly, sequence comparison reveals that Tenp and OCX36 amino acid sequences have only 18% identity with each other (homology climbs to 39% for conservative replacements). OCX36 and Tenp are therefore different egg proteins with significant sequence similarity to BPI proteins, although present at different sites; OCX36 is restricted to the calcified eggshell and to the eggshell membranes. It is likely that OCX36 contributes to natural egg defenses by providing chemical protection for the egg contents, particularly in the lumen of the distal oviduct during eggshell formation in the white isthmus, red isthmus, and uterus, where bacteria could be already present via retrograde movement from the cloaca before eggshell formation.

2.5.3. Cathelicidins, β-Defensins, and Eggshell Antimicrobial Peptides

Numerous low-molecular-weight cationic membrane binding antimicrobial peptides have been identified within avian tissues. Such antimicrobial peptides are present as components of innate immunity and provide protection against invading pathogens. These peptides, expressed by heterophils and epithelial cells of various tissues, demonstrate broad-spectrum antimicrobial activity against Gram-positive and Gram-negative bacteria as well as fungi, protozoa, and enveloped viruses. Expression in the hen reproductive system has been documented. For example, chicken ovary and oviduct possess components with bactericidal activity against Gram-positive and Gram-negative bacteria. These antimicrobial polypeptides were extracted, purified and identified as histones H1 and H2B (Silphaduang et al. 2006).

In another approach, functional genomic information was utilized to study the expression of defensins in chicken (Xiao et al. 2004). Defensins are a large family of antimicrobial peptides that are capable of killing pathogens and thus play a critical role in the first line of host defense. A genomewide screen of the chicken sequence was used for comparative analysis of the defensins among chicken, mouse, and human (Xiao et al. 2004). These proteins all share a consensus defensin motif that was used to screen the chicken genome. Fourteen different chicken defensin genes, designated as Gal 1–14, have been identified. The gene products have been termed *gallinacins*, which correspond to the prepropeptides. On the other hand, the mature peptides are termed *β-defensins* (1–13). Specific primers were designed to study their expression in various chicken tissues or organs, including oviduct and ovary. Expression of Gal 10, 11, and 12, corresponding to gallinacin-8, β-defensin-11, and gallinacin-10, respectively, was specifically detected in oviduct tissues and in uterus (V. Hervé-Crepinet, personal communication) indicating that they could be deposited in the egg or shell to play a role in defense against pathogens. In agreement with this prediction, proteomic studies have detected β-defensin-11 (Gal 11) and gallinacin-8 (Gal 10) in the eggshell matrix (Mann et al. 2006). In addition, β-defensins such as gallinacins-1, -2, and -3 are expressed in the vaginal mucosa of laying hens (Yoshimura et al. 2006). A similar approach was used to identify three potential cathelicidins, also involved in host defense against pathogens (Xiao et al. 2006). These have been termed *fowlicidins*-1, -2, and -3. Cathelicidins are a family of animal antimicrobial peptides with hallmarks of a highly preserved prosequence (cathelin domain) and an extremely variable, antibacterial sequence at the *C*-terminus. Fowlicidin-2 is identical to CMAP27, a chicken myeloid antimicrobial peptide, which is expressed at low levels in skin tissue and in higher levels in the uropygial gland (van Dijk et al. 2005). The uropygial gland is a sebaceous gland located at the base of the tail of birds; it secretes a holocrine solution of waxes and oils possessing antimicrobial properties and serves to maintain feather condition (van Dijk et al., 2005). It is probable that these secretions come in contact with the surface of the egg and eggshell during incubation and may thereby provide some protective value.

These antimicrobials are likely to become incorporated in the egg and eggshell during development and may play roles in antimicrobial defense of the avian egg. Their localization within this structure remains to be determined.

Antimicrobial protection may involve a synergistic effect between antibacterial proteins and peptides. It is not yet clear how they would provide this protection if entrapped within the cuticle or calcitic shell. However, it can be proposed that the chitinase activity of *C*-type lysozyme in the cuticle would help defend against fungi, which break down the cuticle and can facilitate microbial invasion by increasing the number of open pores that allow bacterial ingress (Cook et al. 2003). With respect to the calcified shell, antimicrobial proteins or peptides are expected to require some mobility and water to interact with microbial surface molecules. Trapped in an almost waterless mineral they may be inactive, even if located in the surface regions. However, eggshell dissolution from the calcium reserve body at the bases of the mammillary bodies occurs prior to hatching in response to acid secreted by the chorioallantoic membrane, a tissue that is elaborated by the developing embryo and lines the inner shell membrane (Narbaitz et al. 1995). This process would solubilize OCX36 and other antimicrobial proteins and/or peptides located within this region of the shell. This mechanism would serve to upregulate antimicrobial protection as the eggshell becomes progressively weaker in preparation for hatching.

2.6. CONCLUSION

The proteomic era is maturing, and the majority of eggshell constituents should be identified in the near future. In light of this information, transcriptomic approaches should lead to a better understanding of the regulation of the process of eggshell formation. This knowledge will help establish clusters of genes involved in a particular function and identify candidate genes that explain genetic variability in eggshell quality. However, a great challenge will be to fully determine the function of the identified eggshell matrix proteins. Two functional roles have been proposed: (1) regulation of eggshell mineralization and (2) antimicrobial protection of the egg and its contents. The availability of nucleotide and amino acid sequences, combined with bioinformatic approaches, will contribute to identification of putative function of proteins. However, studies with purified native or recombinant proteins to screen activities using various *in vitro* tests will be necessary to gain insight into the role of each component. One important goal will be to determine the impact and importance of posttranslational modification of matrix components (glycosylation, glycanation, phosphorylation, etc.), which could greatly alter their properties and interactions.

ACKNOWLEDGMENTS

The assistance of Yung-Ching Chien and Line Mongeon with scanning electron microscopy is gratefully acknowledged (MDM). Financial support was provided by NSERC via the Discovery (MTH) and Collaborative Research Opportunities (MTH, MDM, YN) Programs, and in part by the Centre for Biorecognition and Biosensors (MDM).

REFERENCES

Abatangelo G, Daga-Gordini D, Castellani I, Cortivo R (1978). Some observations on the calcium ion binding to the eggshell matrix. Calc Tissue Res 26:247–252.

Ahmed AM, Rodriguez A, Vidal ML, Gautron J, Garcia-Ruiz JM, Nys Y (2003). Effect of moult on eggshell quality. Br Poultry Sci 44:782–793.

Ahmed AM, Rodriguez-Navarro AB, Vidal ML, Gautron J, Garcia-Ruiz JM, Nys Y (2005). Changes in eggshell mechanical properties, crystallographic texture and in matrix proteins induced by moult in hens. Br Poultry Sci 46:268–279.

Arias JL, Jure C, Wiff JP, Fernandez MS, Fuenzalida V, Arias JL (2002). Effect of sulfate content of biomacromolecules on the crystallization of calcium carbonate. Mater Res Soc Symp Proc 711:243–248.

Arias JL, Nakamura O, Fernandez MS, Wu JJ, Knigge P, Eyre DR, Caplan AI (1997). Role of type X collagen on experimental mineralization of eggshell membranes. Connect Tissue Res 36:21–33.

Arias JL, Fink DJ, Xiao SQ, Heuer AH, Caplan AI (1993). Biomineralization and eggshells: Cell-mediated acellular compartments of mineralized extracellular matrix. Int Rev Cytol 45:217–250.

Arias JL, Carrino DA, Fernandez MS, Rodriguez JP, Dennis JE, Caplan AI (1992). Partial biochemical and immunochemical characterization of avian eggshell extracellular matrices. Arch Biochem Biophys 298:203–302.

Arias JL, Fernandez MS, Dennis JE, Caplan AI (1991). Collagens of the chicken eggshell membranes. Connect Tissue Res 26:37–45.

Arthur M, Depardieu F, Snaith HA, Reynolds PE, Courvalin P (1994). Contribution of VanY D,D,-carboxypeptidase to glycopeptide resistance in Enterococcus faecalis by hydrolysis of peptidoglycan precursors. Antimicrob. Agents Chemother 38(9): 1899–1903.

Beamer LJ, Carroll SF, Eisenberg D (1997). Crystal structure of human BPI and two bound phospholipids at 2.4 angstrom resolution. Science 276(5320):1861–1864.

Belcher AM, Wu XH, Christensen RJ, Hansma PK, Stucky GD, Morse DE (1996). Control of crystal phase switching and orientation by soluble mollusc-shell proteins. Nature 381:56–58.

Bera A, Herbert S, Jakob A, Vollmer W, Gotz F (2005). Why are pathogenic staphylococci so lysozyme resistant? The peptidoglycan O-acetyltransferase OatA is the major determinant for lysozyme resistance of *Staphylococcus aureus*. Mol Microbiol 55:778–787.

Bingle CD, Craven CJ (2004). Meet the relatives: a family of BPI- and LBP-related proteins. Trends Immunol 25:53–55.

Burgess SC (2003). Proteomics in the chicken: Tools for understanding immune responses to avian disease. Poultry Sci 83:552–573.

Burley RW, Vadehra DV (1989). The Avian Egg. Chemistry and Biology. New York: Wiley-Interscience.

Burt DW (2005). Chicken genome: The current status and future opportunities. Genome Res 15:1692–1698.

Chowdhury SD (1990). Shell membrane system in relation to latyrogen and copper deficiency. World's Poultry Sci J 46:153–169.

Cook MI, Beissinger SR, Toranzos GA, Rodriguez RA, Arendt WJ (2003). Trans-shell infection by pathogenic micro-organisms reduces the shelf life of non-incubated bird's eggs: A constraint on the onset of incubation? Proc Roy Soc Lond B 270:2233–2240.

Cortivo R, Castellani I, Martelli M, Michelotto G, Abatangelo G (1982). Chemical characterization of the hen eggshell matrix: isolation of an alkali-resistant peptide. J Chromatogr 237:127–135.

Corzo A, Kidd MT, Pharr GT, Burgess SC (2004). Initial mapping of the chicken blood plasma proteome. Int J Poultry Sci 3:157–162.

Deeming DC (2002). Avian Incubation: Behaviour, Environment, and Evolution. New York: Oxford Univ Press.

Dennis JE, Xiao S-Q, Agarwal M, Fink DJ, Heuer AH, Caplan AI (1996). Microstructure of matrix and mineral components of eggshells from white leghorn chickens (*Gallus gallus*). J Morphol 228:287–306.

De Reggi M, Gharib B (2001). Protein-X, pancreatic stone-, pancreatic thread-, reg-protein, P19, lithostathine, and now what? Characterization, structural analysis and putative function(s) of the major non-enzymatic protein of pancreatic secretions. Curr Protein Peptide Sci 2:19–42.

Doherty MK, McLean L, Hayter JR, Pratt JM, Robertson DHL, El-Shafei A, Gaskell SJ, Benyon RJ (2004). The proteome of chicken skeletal muscle: Changes in soluble protein expression during growth in a layer strain. Proteomics 4:2082–2093.

Dominguez-Vera JM, Gautron J, Garcia-Ruiz, JM, Nys Y (2000). The effect of avian uterine fluid on the growth behavior of calcite crystals. Poultry Sci 79:901–907.

Dunn IC, Joseph NT, Milona P, Bain M, Edmond A, Wilson PW, Nys Y, Gautron J, Schmutz M, Preisinger R, Waddington D (2007). Polymorphisms in eggshell organic matrix genes are associated with eggshell quality measurements in pedigree Rhode Island Red hens (manuscript submitted).

Ewart KV, Fletcher GL (1993). Herring antifreeze protein: primary structure and evidence for a C-type lectin evolutionary origin. Mol Marine Biol Biotechnol 2(1):20–27.

Fernandez MS, Araya M, Arias JL (1997). Eggshells are shaped by a precise spatiotemporal arrangement of sequentially deposited macromolecules. Matrix Biol 16:13–20.

Fernandez MS, Escobar C, Lavelin I, Pines M, Arias JL (2003). Localization of osteopontin in oviduct tissue and eggshell during different stages of the avian egg laying cycle. J Struct Biol 143:171–180.

Fernandez MS, Moya A, Lopez L, Arias JL (2001). Secretion pattern, ultrastructural localization and function of extracellular matrix molecules involved in eggshell formation. Matrix Biol 19:793–803.

Garcillán DA, Gómez-Esquer F, Diaz-Gil G, Martinez-Arribas F, Delcán J, Schneider J, Palomar MA, Linares R (2005). Proteomic analysis of the *Gallus gallus* embryo at stage-29 of development. Proteomics 5:4946–4957.

Gautron J, Murayama E, Vignal A, Morisson M, McKee MD, Rehault S, Labas V, Belghazi M, Vidal ML, Nys Y, Hincke MT (2007). Cloning of ovocalyxin-36, a novel chicken eggshell protein related to lipopolysaccharide-binding proteins (LPB) bactericidal permeability-increasing proteins (BPI), and PLUNC family proteins. J Biol Chem (Epub ahead of print).

Gautron J, Rodriguez-Navarro AB, Gomez-Morales J, Hernandez-Hernandez MA, Dunn IC, Bain M, Garcia-Ruiz JM, Nys Y (2005). Evidence for the implication of chicken eggshell matrix proteins in the process of shell mineralization. Proc 9th Int Symp Biomineralization (Biom 09). Pucon, Chile.

Gautron J, Hincke MT, Panheleux M, Bain M, McKee MD, Solomon SE, Nys Y (2001a). Ovocalyxin-32, a novel chicken eggshell matrix protein. Isolation, amino acid sequencing, cloning, and immunocytochemical localization. J Biol Chem 276(42): 39243–39252.

Gautron J, Hincke MT, Panheleux M, Garcia-Ruiz JM, Boldicke T, Nys Y (2001b). Ovotransferrin is a matrix protein of the hen eggshell membranes and basal calcified layer. Connect Tissue Res 42(4):255–267.

Gautron J., Hincke MT, Nys Y (1997). Precursor matrix proteins in the uterine fluid change with stages of eggshell formation in hens. Connect Tissue Res 36:195–210.

Gautron J, Bain M, Solomon S, Nys Y (1996). Soluble matrix of hen's eggshell extracts changes in vitro the rate of calcium carbonate precipitation and crystal morphology. Br Poultry Sci 37:853–866.

Ghim S-Y, Choi S-K, Shin B-S, Jeong Y-M, Sorokin A, Ehrlich SD, Park S-H (1998). Sequence analysis of the *Bacillus subtilis* 168 chromosome region between the *ssp*C and *odh*A loci (184°–180°). DNA Res 5:195–201.

Guerin-Dubiard C, Pasco M, Molle D, Desert C, Croguennec T, Nau F (2006). Proteomic analysis of hen egg white. J Agric Food Chem 54(11):3901–3910.

Hagiwara K, Kikuchi T, Endo Y, Huqun Usui K, Takahashi M, Shibata N, Kusakabe T, Xin H, Hoshi S, Miki M, Inooka N, Tokue Y, Nukiwa T (2003). Mouse SWAM1 and SWAM2 are antibacterial proteins composed of a single whey acidic protein motif. J Immunol 170:1973–1979.

Hamilton RMG (1986). The microstructure of the hens egg-shell—a short review. Food Microstruct 5:99–110.

Hincke MT, Gautron J, Mann K, Panhéleux M, McKee MD, Bain M, Solomon SE, Nys Y (2003). Purification of ovocalyxin-32, a novel chicken eggshell matrix protein. Connect Tissue Res 44(1):16–19.

Hincke MT, Gautron J, Panhéleux M, Garcia-Ruiz J, McKee MD, Nys Y (2000a). Identification and localization of lysozyme as a component of eggshell membranes and eggshell matrix. Matrix Biol 19:443–453.

Hincke MT, St. Maurice M, Nys Y, Gautron J, Panheleux M, Tsang CPW, Bain MM, Solomon SE, McKee MD (2000b). Eggshell proteins and shell strength: Molecular biology of eggshell matrix proteins and industry applications. In: Sim JS, Nakai S, Guenter W, eds. Egg Nutrition and Newly Emerging Ovo-Bio Technologies. New York: CABI Publishing; pp 447–462.

Hincke MT, St. Maurice M (2000). Phosphorylation-dependent modulation of calcium carbonate precipitation by chicken eggshell matrix proteins. In: Goldberg M, Boskey A, Robinson C, eds. Chemistry and Biology of Mineralized Tissues. Rosemont, IL: American Academy of Orthopaedic Surgeons; pp 13–17.

Hincke MT, Gautron J, Tsang CPW, McKee MD, Nys Y (1999). Molecular cloning and ultrastructural localization of the core protein of an eggshell matrix proteoglycan, Ovocleidin-116. J Biol Chem 274:32915–32923.

Hincke MT (1995). Ovalbumin is a component of the chicken eggshell matrix. Connect Tissue Res 31:227–233.

Hincke MT, Tsang CPW, Courtney M, Hill V, Narbaitz R (1995). Purification and Immunochemistry of a soluble matrix protein of the chicken eggshell (Ovocleidin 17). Calc Tissue Int 56:578–583.

Hincke MT, Bernard A-M, Lee ER, Tsang CPW, Narbaitz R (1992). Soluble protein constituents of the domestic fowl's eggshell. Br Poultry Sci 33:505–516.

Huang S-Y, Lin J-H, Chen Y-H, Chuang C-K, Chiu Y-F, Chen M-Y, Chen H-H, Lee W-C (2006). Analysis of chicken serum proteome and differential protein expression during development in single-comb White Leghorn hens. Proteomics 6:2217–2224.

Hubacek JA, Buchler C, Aslanidis C, Schmitz G (1997). The genomic organization of the genes for human lipopolysaccharide binding protein (LBP) and bactericidal permeability increasing protein (BPI) is highly conserved. Biochem Biophys Res Commun 236(2):427–430.

Hunter GK, Szigety SK (1992). Effects of proteoglycan on hydroxyapatite formation under non-steady-state and pseudo-steady-state conditions. Matrix 12:362–368.

Hunter GK (1991). Role of proteoglycan in the provisional calcification of cartilage. A review and reinterpretation. Clin Orthop Related Res 262:256–280.

International Chicken Genome Sequencing Consortium (2004). Sequence and comparative analysis of the chicken genome provide unique perspectives on vertebrate evolution. Nature 432:695–716.

Ishihama Y, Oda Y, Tabata T, Sato T, Nagasu T, Rappsilber J, Mann M (2005). Exponentially modified protein abundance index (emPAI) for estimation of absolute protein amount in proteomics by the number of sequenced peptides per protein. Mol Cell Proteom 4:1265–1272.

Jimenez-Lopez C, Rodríguez-Navarro A, Domingez-Vera JM, Garcia-Ruiz JM (2003). Influence of lysozyme on the precipitation of calcium carbonate. Kinetic and morphological study. Geochim Cosmochim Acta 67:1667–1676.

Kawasaki K, Weiss KM (2006). Evolutionary genetics of vertebrate tissue mineralization: the origin and evolution of the secretory calcium-binding phosphoprotein family. J Exper Zool Pt B Mol Devel Evol 306(3):295–316.

Lakshminarayanan R, Kini RM, Valiyaveettil S (2002). Investigation of the role of ansocalcin in the biomineralization in goose eggshell matrix. Proc Nat Acad Sci USA 99:5155–5159.

Lakshminarayanan R, Valiyaveettil S, Roa VS, Kini RM (2003). Purification, characterization, and in vitro mineralization studies of a novel goose eggshell matrix protein, ansocalcin. J Biol Chem 278:2928–2936.

Lakshminarayanan R, Joseph JS, Kini RM, Valiyaveettil S (2005). Structure-function relationship of avian eggshell matrix proteins: A comparative study of two major eggshell matrix proteins, ansocalcin and OC-17. Biomacromolecules 6:741–751.

Liu Q, Yu L, Gao J, Fu Q, Zhang J, Zhang P, Chen J, Zhao S (2000). Cloning, tissue expression pattern and genomic organization of latexin, a human homologue of rat carboxypeptidase A inhibitor. Mol Biol Rep 27:241–246.

Mann K, Olsen JV, Macek B, Gnad F, Mann M (2007). Phosphoproteins of the chicken eggshell calcified layer. Proteomics 7:106–115.

Mann K, Siedler F (2006). Amino acid sequences and phosphorylation sites of emu and rhea eggshell C-type lectin-like proteins. Compar Biochem Physiol Pt B 143:160–170.

Mann K, Maček B, Olsen JV (2006). Proteomic analysis of the acid-soluble organic matrix of the chicken calcified eggshell layer. Proteomics 6:3801–3810.

Mann K (2004). Identification of major proteins of the organic matrix of the emu (*Dromaius novaehollandiae*) and rhea (*Rhea americanus*) eggshell calcified layer. Br Poultry Sci 45:483–490.

Mann K, Siedler F (2004). Ostrich (Struthio camelus) eggshell matrix contains two different C-type lectin-like proteins. Isolation, amino acid sequence and post-translational modifications. Biochim Biophys Acta 1696:41–50.

Mann K, Gautron J, Nys Y, McKee MD, Bajari T, Schneider WJ, Hincke MT (2003). Disulfide-linked heterodimeric clusterin is a component of the chicken eggshell matrix and egg white. Matrix Biol 22(5):397–407.

Mann K (1999). Isolation of a glycosylated form of the eggshell protein ovocleidin and determination of the glycosylation site. Alternative glycosylation/phosphorylation at an N-glycosylation sequon. FEBS Lett 463:12–14.

Mann K, Siedler F (1999). The amino acid sequence of ovocleidin-17, a major protein of the avian eggshell calcified layer. Biochem Mol Biol J 47:997–1007.

Mann M, Ong S, Grønborg M, Steen H, Jensen ON, Pandey A (2002). Analysis of protein phosphorylation using mass spectrometry: deciphering the phosphoproteome. Trends Biotechnol 20:261–268.

Mann M, Hendrickson RC, Pandey A (2001). Analysis of proteins and proteomes by mass spectrometry. Annu Rev Biochem 70:437–473.

McKee MD, Nanci A (1996). Osteopontin at mineralized tissue interfaces in bone, teeth, and osseointegrated implants: Ultrastructural distribution and implications for mineralized tissue formation, turnover, and repair. Microsc Res Tech 33:141–164.

McKee MD, Nanci A (1995). Osteopontin and the bone remodeling sequence. Ann NY Acad Sci 760:177–189.

McKee MD, Nanci A, Khan SR (1995). Ultrastructural immunodetection of osteopontin and osteocalcin as major matrix components of renal calculi. J Bone Mineral Res 10:1913–1929.

Mine Y, Oberle C, Kassaify Z (2003). Eggshell matrix proteins as defence mechanism of avian eggs. J Agric Food Chem 51:249–253.

Narbaitz R, Bastani B, Galvin NJ, Kapal VK, Levine DZ (1995). Ultrastructural and immunocytochemical evidence for the presence of polarised plasma membrane H(+)-ATPase in two specialised cell types in the chick embryo chorioallantoic membrane. J Anat 186(Pt 2):245–252.

Nys Y, Gautron J, Garcia-Ruiz JM, Hincke MT (2004). Avian eggshell mineralization: Biochemical and functional characterization of matrix proteins. CR Paleovol 3:549–562.

Nys Y, Gautron J, McKee MD, Garcia-Ruiz JM, Hincke MT (2001). Biochemical and functional characterisation of eggshell matrix proteins in hens. World's Poultry Sci J 57:401–413.

Nys Y, Hincke M, Arias JL, Garcia-Ruiz JM, Solomon S (1999). Avian eggshell mineralization. Poultry Avian Biol Rev 10:143–166.

Nys Y, Zawadzki J, Gautron J, Mills AD (1991). Whitening of brown-shelled eggs: Mineral composition of uterine fluid and rate of protoporphyrin deposition. Poultry Sci 70:1236–1245.

Panheleux M, Nys Y, Williams J, Gautron J, Boldicke T, Hincke MT (2000). Extraction and quatification of eggshell matrix proteins (ovocleidins, ovalbumin, ovotransferrin) in shell from young and old hens. Poultry Sci 79:580–588.

Panheleux M, Bain M, Fernandez MS, Morales I, Gautron J, Arias JL, Solomon SE, Hincke M, Nys Y (1999). Organic matrix composition and ultra-structure of eggshell: A comparative study. Br Poultry Sci 40:240–252.

Parada C, Gato A, Aparicio M, Bueno D (2006). Proteome analysis of chick embryonic cerebrospinal fluid. Proteomics 6:312–320.

Parsons AH (1982). Structure of the eggshell. Poultry Sci 61:2013–2021.

Pines M, Knopov V, Bar A (1994). Involvement of osteopontin in egg shell formation in the laying chicken. Matrix Biol 14:765–771.

Reyes-Grajeda JP, Moreno A, Romero A (2004). Crystal structure of ovocleidin-17, a major protein of the calcified *Gallus gallus* eggshell: Implications in the calcite mineral growth pattern. J Biol Chem 279:40876–40881.

Robey PG (1996) Vertebrate mineralized matrix proteins: Structure and function. Connect Res Tissue 35(1–4):131–136.

Silphaduang U, Hincke MT, Nys Y, Mine Y (2006). Antimicrobial proteins in chicken reproductive system. Biochem Biophys Res Commun 340:648–655.

Sparks NHC (1994). Shell accessory materials: Structure and function. In: Board RG, Fuller R, eds. Microbiology of the Avian Egg. London: Chapman & Hall.

Sparks NHC, Board RG (1984). Cuticle, shell porosity and water intake through hen's eggshells. Br Poultry Sci 25:267–276.

Steen H, Mann M (2004). The ABC's (and XYZ's) of peptide sequencing. Nat Rev/Mol Cell Biol 5:699–711.

Uratani Y, Takiguchi-Hayashi K, Miyasaka N, Sato M, Jin M, Arimatsu Y (2000). Latexin, a carboxypeptidase A inhibitor, is expressed in rat peritoneal mast cells and is associated with granular structures distinct from secretory granules and lysosomes. Biochem J 346:817–1826.

Vadehra DV, Baker RC, Naylor HB (1972). Distribution of lysozyme activity in the exteriors of eggs from *Gallus gallus*. Compar Biochem Biophys B 43:503–508.

Valenti P, De Stasio A, Mastromerino P, Seganti L, Sinibaldi L, Orsi N (1981). Influence of bicarbonate and citrate on the bacteriostatic action of ovotransferrin towards staphylococci. FEMS Microbiol Lett 10:77–79.

van Dijk A, Veldhuizen EJA, van Asten AJAM, Haagsman HP (2005). CMAP27, a novel chicken cathelicidin-like antimicrobial protein. Vet Immunol Immunopathol 106:321–327.

von Hunolstein C, Ricci ML, Valenti P, Orefici G (1992). Lack of activity of transferrins towards Streptococcus spp. Med Microbiol Immunol 181:351–357.

Watanabe H, Yamada Y, Kimata K (1998). Roles of aggrecan, a large chondroitin sulfate proteoglycan, in cartilage structure and function. J Biochem 124:687–693.

Wellman-Labadie O, Picman J, Hincke MT (2007a). Antimicrobial activity of cuticle and outer eggshell protein extracts from three species of domestic fowl. Br Poult Sci (in revision).

Wellman-Labadie O, Picman J, Hincke MT (2007b). Antimicrobial activity of the anseriform outer eggshell and cuticle (submitted).

Whittow GC (2000). Sturkie's Avian Physiology, 5th ed. San Diego: Academic Press.

Wong M, Hendrix MJC, Von der Mark K, Little C, Stern R (1984). Collagen in the egg shell membranes of the hen. Devel Biol 104:28–36.

Xiao Y, Cai Y, Bommineni YR, Fernando SC, Prakash O, Gilliland SE, Zhang G (2006). Identification and functional characterization of three chicken cathelicidins with potent antimicrobial activity. J Biol Chem 281:2858–2867.

Xiao Y, Hughes AL, Ando J, Matsuda Y, Cheng J-F, Skinner-Noble D, Zhang G (2004). A genome-wide screen identifies a single β-defensin gene cluster in the chicken: Implications for the origin and evolution of mammalian defensins. BioMed Central Genom 5:56.

Xing J, Wellman-Labadie O, Gautron J, Hincke MT (2007). Recombinant eggshell ovocalyxin-32: Expression, purification and biological activity of the glutathione S-transferase fusion protein. Compar Biochem Physiol (in press).

Yang KT, Lin CY, Liou JS, Fan YH, Chiou SH, Huang CW, Wu CP, Lin EC, Chen CF, Lee YP, Lee WC, Ding ST, Cheng WT, Huang MC (2007) Differentially expressed transcripts in shell glands from low and high egg production strains of chickens using cDNA microarrays. Anim Reprod Sci (Epub ahead of print Sept 9, 2006).

Yoshimura Y, Ohashi H, Subedi K, Nishibori M, Isobe N (2006). Effects of age, egg-laying activity, and salmonella-inoculatioon on the expressions of gallinacin mRNA in the vagina of the hen oviduct. J Reprod Devel 52:211–218.

3

BIOAVAILABILITY AND PHYSIOLOGICAL FUNCTION OF EGGSHELLS AND EGGSHELL MEMBRANES

Yasunobu Masuda and Hajime Hiramatsu
R&D Division, Q.P. Corporation, Tokyo, Japan

3.1. INTRODUCTION

Approximately 2,500,000 tons of eggs are consumed in Japan every year, which poses a significant challenge regarding disposal of the eggshells. Rather than discarding them, practical utilization of eggshells and their membranes has been sought. The main objective of this chapter is to describe the research on the bioavailability and physiological functions of eggshell components and to identify potential uses for them.

3.2. EGGSHELL STRUCTURE

Eggshells are composed of a matrix consisting of interwoven protein fibers and spherical masses with interstitial calcium carbonate crystals. The proportion of matrix to crystalline material is about 1:50. The surface of the calcified eggshell is covered by a protein cuticle and has over 10,000 funnel-shaped pores forming passages between the cuticle and the shell membrane.

Egg Bioscience and Biotechnology Edited by Yoshinori Mine

3.3. EGGSHELL POWDERS

3.3.1. Manufacturing and Composition

To be added to foods, eggshells must be refined; Figure 3.1 shows the manufacturing process for producing eggshell powders. First, eggs are washed and cracked and the liquid egg is removed. The eggshells are then collected, ground, and washed with water to separate the membrane. Next, the eggshells are dried, pasteurized, and ground into a fine powder. This powder is also referred to as *eggshell calcium*.

Table 3.1 shows the composition of eggshell calcium. The major component is calcium carbonate, with 38.0% of the powder as pure calcium. Because of the high calcium content, only a small amount of eggshell powder would be required to adequately fortify foods. Eggshell powder also contains magnesium, sodium, potassium, and a small amount of protein.

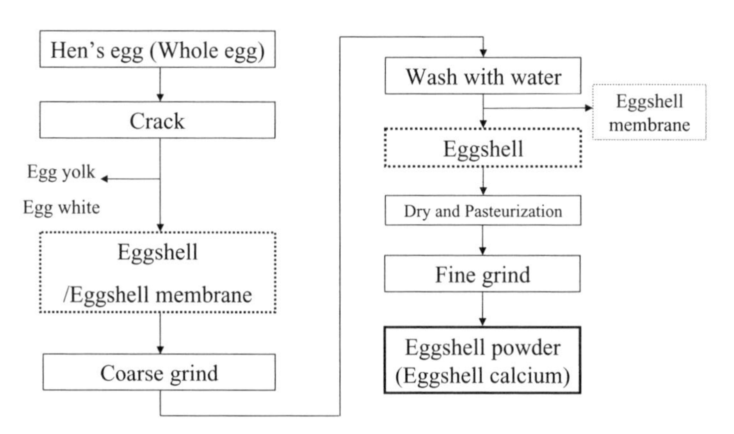

Figure 3.1. Manufacturing process for eggshell powder.

TABLE 3.1. Chemical Composition[a] of Eggshell Powder

Moisture	0.5 g
Protein	2.1 g
Ash	96.9 g
Calcium	38.0 g
Potassium	41.6 mg
Sodium	87.0 mg
Phosphorus	99.3 mg
Ferrum	0.5 mg
Magnesium	375.0 mg

[a]Analytical values per 100 g.

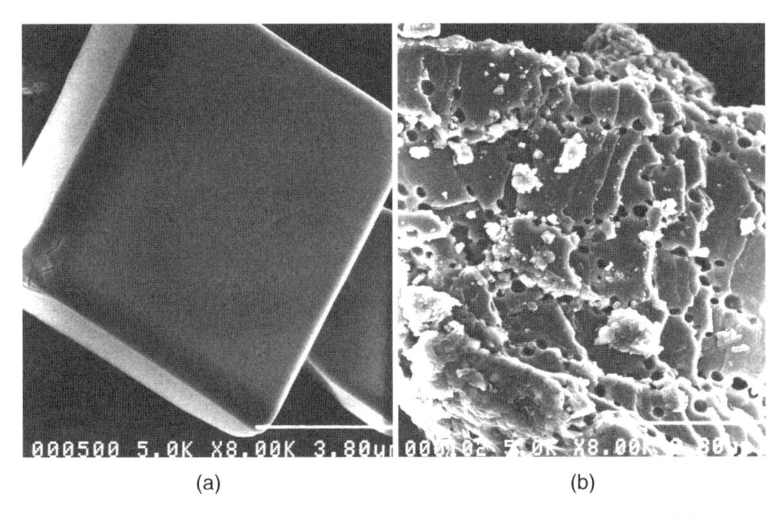

Figure 3.2. Electron photomicrographs (×8000) of calcium carbonate (a) and eggshell powder (b).

The physical structure of the eggshell powder is very porous, while manufactured calcium carbonate has a smooth surface. Figure 3.2 shows electron photomicrographs of both synthetic calcium carbonate and eggshell powder.

3.3.2. Calcium Intake in Japan

Japan has a high rate of osteoporosis among its elderly population, which has been linked to insufficient calcium intake. This is a serious nutritional problem. The Japanese adequate intake (AI) of calcium for male and female adults (over 18 years) is >650 mg and >600 mg, respectively (Ministry of Health, Labour and Welfare 2005). However, the latest National Health and Nutrition Survey, conducted in 2003 (Ministry of Health, Labour and Welfare 2006), states that the actual calcium intake was only 536 mg; this value has remained unchanged since 1975. The AI of calcium in Japan is also much lower than that of other countries. For example, the recommended dietary allowance for Canada and the United States is 800 and 1000 mg, respectively. This may suggest that fulfilling the AI requirement of calcium from the typical Japanese diet is extremely difficult. Calcium fortification with eggshell powder may be the solution to this problem.

The ideal dietary calcium : phosphorus ratio is suggested to be between 1:1 and 1:2 (Hirota and Hirota 2001). However, studies show that phosphorus intake greatly exceeds that of calcium. This is due not only to diets low in calcium but also to the fact that many calcium-fortifying agents such as fish or animal bones contain high levels of phosphorus. Fortifying with eggshell powder, which contains only 0.10% phosphorus, would help restore the proper dietary balance between calcium and phosphorus.

TABLE 3.2. Solubility of Two Calcium Sources in the Stomach

	Percent Calcium Solubilized	
Time, min	Eggshell	$CaCO_3$
30	63.2 ± 21.8	28.8 ± 24.9 ($p < .05$)
60	88.6 ± 15.7	51.8 ± 28.1 ($p < .01$)
90	88.8 ± 14.6	73.1 ± 20.0 ($p < .05$)
120	95.2 ± 6.7	69.6 ± 26.7 ($p < .01$)

Source: Kusumi et al. (1999).

3.3.3. Solubility of Eggshell Powder

It has been stated that the digestibility of calcium is directly related to its solubility in the stomach. A study by Kusumi et al. (1999) investigated this hypothesis. Human subjects (elderly patients suffering from cerebrovascular disease) were randomly assigned to a liquid diet containing either eggshell powder or synthetic calcium carbonate. The contents of their stomachs were periodically removed and the calcium solubility analyzed. Table 3.2 shows that the eggshell calcium was significantly more soluble than synthetic calcium carbonate. This result suggests a rapid solubility of eggshell powder in the stomachs of the aged. It is thought that eggshell powder is more easily dissolved with gastric juice because of its porous structure.

3.3.4. Absorption Rate of Eggshell Powder

Figure 3.3 shows the results of a study conducted by Goto et al. (1981), who compared calcium absorption rates on administration of three different sources of calcium in male rats (i.e., calcium carbonate, coarse eggshell powder, and fine eggshell powder).

The calcium absorption rates were assessed during both the growth and the maturation periods. No significant differences were observed during the rat growth period. On the other hand, during the maturation period the absorption rates obtained on ingestion of eggshell powders were significantly higher than the rate observed after calcium carbonate administration. More specifically, the fine eggshell powder absorption rate (34.8%) was greater than that of the coarse powder (21.3%) and twice as high as that of calcium carbonate (13.7%). A subsequent study reported similar results in pregnant rats (Niiyama and Sakamoto 1984).

3.3.5. Effect of Eggshell Powder on Bone Mineral Density

It has been reported that eggshell powder is effective in increasing bone mineral density (BMD) in osteoporotic model rats (Omi and Ezawa 1998). In this study, the effects of eggshell powder and calcium carbonate were compared by measuring bone metabolism in oophorectomized rats. Results showed no significant differences in body weight gain, food intake, or food efficiency

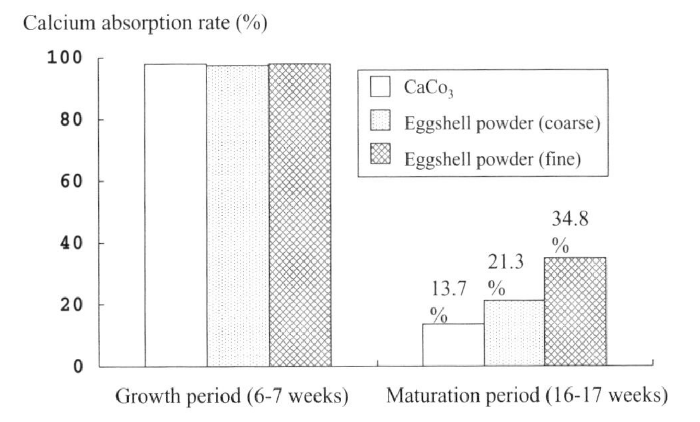

Figure 3.3. Comparison between eggshell powder and calcium carbonate for absorptivity in male rats (Goto et al. 1981).

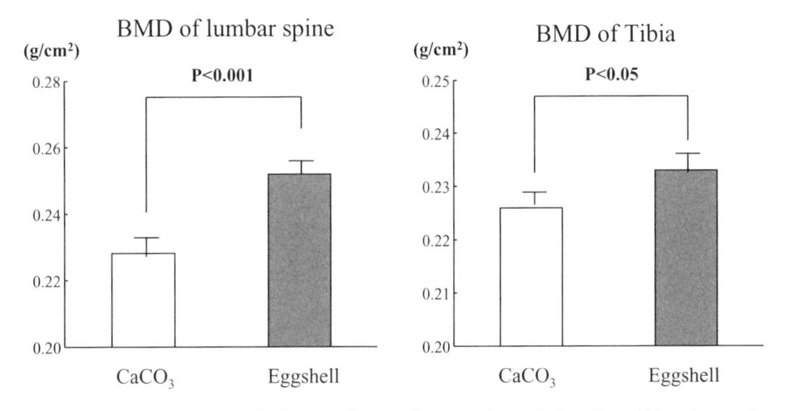

Figure 3.4. The effect of eggshell powder on bone mineral density of lumbar spine and tibia in oophorectomized rats (Omi and Ezawa 1998).

between the two groups. The serum levels of calcium and phosphorus were also normal for both groups. However, significant differences were observed between the two groups in the BMD values. Figure 3.4 shows that the BMD value of the lumbar spine and the tibia for the eggshell calcium group was significantly greater than that of the calcium carbonate group. These data suggest that eggshell calcium could be effective in preventing bone mineral loss following an oophorectomy. A second study (Kimura et al. 1999) has shown that the BMD value of the lumbar spine for the eggshell calcium treatment was greater than those of other calcium sources (shellfish or bone marrow) in oophorectomized rats (Fig. 3.5).

In an attempt to elucidate the mechanism underlying the effects of eggshell powder on bone, the biochemical markers of bone turnover were measured

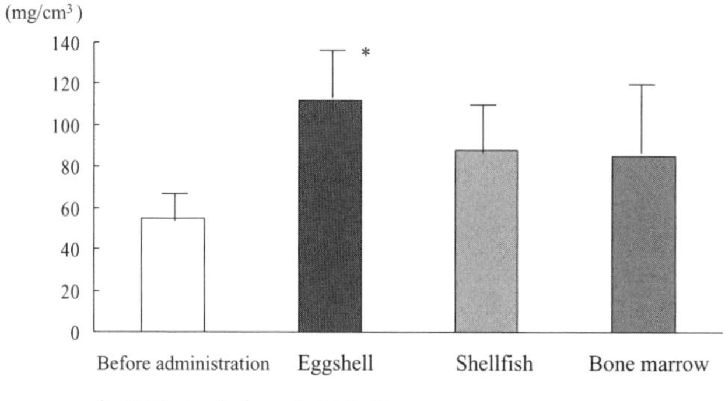

Figure 3.5. The effect of natural calcium sources on bone mineral density of lumbar spine in oophorectomized rats (Kimura et al. 1999).

TABLE 3.3. Effects of Eggshell Administration on Calcium Metabolism and Biological Markers of Bone Turnover (Mean ± SE)

Mineral or Compound	Before $CaCO_3$ (*A*)	After $CaCO_3$ (*A*)	After Eggshell (*B*)	Rate of Change (*B*–*A*)/*A*
Ca (mg/dL)	9.2 ± 0.4	9.1 ± 0.3	8.7 ± 0.4[a]	–4.2
#Ca[b] (mg/dL)	9.6 ± 0.3	9.5 ± 0.4	9.0 ± 0.3[a]	–5.3
P[c] (mg/dL)	3.8 ± 0.4	3.4 ± 0.6	3.5 ± 0.4	3
ALP[d] (IU)	181 ± 51	165 ± 69	51 ± 60	–7.6
ALP3 (IU)	76.7 ± 28.3	71.6 ± 43.7	63.6 ± 32.4	–6.3
Urine Ca/Cr[e]	0.20 ± 0.14	0.19 ± 0.14	0.15 ± 0.10	–8.8
Intact PTH[f] (pg/mL)	35.6 ± 26.8	35.9 ± 22.2	26.6 ± 23.5[a]	–28.2
DPD[g] (nm/mmCr)	13.5 ± 6.6	12.2 ± 5.9	10.6 ± 6.1	–14.4

[a]$p < .05$: vs. (*A*).
[b]Compensate calcium.
[c]Phosphorus.
[d]Alkaline phosphatase.
[e]Creatinine.
[f]Parathyroid hormone.
[g]Deoxypyridinoline.
Source: Masaki et al. (2000).

in a group of elderly women. Table 3.3 summarizes the results obtained after administration of either eggshell powder or calcium carbonate in this group of subjects. The ingestion of eggshell powder led to a significant decrease in serum parathyroid hormone (PTH) concentration, and lower levels of urine deoxypyridinoline (DPD) (Masaki et al. 2000). Although the bioactive mechanisms underlying the beneficial effects of eggshell calcium remain unclear, it is believed that their porous structure, as well as the presence of trace amount of proteins, may be involved. It has also been reported that eggshell matrix proteins could enhance calcium transport across the human intestinal epithelial cells.

3.3.6. Use of Eggshell Powder as a Phosphate Binder

Eggshell powder has also been investigated as an orally administered form of phosphate binder. Patients suffering from renal failure often require hemodialysis. However, this is an insufficient method of removing serum phosphorus, and hyperphosphatemia often occurs in these patients. One way to prevent this is a low-phosphate diet. Another is the use of phosphate binders to obstruct intestinal absorption. Calcium carbonate is a commonly used phosphate binder (Slatopolsky et al. 1986). However, a study by Ogihara et al. (1996) has shown eggshell powder to be more efficient than calcium carbonate in lowering the serum phosphorus levels without causing the serum calcium concentration to rise.

For some patients taking calcium carbonate, the serum phosphorus concentration is insufficiently reduced. This often occurs in patients using gastric inhibitory agents such as H_2-blockers. However, improvements were observed in these patients when eggshell powder was administered instead of calcium carbonate. Table 3.4 shows the effects of eggshell powder as the phosphate binder in H_2-blocker-treated patients. In non-H_2-blocker patients, both eggshell powder and calcium carbonate were equally effective in decreasing serum phosphorus. However, in H_2-blocker-treated patients, the administration of calcium carbonate produced only a slight decrease in serum phosphorus concentrations while ingestion of eggshell calcium decreased phosphorus concentrations down to levels seen in non-H_2-blocker-treated patients (Ogihara et al. 1996).

3.3.7. Effect of Eggshell Powder on Fat Absorption

The effects of eggshell powder on the absorption of fat from chocolate in Japanese men have been reported. The results were as follows: (1) fecal levels of total lipids were higher in the eggshell-treated group than in the control group, (2) fecal levels of palmitic and stearic acids were higher in the eggshell-treated group, (3) there was a marked correlation between the concentration of fecal calcium and fecal total lipids, and (4) no change was observed in serum lipid or calcium concentrations (Murata et al. 1998). The increase in fat excretion is likely to result from an interaction between calcium and saturated fatty

TABLE 3.4. Effects of Eggshell as a Phosphate Binder under H_2-Blocker Administration

	Non-H_2-Blocker Patients		H_2-Blocker Patients	
Modification	$CaCO_3$	Eggshell	$CaCO_3$	Eggshell
		(%)		
Change of serum P	−15.6	−18.5	−5.3	−17.7
Change of serum Ca	+4.3	+2.1	+6.3	+3.4

Notes: *Patients*—non-H_2-blocker (n = 14), H_2-blocker (n = 24); *phosphate binder*—$CaCO_3$ or eggshell (3.0 g/day); *experimental period*—4 weeks.

Source: Ogihara et al. (1996).

TABLE 3.5. Clinical Features and Results of Oral Fine Eggshell Powder Challenge in Subjects with Egg Hypersensitivity

Initials	Age, (years)	Sex	Serum IgE (IU/mL)	CAPRAST Score to Eggwhite	Whole-Egg Challenge	Eggshell Challenge
SW	3	F	414	3	Positive	Negative
SO	4	M	431	3	Positive	Negative
YH	2	M	932	6	Positive	Negative
SM	3	M	14,900	6	Positive	Negative
MS	2	M	271	4	Positive	Negative
SM	6	M	4440	4	Positive	Negative

Source: Ebisawa et al. (2005).

acids, leading to the formation of insoluble complexes and hence a reduction in fat absorption. More recently, it has been reported that intake of dietary calcium may play important regulatory functions in energy metabolism (Zemel and Miller 2004).

3.3.8. Allergic Response to Fine Eggshell Powder

There are few data regarding the allergenicity of eggshell powder. To test the possibility of egg allergen contamination in fine noncalcinated eggshell powder (FNCEP) (not manufactured at very high temperature), samples were analyzed by both *in vitro* methodologies (SDS-PAGE, Western blotting, inhibition ELISA, and sandwich ELISA) and by a single-blind challenge on six egg-hypersensitive patients. The allergenic activity of FNCEP was negligible compared to egg white evaluated by *in vitro* methods. Furthermore, the six patients with egg hypersensitivity were unresponsive to oral ingestion of FNCEP (Table 3.5). These results indicate that there is little contamination of egg white in FNCEP, and that its allergenic activity is equivalent to that of calcinated eggshell powder (Ebisawa et al. 2005).

3.3.9. Nutraceutical Uses of Eggshell Powder

Eggshell powder represents a natural and health promoting source of calcium. Its incorporation to a variety of food products has therefore been encouraged. Commonly found eggshell powder-supplemented products include rice crackers, rice condiments, confectionaries, and fruit juices.

3.4. EGGSHELL MEMBRANE

3.4.1. Eggshell Membrane as a Biological Dressing

The eggshell membrane has long been used as a treatment for skin injuries. More recently, animal experiments and clinical usage of eggshell membrane have shown satisfactory results. Maeda et al. (1981) reported on the successful use of eggshell membrane as a biological dressing for covering burns and deepithelialized donor areas.

3.4.2. Effects of Hydrolyzed Eggshell Membrane on Growth of Human Skin Fibroblasts and Their Production of Type III Collagen

In a study by Suguro et al. (2000), the effects of hydrolyzed eggshell membrane (HEM) on cell growth were investigated. Culture dishes were prepared by pouring a solution of suspended HEM (concentrations of 0.1, 1, 10, and 100 g/cm^2) into them and allowed to dry. Normal human skin fibroblasts (HSF) were cultured into untreated dishes, tissue culture dishes, and the HEM-coated dishes. Untreated dishes did not support HSF growth. However, they did grow in both the tissue culture dishes and the dishes coated with HEM (Fig. 3.6). Growth rates appeared to be related to the HEM concentration, with greater cell growth at higher concentrations.

HSF also produce collagen. Type III collagen, which is rich in the skin of infants, is believed to soften skin. The quantities of collagen formed in the test were compared between the HEM-coated dishes and tissue culture dishes (Suguro et al. 2000). Table 3.6 shows that the collagen ratio using HEM-coated dishes was around 9.3%, while the ratio using tissue culture dishes was only about 7.6%. These results suggest that HEM tends to allow growth of HSF and may facilitate their production of type III collagen.

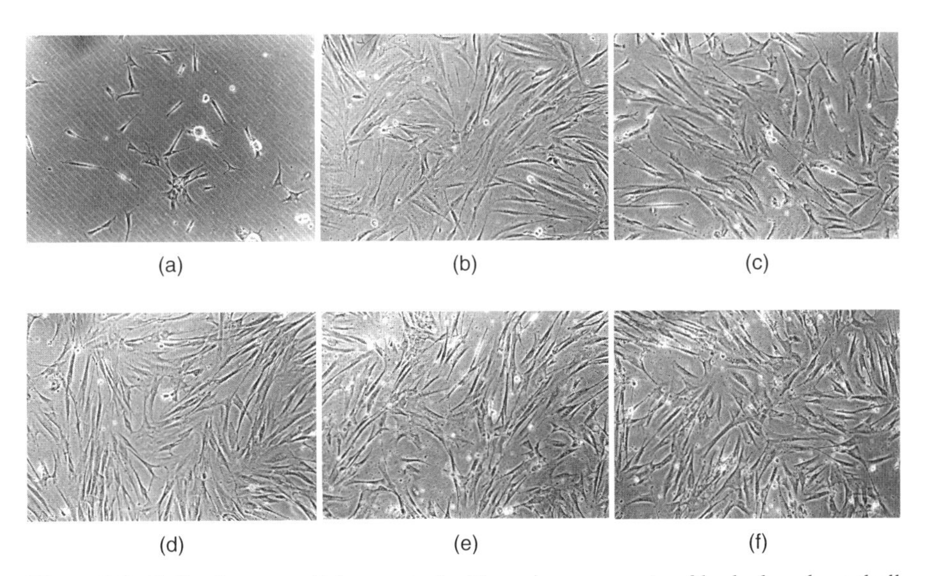

Figure 3.6. Cell culture on dishes coated with various amounts of hydrolyzed eggshell membrane: (a) nontreated dish; (b) tissue culture dish; (c) hydrolyzed eggshell membrane, 0.1 μg/cm^2; (d) hydrolyzed eggshell membrane, 1 μg/cm^2; (e) hydrolyzed eggshell membrane, 10 μg/cm^2; (f) hydrolyzed eggshell membrane, 100 μg/cm^2 (Suguro et al. 2000).

TABLE 3.6. Amount of Collagen Produced by Human Fibroblasts

Medium	Total Collagen (I + III) (cpm/10^5cells)	Type III Collagen Ratio (III : I + III) (%)
Egg membrane–coated dish	4039 ± 209	9.28 ± 0.65
Tissue culture dish	3932 ± 25	7.59 ± 0.26

Source: Suguro et al. (2000).

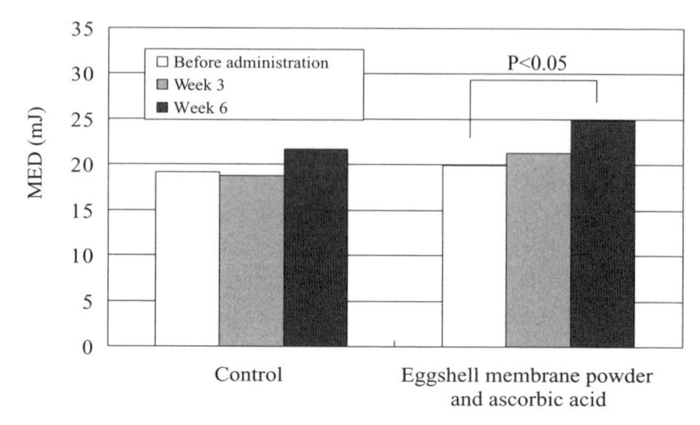

Figure 3.7. The effect on MED by taking eggshell membrane powder with ascorbic acid (Tanino 2003).

3.4.3. Dermatologic Effects of Eggshell Membrane Powder Combined with Ascorbic Acid

Tanino (2003) studied the effects of ingesting eggshell membrane on the reaction of skin to ultraviolet light. Twenty volunteers were evenly divided into two groups, with the first group taking 2.5 g of eggshell membrane powder and 300 mg of ascorbic acid daily for 6 weeks. Subjects of the second (control) group took dextrin daily for 6 weeks. After 0, 3, and 6 weeks, the minimal erythemal dose (MED) and minimal persistent pigmenting dose (MPPD) for each subject was determined. The MED is defined as dosage of UV light irradiation sufficient to produce a minimal, perceptible erythema on untreated skin. The MPPD is the UVA (ultraviolet A) dose necessary to induce a minimal, perceptible pigmentation, at a specified observation time. The MED and MPPD of the treatment group both increased after 6 weeks. This suggests that taking eggshell membrane powder with ascorbic acid tends to prevent UV-induced erythema and pigmentation of skin (Figs. 3.7 and 3.8).

3.4.4. Metal Ion Adsorptive Function of Eggshell Membrane

A study has shown eggshell membrane to rapidly accumulate and eliminate various heavy-metal ions from dilute aqueous solution. Under certain condi-

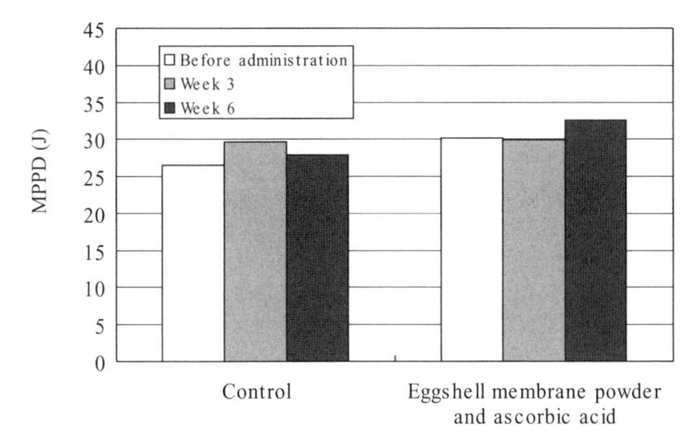

Figure 3.8. The effect on MPPD by taking eggshell membrane powder with ascorbic acid (Tanino 2003).

tions, the accumulation level of the ions, Au, Pt, Pd, and U, approached 55%, 25%, 22%, and 30% of dry weight of eggshell membrane, respectively (Suyama et al. 1994). This suggests that eggshell membrane could potentially be used for the removal or recovery of metals and for water pollution control.

3.5. CONCLUSIONS

Currently, research is being done to help elucidate the bioactive mechanisms and physiological functions of eggshells and their membranes. Once these mechanisms and functions are better understood, utilization of these by-products may advance areas of health and medical care while promoting environmental sustainability by reducing the amount of waste materials.

ACKNOWLEDGMENT

We thank Karen Moss of Henningsen Foods, Inc. for useful advice in the preparation of this manuscript.

REFERENCES

Ebisawa M, Tachimoto H, Ikematsu K, Sugizaki C, Masuda Y, Kimura M (2005). Allergenic activity of non-calcinated egg shell calcium. Jpn J Allergol 54:471–477.

Goto S, Suzuki K, Kanke Y, Kokubu T, Kurokawa T (1981). The utilization of egg shell as calcium source. In: Abstracts 35th Annual Mtg Japanese Society of Food and Nutrition, Tokushima; p 124.

Hirota T, Hirota K (2001). The diet therapy in osteoporosis. Clin Nutr 99:290–297.

Kimura M, Hayashi N, Takizawa K, Suguro N, Kunou M (1999). The effect of several natural calcium sources on bone mineral density in ovariectomized rats. In: Abstracts 1st Annual Mtg Japan Osteoporosis Society, Kurashiki; p 64.

Kusumi N, Nakamura M, Tando Y, Suda T, Kudo K (1999). Egg-shell calcium solubility in stomach. Jpn J Nutr Assess 16:291–296.

Maeda K, Sasaki Y (1981). An experience of hen-egg membrane as a biological dressing. Burns 8:313–316.

Masaki H, Nakatsuka Y, Miki T, Takamoto K, Ohnishi T, Suguro N, Kunou M, Kawamura M, Nishizawa Y, Morii H (2000). Inhibitory effects of eggshell calcium on bone resorption in elderly subjects—comparison with calcium carbonate. Osteoporosis (Jpn) 8:245–247.

Ministry of Health, Labour and Welfare (2005). Dietary Reference Intakes for Japanese, 2005. Tokyo: Daiichi Publications; pp 135–143.

Ministry of Health, Labour and Welfare (2006). The National Health and Nutrition Survey in Japan, 2003. Tokyo: Daiichi Publications; pp 58–70, 310.

Murata T, Kuno T, Hozumi M, Tamai H, Takagi M, Kamiwaki T, Itoh Y (1998). Inhibitory effects of calcium (derived from eggshell)-supplemented chocolate on absorption of fat in human males. J Jpn Soc Nutr Food Sci 51:165–171.

Niiyama Y, Sakamoto S (1984). Calcium utilization in pregnant rats fed soy protein isolate. Nutr Sci Soy Protein 5:53–58.

Ogihara M, Suzuki T, Umeda H, Nakamura T, Ishibashi K, Nomiya M, Yamaguchi O, Shiraiwa Y, Sasaki S (1996). Change of effects of phosphate binder under histamine H_2-receptor antagonist administration; comparative study between calcium carbonate and egg shell calcium. Kidney Dialysis 41:695–698.

Omi N, Ezawa I (1998). Effect of egg-shell Ca on preventing of bone loss after ovariectomy. J Home Econ Jpn 49:277–282.

Slatopolsky E, Weerts C, Lopez-Hilker S, Norwwod K, Zink M, Windus D, Delmez J (1986). Calcium carbonate as a phosphate binder in patients with chronic renal failure undergoing dialysis. New Engl J Med 315:157–161.

Suguro N, Horiike S, Masuda Y, Kunou M, Kokubu T (2000). Bioavailability and commercial use of eggshell calcium, membrane proteins and yolk lecithin products. In: Sim JS, Nakai S, Guenter W, eds. Egg Nutrition and Biotechnology. CABI Publishing; pp 219–232.

Suyama K, Fukazawa Y, Umetsu Y (1994). A new biomaterial, hen egg shell membrane, to eliminate heavy metal ion from their dilute waste solution. Appl Biochem Biotechnol 45/46:871–879.

Tanino S (2003). Functional materials for food from hen egg components. Tech J Food Chem Chem 19(5):32–35.

Zemel MB, Miller SL (2004). Dietary calcium and daily modulation of adiposity and obesity risk. Nutr Rev 62:125–131.

4

BIOACTIVE COMPONENTS IN EGG WHITE

Yoshinori Mine and Icy D'Silva
Department of Food Science, University of Guelph, Guelph, Ontario, Canada

4.1. INTRODUCTION

Hen eggs are an important source of nutrients and a major source of active molecules. Modification of the nutrient content of the egg by modifying the diet of the hen has led to "designer eggs" that serve as functional foods and nutraceuticals (Stadelman 1999). With a hen laying more than 300 eggs per year, and with the vast infrastructure of the layer industry housing more than 100,000 hens, the hen holds great promise as a low-cost, high-yield bioreactor of active molecules usable by the food, pharmaceutical, and cosmetic industries.

Eggs from all species of poultry consist of 52–58% albumen, 32–35% yolk, and 9–14% shell and its supporting membranes. The egg contains all the essential components, such as proteins, lipids, vitamins, minerals, carbohydrates, and growth factors required by the developing embryo (Mine and Kovacs-Nolan 2004). Egg components have been attributed diverse biological activities, including antimicrobial and antiviral activity, protease inhibitory action, vitamin-binding properties, anticancer activity, and immunomodulatory activity (Li-Chan et al. 1995). In addition, bioactive components of the egg are of importance in advanced research, and represent promising leads for treatment of various diseases. The albumen is a major water reservoir of the egg, which would seem to make it a weak contributor of proteins. However, the greater mass of the albumen makes it an equal contributor of proteins.

Egg Bioscience and Biotechnology Edited by Yoshinori Mine

Hen egg white has a very broad molecular mass range of proteins, extending from 12.7 kDa for cystatin to 8000 kDa for soluble ovomucin; and pI ranges from 3.9 for ovoglycoprotein to 10.7 for lysozyme (Guérin-Dubiard et al. 2006). Albumen proteins identified to date belong to several functional protein families, including serpin, transferrin, protease inhibitors Kazal, glycosyl hydrolases, lipocalin, bacterial permeability-increasing protein, clusterin, cysteine protease inhibitor, VMO-1, and uPAR/CD59/Ly6/snake neurotoxin folate receptor (Guérin-Dubiard et al. 2006). Proteins in the egg white occur both as free and conjugated forms. The latter are formed when the protein is modified by prosthetic groups such as lipids and carbohydrates to yield lipoproteins and glycoproteins, respectively (Kovacs-Nolan et al. 2005). The major egg white proteins include ovalbumin, ovotransferrin (conalbumin), ovomucoid, and ovomucin. Other egg white proteins include lysozyme, ovoinhibitor, ovomacroglobulin (ovostatin), cystatin, and avidin (Sugino et al. 1997). Specific binding activities of egg white proteins, such as ovoflavoprotein (Farrell et al. 1969), thiamin-binding protein (Muniyappa and Adiga 1979), ovoglycoprotein (Ketterer 1965), ovofactor 1 (Nakamura et al. 1995), ovoglobulins (Longsworth et al. 1940), and minor glycoproteins (Itoh et al. 1993) have been identified.

Proteomic analysis of hen egg albumen using two-dimensional polyacrylamide gel electrophoresis (2D-PAGE) and mass spectrometry has revealed the presence of 3 yet-to-be identified proteins, and 16 known proteins, 2 of which were not previously detected, namely, a protein called transiently expressed in neural precursors (Tenp), with strong homology to a bacterial permeability-increasing protein family, and VMO-1, a vitelline membrane protein (Guérin-Dubiard et al. 2006). The presence of two unidentified proteins FLJ10305 and Fatso proteins in the proteome was confirmed by Raikos et al. (2006). With the sequencing of the 1200-Mb (mega-base-pair) genome of the domesticated chicken (*Gallus gallus*) (International Chicken Gene Sequencing Consortium 2004) and the availability of powerful tools for proteomics, the study of known and yet-to-be identified bioactive egg white proteins will be further facilitated.

Many of the properties of food proteins, including egg white proteins, with physiological significance beyond the nutritional requirements of nitrogen for growth and maintenance, are attributed to protein-encrypted physiologically active peptides (Korhonen and Pihlanto 2003) that are released during *in vivo* gastrointestinal digestion or *in vitro* enzyme treatment and processing as bioactive peptides or "food hormones." A number of peptides with different biological activities have been described since the discovery in 1970 of bradykinin-potentiating peptides possessing angiotensin 1–converting enzyme-(ACE) inhibitory activity (Ferreira et al. 1970; Rutherfurd-Markwick and Moughan 2005). Functional peptides can be generated through proteolytic digestion of a parent protein and may have a different physiological function in the organism, suggesting that multifunctionality is an intrinsic property of most proteins, and that bioactive peptides can be "tailored and modeled" to achieve a desired function (Pellegrini 2003). Short sequences; hydrophobic amino acids in addition to proline, lysine, or arginine groups; and resistance to digestive peptidases are some of the common structural properties of bioactive peptides (Kitts and Weiler 2003).

Differing viscosities and pH values in the egg white inhibit bacterial proliferation. These antimicrobial effects may be attributed to several mechanisms, including bacterial cell lysis, metal binding, and vitamin binding (Li-Chan and Nakai 1989). Antimicrobial peptides represent an important component of innate immunity and nutritional immunity and offer the advantage of safety for use in medicine and in the food industry. The adhesion of microorganisms to host tissues is the first step in the infection process. In many cases the adhesion is mediated by an interaction between components on the surface of the microorganism and carbohydrates on the mucosal surface of the host (Sharon and Ofek 2002). It has been suggested that glycoproteins and glycolipids on the mucosal surface of the intestine competitively inhibit microorganism—carbohydrate adhesion, thereby preventing microbial infection (Kobayashi et al. 2004).

Proteases play key roles in several physiological processes, including intracellular protein degradation, bone remodeling, and antigen presentation, and their activities are increased in pathophysiological conditions such as cancer metastasis and inflammation. Microbial proteases are involved in the mechanism of penetration of tissues by bacteria, in the proteolytic cleavage of precursor proteins for virus replication, and in the facilitation of host invasion by parasites (Henskens et al. 1996). Therefore, protease inhibitors represent an important class of compounds of therapeutic significance. Four protease inhibitors have been identified in egg white: cystatin, ovomucoid, ovomacroglobulin, and ovoinhibitor (Li-Chan and Nakai 1989).

The immune system responds to antigenic stimulation with a complex array of molecular events involving antigen-presenting cells, B cells, T cells, and phagocytes. Cytokines play a significant role in regulating such immune responses (Wahn 2003). Cytokines and growth factors mediate a wide range of physiological processes, including hematopoiesis, immune responses, wound healing, and general tissue maintenance. They are concomitantly involved in the pathology of a wide range of diseases and have potential use in replacement and immunomodulatory therapy (Mire-Sluis 1999). Several albumen proteins and peptides have demonstrated immunomodulating activity.

It has been reported that certain egg white–derived peptides can play a role in controlling the development of hypertension by exerting vasorelaxing effects (Davalos et al. 2004). Improvements in antihypertensive activity have been reported to result from incorporation of peptides into liposomes to protect them through the gastrointestinal system, modification of the structure of the peptides by cyclization (Chen et al. 2003), genetic modification to incorporate peptide sequences, incorporation of antihypertensive egg peptide sequences into soybean protein using controlled mutagenesis, and modification of residues close to the active peptide to facilitate its release *in vivo* (Matoba et al. 2001a, 2001b, 2003; Onishi et al. 2004).

Reactive-oxygen species and other free radicals cause oxidative damage to DNA, proteins, and other macromolecules. They have also been implicated in a number of multifactorial degenerative diseases, including diabetes, cancer, and cardiovascular disease (Ames 1993). Davalos et al. (2004) reported that the enzymatic hydrolysis of crude albumen proteins with pepsin resulted in the production of peptides with strong antioxidant activities. The egg white peptide Tyr–Ala–Glu–Glu–Arg–Tyr–Pro–Ile–Leu, which was shown previously to possess ACE-inhibitory activity, also exhibited a high radical-scavenging activity. These results would suggest that the combined antioxidant and ACE-inhibitory properties of albumen hydrolysates, or the corresponding peptides, would make a useful multifunctional preparation for the control of cardiovascular diseases, in particular hypertension.

Hen egg white proteins are a major cause of antigenicity and allergenicity (Mine and Zhang 2002a, 2002b) with almost two-thirds of children diagnosed with food allergies being reactive to egg white (Sampson and Ho 1997). Ovomucoid (Gal d1), ovalbumin (Gal d2), ovotransferrin (Gal d3), and lysozyme (Gal d4) have been clearly identified as egg white allergens causing IgE-mediated reactions (type I) known as *immediate* (within one minute to a few hours) *hypersensitivity reactions*.

There is increasing commercial interest in the production of bioactive peptides with minimal destruction of activity, as well as with enhanced activity for various purposes, such as therapeutics, functional foods, nutrigenomics, and specific nutraceutical applications (Korhonen and Pihlanto 2003). Furthermore, bioactive peptides can have several clinical advantages over traditional small-molecule chemotherapeutics, including specificity, selectivity, and potency (Brissette et al. 2006). Industrial-scale production of such peptides involves their separation by automated and continuous systems but is hampered by the lack of suitable technologies. However, selective column chromatography methods are replacing batch methods of salting out. At the laboratory scale for biochemical, biological, and biophysical studies, ion exchange chromatography is widely used because of the low resulting denaturation, release of a nonaltered by-product, and easy scale up, as reported for albumen (Guérin-Dubiard et al. 2005, 2006). Ion exchange electrochromatography with an oscillatory transverse electric field is a promising technique for high-capacity purification of proteins (Yuan et al. 2006). Phage display is a well-established approach for the rapid generation of peptide libraries that can be "panned" for specific activities leading to "targeted therapeutics," a current clinical paradigm of choice in the field of oncology (Brissette et al. 2006).

This chapter is a description of the biological activities of bioactive components of the hen egg white. Table 4.1 is a summary of the components' biological activities.

TABLE 4.1. Biological Activities of Egg White Components

Component	Biological Activity	Reference
Ovalbumin	Anticancer	
	Highly antimutagenic	Hosono et al. 1988; Vis et al. 1998
	Antihypertensive	
	Ovokinin (FRADHPFL)	Fujita et al. 1995a
	Ovokinin (2–7) (RADHPF)	Matoba et al. 1999; Scruggs et al. 2004
	Synthetic peptides RPLKPW and RPFHPF	Matoba et al. 2001b; Yamada 2002
	Dipeptide LW	Fujita et al. 2000
	Peptides RADHPFL and YAEERYPIL	Miguel et al. 2007
	Antimicrobial	
	Trypsin-digested peptides SALAM, SALMVY, YPILPEULQ, ELINSW, NVLQPSS	Pellegrini 2004
	Chymotrypsin-digested peptides AEERYPILPEYL, GIIRN, TSSNVMEER	Pellegrini 2004
	Antioxidant	
	Restricts lipid oxidation	Nakamura and Kato 2000
	Immunomodulating	
	Releases TNF-α *in vitro*	Fan et al. 2003
	Peptides increase phagocytic activity of macrophages	Tezuka and Yoshikawa 1995
Ovotransferrin	Antimicrobial	
	Bacteriostatic and bactericidal	Tranter and Board 1984
	Peptide OTAP-92 kills Gram-negative bacteria	Ibrahim et al. 2000
	Antiviral against Marek's disease virus	Giansanti et al. 2002
	Antifungal against *Candida* sp.	Valenti et al. 1985
	Iron-chelating	Abdallah and Chahine 1999
	Immunomodulating	
	Modulates macrophage and heterophil function	Xie et al. 2002
	Inhibits proliferation of mouse spleen lymphocytes	Otani and Odashima 1997

TABLE 4.1. (*Continued*)

Component	Biological Activity	Reference
Ovomucoid	Biospecific ligand	
	Overcomes degradation of protein drugs	Agarwal et al. 2001b
	Targets drugs to the blood	Plate et al. 2002
	Protease inhibition	
	Inhibits trypsin and chymotrypsin	Kato et al. 1987; Saxena and Tayyab 1997
	Improves oral delivery of protein/ peptide therapeutics	Shah and Khan 2004
	Model for the design of therapeutic inhibitory peptides	Hilpert et al. 2003
	Immunomodulating	
	Synthetic peptides induce secretion of cytokines	Holen et al. 2001
Ovomucin	Antiadhesive	
	Glycopeptides protective against *E. coli*	Kobayashi et al. 2004
	Anticancer	
	70-kDa glycosylated peptide inhibits tumor growth	Oguro et al. 2001
	Antimicrobial	
	Maintains structure and viscosity of egg albumen	Tsuge et al. 1997a, 1997b
	Antiviral	Watanabe et al. 1998a
	Hypercholesterolemic	
	Attenuates hypercholesterolemia	Nagaoka et al. 2002
	Immunomodulating	
	Sulfated glycopeptides are macrophage-stimulating *in vitro*	Tanizaki et al. 1997
Ovoglobulin G2 and G3	Contributes to the foaminess of egg white	Nakamura et al. 1980
Lysozyme	Anticancer	
	Inhibits tumor growth when administered orally	Pacor et al. 1999; Sava et al. 1991
	Antimicrobial	
	Peptides 98–112, 98–108, 15–21 antibacterial against *E. coli* and *Staphylococcus aureus*	Pellegrini et al. 2000; Mine et al. 2004
	Synthetic peptides damage bacterial membranes and inhibit DNA and RNA synthesis	Pellegrini et al. 2000; During et al. 1999
	Antiviral against HIV, chickenpox, and herpes simplex	Sava 1996
	Antioxidant	
	Suppresses reactive-oxygen species and oxidative stress genes	Liu et al. 2006
	Immunomodulating	
	Enhances IgM production in hybridomas and lymphocytes	Sugahara et al. 2000

TABLE 4.1. (*Continued*)

Component	Biological Activity	Reference
Ovoinhibitor	Protease inhibition	
	Inhibits a wide spectrum of proteinases	Laskowski and Kato 1980
	Inhibits formation of active-oxygen species	Frenkel et al. 1987
Ovoglycoprotein	Used to separate drug enantiomer	Sadakane et al. 2002
Ovoflavoprotein	Functions as a sweetness-suppressing protein	Maehashi et al. 2006
Ovomacroglobulin	Protease inhibition	
	Broad-spectrum activity against proteases	Kitamoto et al. 1982; Molla et al. 1987
	Strong anticollagenase activity in egg white	Nagase et al. 1983
Cystatin	Anticancer	
	Inhibits tumor-associated activity of intracellular proteases	Cegnar et al. 2004
	Antimicrobial	
	Inhibits growth of bacteria	Bjorck 1990
	Immunomodulating	
	Induces cytokines that upregulate nitric oxide and reduce parasite numbers	Das et al. 2001
	Protease inhibition	
	Inhibits proteases	Ogawa et al. 2002
Avidin	Anticancer	
	Used in cancer treatment to localize and image cancer cells and to pretraget drugs to tumors	Corti et al. 1998
	Antimicrobial	
	Inhibits growth of biotin-requiring microbes	Eakin et al. 1941; Green 1975
	Biospecific ligand	
	Facilitates delivery of therapeutics to the brain	Bickel et al. 2001
	Immunomodulating	
	May function as a host defense factor	Korpela 1984
Ovofactor-1	Stimutales DNA synthesis and cultured cells from chicken embryos	Nakamura et al. 1995
Extracellular fatty acid binding protein	Involved in heart development, fatty acid transport, and lipid metabolism	Gentili et al. 2005
Chondrogenesis-associated lipocalins β and γ	Acts synergistically with ex-FABP in bone formation in chicken embryos	Pagano et al. 2003
Tenp	Binds to lipid A of Gram-negative lipopolysaccharide	Bingle and Craven 2004

TABLE 4.1. (*Continued*)

Component	Biological Activity	Reference
Clusterin	Marker for follicular atresia and resorption Carrier for receptor-mediated endocytosis into oocytes during embryonic development	Mahon et al. 1999
	Prevents premature aggregation and precipitation of eggshell matrix components before and during their assembly into the rigid protein scaffold necessary for ordered mineralization	Mann et al. 2003
Hep21	Belongs to the multifunctional uPAR/CD59/Ly6/snake neurotoxin superfamily, but biological function not yet known	Nau et al. 2003
VMO-1 protein	Along with lysozyme, binds to ovomucin	Schäfer et al. 1998
Thiamin binding protein	Exhibits specificity for riboflavin-binding protein	Muniyappa and Adiga 1979
Minor glycoproteins	Globulin-like properties	Itoh et al. 1993

4.2. EGG WHITE PROTEINS AND THEIR BIOLOGICAL ACTIVITIES

4.2.1. Ovalbumin

Ovalbumin (OVA) constitutes over half of the total egg white proteins (Ibrahim 1997). Despite being one of the first proteins to be isolated in a pure form, its biological function still remains unknown (Huntington and Stein 2001). It has been suggested that ovalbumin may serve as a source of amino acids for the developing embryo (Ibrahim 1997).

4.2.1.1. Anticancer Activity

Heat-denatured ovalbumin was found to be highly antimutagenic against *N*-methyl-*N*′-nitro-*N*-nitrosoguanidine and pepper-induced mutagenicity (Hosono et al. 1988; Vis et al. 1998).

4.2.1.2. Antihypertensive Activity

Ovokinin, an antihypertensive and vasorelaxing octapeptide (FRADHPFL) corresponding to fragments 358–365 of ovalbumin, was isolated by peptic digestion. Its effect was mediated by B_1 receptors that stimulated the release of prostacyclin (Fujita et al. 1995a). Oral administration of ovokinin in high doses in the form of an emulsion of egg yolk to spontaneously hypertensive rats (SHR) was highly effective, possibly because of the yolk phospholipids that improved intestinal absorption and protected the peptide from intestinal peptidases. Ovokinin(2–7), a peptide produced by chymotrypsin digestion, and corresponding to OVA 359–364 (RADHPF), was also found to possess vaso-relaxing activity mediated by nitric oxide (Matoba et al. 1999) and bradykinin B_2 vascular receptors (Scruggs et al. 2004) in SHR rats, but not in normotensive Wistar–Kyoto (WKY) rats. Both peptides were found to significantly lower the systolic blood pressure in SHR rats, when administered orally (Fujita et al. 1995b); however, ovokinin(2–7) was more effective at 10 times smaller doses than were effective doses of ovokinin. Administration of very high doses of ovokinin(2–7) resulted in only a slight decrease in arterial blood pressure. Technological and physiological factors have been reported to condition the activity and bioavailability of the peptides. Synthetic peptides RPLKPW and RPFHPF designed by the replacement of amino acids in the ovokinin(2–7) peptide have resulted in enhanced antihypertensive activity, in turn resulting in 100-fold and 10-fold more potent anti-hypertensive activity, respectively (Matoba et al. 2001b; Yamada 2002).

However, neither ovokinin(2–7) nor its derivatives inhibited ACE *in vitro*. RPLKPW was genetically introduced into homologous sequences in soybean β-conglycinin α′ subunit, and the recombinant product was expressed in *E. coli*. Oral administration of RPLKPW-incorporated α′ subunit in SHR resulted in antihypertensive activity, making it the first example of a genetically modified food protein possessing physiological activity based on a bioactive peptide (Matoba et al. 2003). Two ACE-inhibitory peptides, corresponding to ovalbumin fragments 183–184 and 200–218, were also identified by peptic and tryptic digestions, respectively (Yoshikawa and Fujita 1994). Six ACE-inhibitory peptides were isolated from a pepsin digest of ovalbumin with IC_{50} values in the range of 0.4–15 μmol/L, but only one dipeptide, Leu–Trp (LW), showed antihypertensive activity in SHR rats (Fujita et al. 2000). Fujita et al. (2000) also showed that hydrolyzates obtained from ovalbumin using pepsin and thermolysin exhibited ACE-inhibitory activity.

Miguel et al. (2004) examined peptides with ACE-inhibitory properties produced by enzymatic hydrolysis of crude egg white. Peptide sequences RADHPFL, YAEERYPIL, and IVF with IC_{50} values of 4.7, 6.2, and 33.11 μmol/L, respectively, derived after a 3 h incubation with pepsin followed by ultrafiltration exhibited potent ACE-inhibitory activity and antihypertensive activity (Miguel et al. 2005). Among these peptides, two novel peptides with potent ACE-inhibitory activity were found, with amino acid sequences RADHPFL and YAEERYPIL. Studies simulating gastrointestinal digestion indicated that the sequences YAEERYPIL and RADHPFL hydrolyze when administered orally (Miguel and Aleixandre 2006; Miguel et al. 2006a) and that these hydrolyzates and the IVF sequence possibly act directly in attenuating the onset of arterial hypertension (Miguel et al. 2006b). Termination of oral administration resulted in arterial blood pressure values similar to those in untreated rats. RADHP and YPI, the end products of gastrointestinal digestion of RADHPFL and YAEERYPIL, reduced vascular resistance and could be used as functional food ingredients in the prevention and treatment of hypertension (Miguel et al. 2007).

4.2.1.3. Antimicrobial Activity

Unlike full-length ovalbumin, which is not antibacterial, peptides produced by digestion of ovalbumin with gastrointestinal proteases trypsin (SALAM 36–40, SALMVY 36–41, YPILPEULQ 111–119, ELINSW 143–148, and NVLQPSS 159–165) and chymotrypsin (AEERYPILPEYL 127–138, GIIRN 155–159, TSSNVMEER 268–276), as well as synthetic peptide counterparts, were strongly active against *Bacillus subtilus* and active to a lesser extent against *E. coli, Bordetella bronchiseptica, Pseudomonas aeruginosa*, and *Serratia marcescens*, as well as to the fungus *Candida albicans* (Pellegrini 2004). The charge state and hydrophobic properties of these peptides did not always correlate with their activities, thus challenging existing models for the mode of action, that positive charge and hydrophobic properties are important for bactericidal activity of the peptide. Compared to other antimicrobial peptides, peptides derived from ovalbumin have a reduced bactericidal potency compensating for the quantity of ovalbumin ingested. Biziulevičius et al. (2006) have shown that ovalbumin hydrolysates such as hydrolysates of serum albumin, β-lactoglobulin, α-lactalbumin and casein, generated by treatment with trypsin, pepsin, chymotrypsin, and pancreatin, respectively, stimulated their antimicrobial and immunostimulatory activities.

4.2.1.4. Antioxidant Activity

Ovalbumin restricts lipid oxidation, and this property can be further improved by increasing its lipid affinity by exposing active sulfhydryl groups in the molecule through conjugation with polysaccharide, such as galactomannan (Nakamura and Kato 2000).

4.2.1.5. Immunomodulating Activity

Ovalbumin may possess some immunomodulatory activity, as it was found to induce the release of tumor necrosis factor (TNF) α in a dose-dependent manner *in vitro*, when modified with dicarbonyl methylglyoxyl (Fan et al. 2003), and immunogenic ovalbumin peptides have been used to enhance immune responses for cancer immunotherapy (Vidovic et al. 2002; Goldberg et al. 2003; He et al. 2003). Tezuka and Yoshikawa (1995) found that the phagocytic activity of macrophages was increased by the addition of ovalbumin peptides, residues 77–84 and 126–134 derived from peptic and chymotryptic digestions, respectively.

4.2.2. Ovotransferrin

Ovotransferrin is a 78–80-kDa monomeric 686-residue polymorphic glycoprotein, belonging to the transferrin family, a group of iron-binding proteins widely distributed in various biological fluids, and having the capacity to reversibly bind two iron ions per molecule concomitantly with two bicarbonate anions (Ibrahim 2000; Ahlborn et al. 2006) and thus conferring nutritional immunity, rise in pasteurization temperature, and resistance to proteolysis. Ovotransferrin differs from animal transferrin only in the carbohydrate moiety.

4.2.2.1. Antimicrobial Activity

The high affinity of ovotransferrin for iron ($10^{30} M^{-1}$) implies that in the presence of unsaturated ovotransferrin, iron will be sequestered and rendered unavailable for the growth of microorganisms. Ovotransferrin is an essential component of the egg's antimicrobial defense system, and possesses both bacterial growth inhibition (static or nutritional immunity) and the iron withholding–independent (cidal) antimicrobial actions at alkaline pH (9.5) and at near 40°C, the physiological temperature of birds (Tranter and Board 1984). However, many microorganisms are able to acquire iron bound to ovotransferrin by the production of high-affinity iron chelators called *siderophores* (Lee et al. 1996; Lim et al. 1998) and by expressing a receptor for the transferrin itself along with using outer membrane and periplasmic iron transport proteins (Modun et al. 1994). A distinct antibacterial domain within the *N*-lobe of ovotransferrin has been identified. The 92–amino acid ovotransferrin peptide (OTAP92) was found to be capable of killing Gram-negative bacteria by crossing the bacterial outer membrane by self-promoted uptake, damaging the cytoplasmic membrane (Ibrahim et al. 2000). It has also been shown that ovotransferrin possesses both antiviral activity against Marek's disease virus in chicken embryo fibroblasts (Giansanti et al. 2002) and antifungal activity against species of *Candida* (Valenti et al. 1985). OTAP92 located in the bactericidal domain showed more than six $\log_{10}$ orders of killing against *Staphylococcus aureus* and more than one $\log_{10}$ order of killing against *E. coli*. It can permeate the *E. coli* outer membrane, reaching the inner membrane, where it selectively causes permeation of ions, resulting in dissipation of the electrical potential without affecting the pH gradient. Similar results were obtained using artificial liposomes, suggesting a direct action of the proteins on the lipid bilayer that was mediated by detectable conformational changes in their structures (Aguilera et al. 2003), thus functioning as an iron scavenger, preventing iron uptake by microorganisms, and acting as an iron delivery agent (Abdallah and Chahine 1999). Ovotransferrin appears to be a key factor in overcoming cephalosporin resistance, through its iron chelating activity, which has been shown to increase the stimulation by an inhibitor of cAMP β-lactamase of some antibiotics that are efficient against β-lactamase producing bacteria (Babini and Livermore 2000). It has been found to exert antibacterial activity against a wide spectrum of bacteria, including *Pseudomonas* spp., *Escherichia coli, Streptococcus mutans* (Valenti et al. 1983), *Staphylococcus aureus, Bacillus cereus* (Abdallah and Chahine 1999), and *Salmonella enteritidis* by permeating bacterial outer membranes, reaching the inner membrane and causing the selective permeation of ions and dissipation of electrical potential (Aguilera et al. 2003).

4.2.2.2. Immunomodulating Activity

Ovotransferrin is an acute-phase protein in chickens, the serum levels of which increase in inflammation and infections. It acts as an immunomodulator, modulating macrophage and heterophil functions *in vitro* and in chickens (Xie et al. 2002), inhibiting proliferation of mouse spleen lymphocytes (Otani and Odashima 1997) and enhancing phagocytic response of peripheral blood mononuclear cells and polymorphonuclear cells in dogs (Hirota et al. 1995). Ovotransferrin was also found to facilitate the recovery of chick eyes from induced myopia (Rada et al. 2001).

4.2.3. Ovomucoid

Ovomucoid is a 27–35-kDa polymorphic glycoprotein with approximately 20–25% carbohydrate content and accounting for roughly 10% of avian egg white proteins.

4.2.3.1. Biospecific Ligand Activity

Like avidin, ovomucoid represents a possible source of biospecific ligand peptides. The incorporation of ovomucoid into polymeric microparticles, to overcome the degradation of protein drugs by proteolytic enzymes, has been examined. Agarwal et al. (2001b) found that when ovomucoid was used, the stability of insulin in polymethacrylate-based microparticulates increased significantly. Inclusion of ovomucoid also resulted in targeting of drugs to the blood, by acting as a biospecific ligand to lectins on the walls of the gastrointestinal tract (Plate et al. 2002). The presence of ovomucoid was found to enhance insulin flux across rat jejunum (Agarwal et al. 2001a), suggesting the use of ovomucoid to enhance the oral delivery of insulin. Using a rat model of experimental pancreatitis, intravenous ovomucoid was found to decrease the trypsin-like activity to the level of intact rats, and reduce primary pancreas destruction (Valueva et al. 1992).

4.2.3.2. Protease Inhibition

Ovomucoid is also one of four egg white proteinase inhibitors inhibiting trypsin and chymotrypsin, and belonging to the group of serine proteinase inhibitors (Kato et al. 1987; Saxena and Tayyab 1997). Because of its allergenic nature, however, ovomucoid has limited biological and medical applications.

Ovomucoid has been shown to be particularly useful for the oral delivery of protein/peptide therapeutics, its application often limited owing to extensive proteolytic degradation in the gastrointestinal tract (Shah and Khan 2004). Because avian egg white ovomucoid inhibits digestive enzymes, such as trypsin, β-chymotrypsin, and elastase, it has been found to improve the oral delivery of insulin (Agarwal et al. 2001a, 2001b) and has been examined for coadministration with calcitonin, a polypeptide associated with calcium homeostasis and bone remodeling, which is often used in the management of osteoporosis (Shah and Khan 2004). In addition, ovomucoid has been used as a model for the design of therapeutic inhibitory peptides (Hilpert et al. 2003). Replacement of Asp→Tyr at the P2′ site of ovomucoid third domain (OVMD3-P1Met) resulted in conversion to a 35,000-fold more effective inhibitor of chymotrypsin with an inhibitor constant (K_i) of 1.17×10^{-11} M. Similarly, enhanced inhibition of trypsin was observed when the Asp→Tyr replacement was introduced into the P2′ site of OVMD3-P1Lys. These results indicated that the P1 site residue, the electrostatic properties of the amino acid residues around the reactive site of the protease inhibitor, and the type of amino acids determine the strength of the interaction of ovomucoid with proteases (Kojima et al. 1999).

The Kazal family of standard mechanism serine proteinase inhibitors, specifically the third domain of avian ovomucoid, have a canonical conformation in the reactive site binding loop with a high degree of superimposition of backbone residues from P3 to P3′ (Krowarsch et al. 2003). A consensus set of 12 inhibitor residues contact the serine proteinase in the enzyme–inhibitor complex (Figs. 4.1a and 4.1b), with Cys16 (P3) and Asn33 (P15′) remaining nearly unvaried. When compared to eglin C (a potato family I inhibitor), turkey ovomucoid third domain (OMTKY3) was a millionfold less effective in inhibiting α-lytic proteinase (a serine proteinase secreted by the soil bacterium *Lysobacter enzymogenes*), despite both the inhibitors having a common P1 Leu (Qasim et al. 2006). However, this study enabled the scientists to design an OMTKY3 that is 10^4 times more powerful as an inhibitor than wild-type OMTKY3.

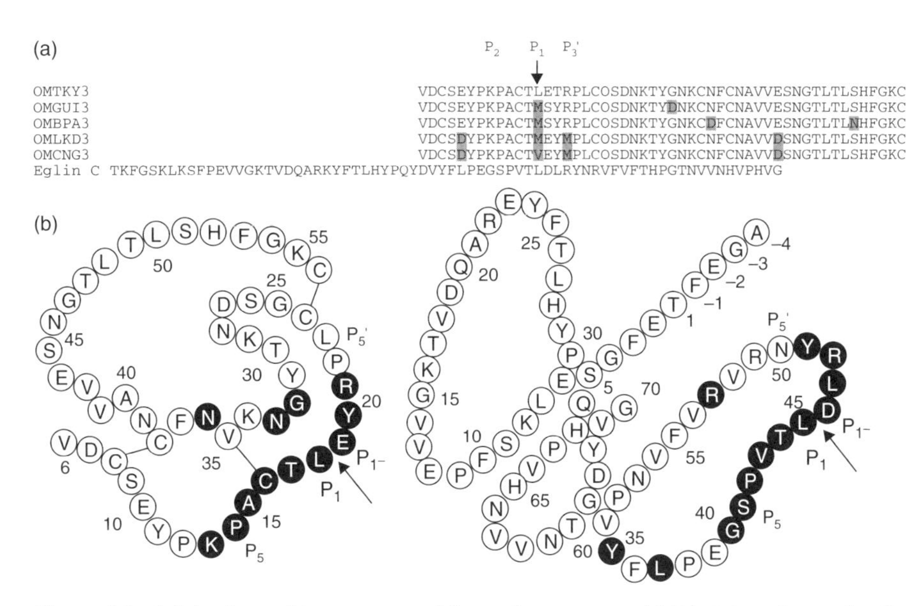

Figure 4.1. (a) Amino acid sequences of five avian ovomucoid third domains and eglin C; (b) covalent structures of turkey ovomucoid third domain, OMTKY3, and eglin C (Qasim et al. 2006). (See insert for color representation.)

4.2.3.3. Immunomodulating Activity

IgE epitopes, each comprising 5–16 amino acids, have been identified in hen egg ovomucoid (Mine and Zhang 2002a, 2002b). Mutational analysis of the epitopes has indicated that charged amino acids, some hydrophobic amino acids, and polar amino acids are important for antibody binding. The study provides a better understanding of designing molecules with reduced allergenicity. Synthetic ovomucoid peptides have also demonstrated immunomodulating activity, inducing T-cell secretion of cytokines interleukin-(IL) 4, IL-10, IL-13, interferon-gamma (IFN-γ), and IL-6 (Holen et al. 2001).

4.2.4. Ovomucin

Egg white ovomucin, which accounts for about 3.5% of the protein in egg white, is a macromolecular and heavily glycosylated glycoprotein, made up of a soluble and an insoluble fraction. Ovomucin, unlike mammalian mucins, does not contain a PTS (proline–threonine–serine) domain, but is located in the same locus as other gel-forming mucins, providing strong support that the mucins are evolutionarily related (Lang et al. 2006).

4.2.4.1. Antiadhesive Activity

Kobayashi et al. (2004) found that pronase digestion of the highly glycosylated ovomucin produced ovomucin glycopeptides with *E. coli* O157:H7-specific binding sites containing sialic acid that may be protective against *E. coli* O157:H7 infection *in vivo*, and is a potentially novel probe for the detection of bacteria in food.

4.2.4.2. Anticancer Activity

Pronase-prepared glycopeptides of ovomucin have demonstrated antitumor effects (complete rejection of direct methylcholanthrene-induced fibrosarcoma tumors and slight growth inhibition of distant tumors in syngeneic mice) in a double-grafted tumor system in mice (Watanabe et al. 1998a, 1998b), suggested to be related to the antiangiogenic activity of a 70-kDa highly glycosylated peptide fragment of ovomucin, inhibiting tumor growth (Oguro et al. 2001).

4.2.4.3. Antimicrobial Activity

Ovomucin serves physical functions within the egg, such as maintaining the structure and viscosity of the egg white albumen (Tsuge et al. 1997a, 1997b), thus preventing the spread of microorganisms (Ibrahim 1997). Ovomucin and ovomucin-derived peptides, besides their physical functions such as maintaining the structure and viscosity of the egg white albumen and thus preventing the spread of microorganisms, have demonstrated antiviral activity against Newcastle disease virus, bovine rotavirus, and human influenza virus *in vitro* (Tsuge et al. 1996a, 1996b, 1997a, 1997b; Watanabe et al. 1998a). Ovomucin peptides, produced by treatment with the enzyme pronase, showed increased solubility, compared to its native form, while still retaining virus-binding activity (Tsuge et al. 1997b; Watanabe et al. 1998a).

It has been reported that ovomucin has the greatest ability to inhibit hemagglutination by heated type B (Lee) influenza virus (Lanni et al. 1949). Bovine rotavirus was inhibited by the α and β subunits of ovumucin while hen Newcastle disease virus required only the β subunit (Tsuge et al. 1996b). Binding of ovomucin to Newcastle disease virus was inhibited when *N*-acetylneuraminic acid in the β subunit was destroyed because of temperature and pH. Disulfide bonds in the α and β subunits contributed to the binding of ovomucin to antiovomucin antibodies (Tsuge et al. 1997a).

4.2.4.4. Hypercholesterolemic Activity

Ovomucin was found to inhibit cholesterol uptake *in vitro* by Caco-2 cells, and attenuate hypercholesterolemia in rats by inhibiting both cholesterol absorption in the intestinal cells and ileal reabsorption of bile acids, thus reducing serum cholesterol levels (Nagaoka et al. 2002).

4.2.4.5. Immunomodulating Activity

O-Linked carbohydrate chains, consisting of *N*-acetylgalactosamine, galactose, *N*-acetylneuraminic acid, and sulfate in sulfated glycopeptides of ovomucin, were identified as immunomodulators, showing macrophage-stimulating activity *in vitro* (Tanizaki et al. 1997).

4.2.5. Ovoglobulins G2 and G3

Feeney et al. (1963) reported egg white globulins A1 and A2 with molecular weights of 30 and 45 kDa, respectively, similar to globulins G2 and G3 reported by Longsworth et al. (1940). Later, Nakamura et al. (1980) purified the proteins, and analysis of their amino acids revealed that G2 and G3 were similar to each other with molecular weights of 49 kDa. The globulins were found to contribute to the foaminess of cgg white.

4.2.6. Lysozyme

Lysozyme is a ubiquitous enzyme, present in almost all organisms, including viruses and plants, and belongs to the family 22 of glycosyl hydrolases. Japan leads the world for the variety of industrial uses of lysozyme. The most plentiful source of lysozyme is hen egg white, containing around 0.3–0.4 g of lysozyme per egg (Losso et al. 2000). The protein has 129 amino acids and a molecular weight of 14.5 kDa with four disulfide bonds, an isoelectric point of 10.7, and a distinctly higher extinction coefficient of 26.4 E1% at 280 nm compared to that of human and bovine lysozyme. The lysozyme molecule consists of two domains, α and β domains, linked by a long α-helix between which lies the active site. Lysozyme, discovered by Alexander Fleming in nasal mucus (Fleming 1922), and later identified by Alderton et al. (1945) in hen egg albumen, was the first egg white protein demonstrated to be effective against Gram-positive bacteria. Since then, it has been found to be effective as an antimicrobial agent, as a nutraceutical and value-added product in even minimally processed food products, and as an antiinflamatory and immunomodulatory agent. The shape of the physiological stability curve of lysozyme suggests the existence of only one folded conformation (Younvanich and Britt 2006), and is due to two of its four disulfide bonds.

4.2.6.1. Anticancer Activity

Lysozyme may act as an anticancer agent, inhibiting tumor growth in a number of experimental tumors when administered orally (Das et al. 1992; Pacor et al. 1996, 1999; Sava 1989; Sava et al. 1989, 1991) and enhancing the efficacy of chemotherapy treatments (Sava et al. 1995). It was also found to have a preventive effect when administered to normal mice (Shcherbakova et al. 2002). Evidence suggests that the anticancer effects of orally administered lysozyme may rely heavily on the host-mediated immune response, including activation of the spleen and macrophages (Sava et al. 1991). However, more recent data indicate that lysozyme may also exert action on the tumor cells themselves (Das et al. 1992; Pacor et al. 1996).

4.2.6.2. Antimicrobial Activity

Lysozymes, also known as *N-acetylmuramideglycanohydrolases*, of avian egg white, hydrolyze cell walls of Gram-positive bacteria by hydrolyzing the β-1,4 linkages between *N*-acetylmuraminic acid and *N*-acetylglucosamine of peptidoglycan (Salton 1957). It is effective against most bacteria, as well as food spoilage and pathogenic organisms such as *Listeria monocytogenes* (Gill and Holley 2000) when used in conjunction with other compounds, such as nisin and EDTA. *In vivo*, it has been shown to be effective as an antimicrobial agent in carp infected by a virulent strain of *Edwardsellia tarda* (Nakamura and Kato 2000). Lysozyme has both enzymatic and nonenzymatic microbicidal activity in native and denatured states, respectively, and hence is not limited by heat treatment encountered during food-processing operations. Genetic evidence of antimicrobial activity independent of muramidase activity of lysozyme was first demonstrated by Ibrahim et al. (2001a). Its high specificity enables it to prevent contamination in cheese because of the absence of inhibition on starter and secondary cultures required for the ripening of cheese. Lysozyme is used as a food preservative for fruits, vegetables, meat, and beverages. Poly(L-glutamic acid)/hen egg white lysozyme nanofilms have been shown to inhibit the growth of *Microcccocus luteus* in the surrounding liquid medium, suggesting potential applications of nanofilms for food preservation and coatings for implant devices (Rudra et al. 2006).

Modification of the lysozyme molecule with a hydrophobic moiety, through fatty acylation (Ibrahim et al. 1991, 1993) or by the genetic fusion of hydrophobic peptides to the *C*-terminus of lysozyme (Arima et al. 1997; Ibrahim et al. 1992, 1994a) enhanced the bactericidal activity of lysozyme against

Gram-negative bacteria, possibly by mediating its interaction and insertion into the bacterial membrane (Ibrahim et al. 2002). Enhanced antimicrobial activity against Gram-negative bacteria has also been demonstrated by lysozyme–polysaccharide conjugates (Nakamura et al. 1991, 1996). Perillaldehyde, a naturally occurring phenolic aldehyde, coupled to lysozyme, markedly enhanced the antibacterial activity against both *Staphylococcus aureus* (Gram-positive) and *E. coli* (Gram-negative) (Ibrahim et al. 1994b).

Through the strategy of "tailoring and modeling," a number of short peptides with high bactericidal activity have been developed from the bactericidal domain of lysozyme (Pellegrini 2003). Enzymatic hydrolysis of lysozyme has been found to enhance its activity, by exposing and generating antibacterial portions of the protein (Ibrahim et al. 1991; Pellegrini et al. 1997). Peptides corresponding to amino acid residues 98–112, 98–108, and 15–21 of lysozyme possessed antimicrobial activity against *E. coli* and *S. aureus* (Pellegrini et al. 2000; Mine et al. 2004) and synthetic bactericidal lysozyme polypeptides were found to not only damage bacterial outer membranes but also inhibit DNA and RNA synthesis (Pellegrini et al. 2000; During et al. 1999). Peptide IVS-DGNGMNAWYAWR-NH2 (residues 98–112) derived from chicken lysozyme exhibited antimicrobial activity against *E. coli* and *S. aureus* but interacted very weakly with membrane lipids when compared to peptide RAWVAWR-NH2 derived from human lysozyme (Hunter et al. 2005). This peptide is part of the helix–loop–helix domain (87–114) located at the upper lip of the active-site cleft of lysozyme (Ibrahim et al. 2001b). The rapid passage of peptides into bacteria suggested that the main bacterial site of action may be located inside bacteria (Rezansoff et al. 2005). Not all segments of the peptide were active. Ile98–Met105 was inactive, while Asn106–Arg112 was weakly active. The *C*-terminal segment NAWVAWR showed improved antimicrobial activity on substitution of Asn with Agr. A requirement of two Arg and two Trp was essential to maintain activity. It was also found to prevent antibiotic-induced bacteriolysis and subsequent endotoxin release, while retaining antibiotic efficacy, suggesting its use for the prevention of endotoxemia in Gram-negative sepsis on treatment with antibiotics (Liang et al. 2003).

Lysozyme has also demonstrated antiviral activity reportedly associated with its charge, rather than with its lytic ability (Losso et al. 2000). Oral and topical applications of lysozyme were found to prevent and control a range of viral skin diseases, as well as herpes simplex and chickenpox (Sava 1996), exerting antiinflammatory action (Sava 1996) and possessing activity against HIV type 1 (Lee-Huang et al. 1999). The antibacterial properties of lysozyme have led to its use in oral healthcare products, such as toothpaste, mouthwash, and chewing gum, and to protect against periodontis-causing bacteria and infections in the oral mucosa (Sava 1996; Tenuovo 2002). Gram-negative bacteria, such as *E. coli*, were sensitized by high pressure to a peptide corresponding to amino acid residues 96–116 of lysozyme. This may contribute to the development of more efficient technology for cold high-pressure pasteurization (Masschalck et al. 2001).

4.2.6.3. Antioxidant Activity

Lysozyme is a defensin, and harbors an 18–amino acid domain that binds agents such as advanced glycation end products known to generate reactive oxygen species, and thereby protects against acute and chronic oxidant injury by suppressing reactive-oxygen species and oxidative stress genes (Liu et al. 2006). Lysozyme was found to raise the levels of antioxidant reserves in both transgenic mice expressing hen egg lysozyme and hepatoma cells Hep G2.

4.2.6.4. Immunomodulating Activity

Lysozyme has been shown to act as an immunomodulating agent, playing an important role in the natural defense mechanism. When combined with immunotherapy, lysozyme was effective in improving chronic sinusitis (Asakura et al. 1990) and normalizing humoral and cellular responses in patients with chronic bronchitis (Sava 1996). The antiviral action has been explained by its role in precipitation of viral particles, its immune-enhancing action on the host, and its interaction with pathogens (Sava 1996). Lysozyme was also found to enhance IgM production in hybridomas and lymphocytes, and thus was termed an *immunoglobulin production stimulating factor* (Sugahara et al. 2000), and to regulate and restore the immune responses in immune-depressed patients undergoing anticancer treatment (Sava 1996). Furthermore, it has been suggested that the antibacterial activity of lysozyme might also occur via stimulation of the macrophage phagocytic function, and the hydrolysis products of peptidoglycan may act as an adjuvant or immunomodulator, enhancing immunoglobulin productivity. Posttranslational modifications, such as nitration of tyrosines and modifications of tryptophans induced by antigen-presenting cells in the T-cell contact residues of the peptides DGSTDYGILQINSRWW (residues 48–62), YTGYSLGNWYCAAKFE (residues 20–35), and synthetic peptides with Tyr replaced by nTyr of hen egg white lysozyme, resulted in the recognition of the peptides by $CD4^+$ T cells on binding to class II histocompatibility molecule $I\text{-}A^k$ (Herzog et al. 2005). Lovitch and Unanue (2005) have shown that two T-cell subtypes, types A and B, recognize distinct conformers of the dominant epitope (residues 48–62) of hen egg white lysozyme presented by $I\text{-}A^k$; the type A conformer is formed in the late endosomes on processing of native protein, and the more flexible type B conformer is formed in early endosomes and at the cell surface.

4.2.7. Ovoinhibitor

Like ovomucoid, ovoinhibitor is a serine proteinase inhibitor, but unlike ovomucoid, ovoihibitor inhibits a wide spectrum of proteinases, including proteinases in chicken egg white and blood plasma (Laskowski and Kato 1980) and bacterial and fungal proteinases (Matsushima 1958; Tomimatsu et al. 1966; Vered et al. 1981). Both ovoinhibitor and ovomucoid have multiple domains with a characteristic pattern of disulfide bridges (Laskowski and Kato 1980), and most likely evolved from a common primordial single-domain inhibitor

(Scott et al. 1987). Kinoshita et al. (2004) have demonstrated linkage mapping of ovomucoid and ovoinhibitor to chicken chromosome 13, and may be orthologous to human pancreatic secretory trypsin inhibitor gene (Tan et al. 1988).

4.2.7.1. Protease Inhibition

Chicken egg white ovoinhibitor is a seven-domain Kazal serine protinase inhibitor with a molecular weight of 48 kDa present at about one-tenth the amount of ovomucoid in egg white (Liu et al. 1971), while Japanese quail egg white ovoinhibitor is a 53-kDa glycoprotein. Japanese quail ovoinhibitor inhibits concanavalin A- and phytohemagglutinin-induced proliferative responses of mouse splenocytes and immunoglobulin production, and also suppresses functional responses of mouse and quail B and T lymphocytes via inhibition of enzymatic activity produced from lymphocytes (Odashima et al. 2000). Ovoinhibitor can simultaneously inhibit two molecules of trypsin, two molecules of chymotrypsin, and one molecule of elastase I. The two trypsin binding sites of ovoinhibitor are nonidentical and differ in their reactivity toward porcine and bovine trypsins. One of the two chymotrypsin binding sites can be selectively inactivated by mild oxidation, while the other is capable of binding elastase II; the affinity for chymotrypsin is stronger than that for elastase II (Vered et al. 1981). Like ovomucoid and cystatin, ovoinhibitor is heat-resistant, but its stability decreases at alkaline pH (Matsushima 1958; Nakamura and Matsuda 1984). Ovoinhibitor–galactomannan conjugate is heat-stable and has better emulsifying properties than does untreated ovoinhibitor, suggesting better prospects for industrial application (Begum et al. 2003). Ovoinhibitor was found to bind rotavirus and inhibit their *in vitro* and *in vivo* replication (Yolken et al. 1987).

Ovoinhibitor has been found to prevent the development of rotavirus gastroenteritis in a mouse model of rotavirus infection (Yolken et al. 1987), and to inhibit the formation of active-oxygen species by human polymorphonuclear leukocytes, which are associated with inflammatory diseases, mutagenicity, and carcinogenicity (Frenkel et al. 1987). It has also been used to study models of autoimmune arthritis in mice (Terato et al. 1996).

Proteinases are involved in the regulation of a number of biological processes, and have been implicated as contributors in several diseases, including viral diseases, such as HIV (Maliar et al. 2002) and Alzheimer's (Schimmoller et al. 2002). Therefore, proteinase inhibitors, such as those from egg white, have significant potential for the treatment and prevention of proteinase-mediated diseases.

4.2.8. Ovoglycoprotein

Ovoglycoprotein was isolated and characterized as an acidic glycoprotein with an isoelectric point of 3.9 and a molecular weight of 24.4 kDa (Ketterer 1965). It constitutes about 1% of the egg white proteins, and like many other glycoproteins does not separate from solution on heating at 100°C or precipitate with trichloroacetic acid. It contains 13.6% hexose made up of mannose and galactose in the ratio 2:1, with 13.8% hexosamine as glucoasmine, and 3% sialic acid as *N*-acetylneuraminic acid. Its biological functions are still unclear. The protein is used as a chiral selector to separate drug enantiomer by HPLC or by capillary electrophoresis (Sadakane et al. 2002).

4.2.9. Ovoflavoprotein

Ovoflavoprotein, also known as *flavoprotein* or *riboflavin-binding protein* (RBP) and a member of the folate receptor family, is a phosphoglycoprotein, giving the egg white a yellowish tinge. RBP is a 219–amino acid acidic monomeric protein (pI = 4.2) with a molecular weight of 34 kDa and occurs as nine isoforms (Amoresano et al. 1999) that vary slightly in size owing to loss of amino acids at the *N*-terminus and to variations in posttranslational modifications such as glycosylation and phosphorylation (Norioka et al. 1985; Rohrer and White 1992). The *N*-terminal end has an unusual pyroglutamyl residue, while the *C*-terminal has a peculiar Gln–Gln–Glu–Glu–Gly–Glu–Glu (Farrell et al. 1969). The apoprotein contains a carbohydrate moiety (14%) made up of mannose, galactose, and glucosamines; seven or eight phosphate groups; and eight disulfide bonds. One mole of apoprotein binds one mole of riboflavin, which is essential for embryonic development, but this binding ability is lost at $pH < 4.2$. Electron paramagnetic resonance indicates that a single type II copper site is present in the RBP of hen egg white, suggesting the possibility of RBP in the storage and/or transport of copper (Smith et al. 2006). Interestingly, RBP has been reported to function as sweetness-suppressing protein (Maehashi et al. 2006).

4.2.10. Ovomacroglobulin

Ovomacroglobulin, the second largest glycoprotein next to ovomucin, is a 640-kDa glycoprotein composed of four subunits joined in pairs by disulfide bonds (Kitamoto et al. 1982) and closely related to human serum α_2-macroglobulin (Donovan et al. 1969). It has demonstrated broad-spectrum inhibitory activity against various types of proteases, including serine proteases, cysteine proteases, thiol proteases, and metalloproteases (Kitamoto et al. 1982; Molla et al. 1987).

4.2.10.1. Protease Inhibition

The antimicrobial effects of ovomacroglobulin against the 56-kDa protease of *Serratia marcescens* and elastase and alkaline protease of *Pseudomonas aeruginosa*, owing to its proteinase inhibitory action, have been demonstrated both *in vitro* (Miyagawa et al. 1991a, 1991b, 1991c, 1994; Molla et al. 1987) and *in vivo*, where it was found to reduce corneal destruction in an experimental keratitis model in rabbits, as well as accelerate wound healing (Ijiri et al. 1993; Miyagawa et al. 1991a, 1994). Ovomacroglobulin was also found to enhance periodontal wound healing in rats, by accelerating fibroblast growth, collagen deposition, and capillary formation in tissue (Ofuji et al. 1992).

Bacterial proteases are known to activate the Hageman factor–dependent cascade, resulting in the generation of bradykinin, and hence enhancing vascular permeability and pain (Molla et al. 1989). Ovomacroglobulin inhibits bacterial protease activity and consequently is able to suppress edema and pain induced by the kinin (Miyagawa et al. 1994). The proteinase inhibitory effects of ovomacroglobulin has demonstrated several other biological effects, including the suppression of *P. aeruginosa* and *Vibrio vulnificus* septicemia due to the inhibition of kinin generating proteases (Maeda et al. 1993; Maruo et al. 1998), the *in vitro* stoichiometric inhibition of the inflammatory proteinase medullasin released from invading granulocytes (Ikai et al. 1989), and the suppression of metalloproteinases and vascular permeability in skin tissues, which play a role in tumor metastasis (Wu et al. 2001).

The physical properties of ovomacroglobulin resemble those of a high-molecular-weight (780 kDa) proteinase inhibitor with strong anticollagenase activity found in chicken egg white and termed *ovostatin* (Nagase et al. 1983). Ovostatin inhibits metalloproteinases more effectively than do proteinases of other classes. The inhibitory activity of ovostatin is very similar to that of α_2-macroglobulin. The proteinase hydrolyzes a peptide bond at a specific locus within an ovostatin polypeptide chain and triggers a conformational change in ovostatin that causes it to bind to the enzyme molecule in such a way that the active site of the enzyme remains free to hydrolyze low-molecular-weight substrates but is restricted from reactions with large protein substrates. Kinetic studies of the binding of collagenase to ovostatin indicate that ovostatin serves as good substrate for mammalian collagenase (Nagase and Harris 1983).

In contrast to α_2-macroglobulin or the closely related duck ovostatin, hen egg white ovostatin subunit does not appear to contain an internal βCys–γGlu thiol ester, and no covalent proteinase binding takes place (Feldman and Pizzo 1984). Furthermore, hen ovostatin–proteinase complexes are not recognized by mammalian α_2-macroglobulin receptor. The lack of thiol esters in hen egg ovostatin is attributed to the presence of Asn949 instead of Cys949 in a 53–amino acid stretch of hen egg white ovostatin corresponding to residues 945–957 of human α_2-macroglobulin.

4.2.11. Cystatin

Cystatin from hen egg white was the first protein to be studied in detail from among a number of cystatins in a homologous group of protein inhibitors of peptidases (Abrahamson et al. 2003). Hen egg white cystatin belongs to the type 2 cystatins with a molecular weight of 12.7 kDa and approximately 115 amino acids and two disulfide loops (Barrett 1986). It inactivates lysosomal cysteine proteinases, such as cathepsins B, H, and L as well as several structurally similar plant proteinases, such as papain and actinidin by forming a reversible tight equimolar complex (Auerswald et al. 1995; Saleh et al. 2003) through hydrophobic interactions (Bode et al. 1988; Machleidt et al. 1989). The complex has a low dissociation constant of 20 n*M*–10 f*M* that effectively blocks the active site of proteinases. According to the crystal structure of cystatin and cystatin–papain complex, papain has three contact regions acting as binding sites to form complementary wedge-shaped edges on the inhibitor. These insert swiftly into the active-site cleft of papain with minimal conformational change (Stubbs et al. 1990). The molecule contains two disulfide bonds that link Cys71–Cys81 and Cys95–Cys115. The Cys71–Cys81 links a small segment of α-helical structure to the main β-sheet of protein, and the Cys95–Cys115 bond joins the two carboxy-terminal strands of this sheet. Reduction of the disulfide bonds leads to a drastic loss of the inhibitory activity. Selective reduction of the Cys95–Cys115 bond induced a conformational change and subsequently decreased the inhibitory activity of chicken cystatin. These phenomena suggest that the disulfide bond may play an important role in folding of the molecular structure and the inhibitory activity of chicken cystatin. Cystatin was isolated as two isoforms, phosphorylated (pI 5.6) and nonphosphorylated (pI 6.5) (Laber et al. 1989). Many cDNAs encoding cystatin have been cloned and expressed in *E. coli*. Most of these recombinants were found to be inert and/or soluble inclusion bodies. However, expression of cystatin in *Pichia pastoris* resulted in recombinant cystatin and Asn106-glycosylated cystatin secreted in the broth and was similar to wild-type cystatin in having K_i of 0.08 nM compared to 0.05 nM, and increased freezing stability of the glycosylated recombinant (Jiang et al. 2002).

4.2.11.1. Anticancer Activity

Several proteolytic enzymes, including cysteine proteases, are believed to play an important role in cancer invasion and metastasis (Kennedy 1993). Increased levels of cysteine proteases, and concomitant decreases in cystatin, have been observed in various cancers (Konduri et al. 2002; Nagai et al. 2003). Cystatins inhibited the tumor-associated activity of intracellular cysteine proteases, and have been suggested as potential anticancer drugs (Cegnar et al. 2004). Cystatin inhibited tumor invasion in *ras*-transformed breast epithelial cells (Premzl et al. 2001), and was found to reduce the activity of the key proteolytic enzymes responsible for the growth of gastric cancer *in vitro*. Multifunctional inhibitors, composed of chicken cystatin in conjunction with other protease inhibitors, have been suggested for the therapy of solid tumors (Kennedy 1993; Krol et al. 2003a, 2003b; Muehlenweg et al. 2000).

4.2.11.2. Antimicrobial Activity

Egg white cystatin has been shown to possess antibacterial activity, preventing the growth of group A *Streptococcus* (Bjorck 1990), *Salmonella typhimurium* (Nakai 2000), and the periodontis-causing *Porphyromonas gingivalis* (Blankenvoorde et al. 1996). Blankenvoorde et al. (1998) found that peptides derived from cystatin were also capable of inhibiting the growth of *P. gingivalis*. Cystatin has been shown to alter viral protein cleavages in infected human cells (Korant et al. 1986, 1995).

4.2.11.3. Immunomodulating Activity

A relationship between cystatins, cytokines, and immune response has also been suggested. Cystatins may also be involved in inflammation and immune responses through the cytokine network, through mechanisms unrelated to the known protease inhibitory regions of the molecule. At physiological concentrations, it stimulates interleukin production by human gingival fibroblast cell lines and murine splenocytes (Kato et al. 2000). Verdot et al. (1996, 1999) found that chicken cystatin induced the synthesis of TNF-α and IL-10, resulting in an upregulation of nitric oxide *in vitro* using mouse peritoneal macrophages, as well as *in vivo*, greatly reducing parasite numbers in a mouse model of visceral leishmaniasis (Das et al. 2001). Cystatin was also found to upregulate the production of IL-6 by human gingival fibroblast cells and murine splenocytes, and the IL-8 production of gingival fibroblasts. Production of IL-6 was upregulated in spite of complete saturation of cystatin by papain (Kato et al. 2000). Oral administration of cystatin prevented the development of human rotavirus–indcued diarrhea in suckling mice by inhibiting proteases required for viral replication (Ebina and Tsukada 1991).

4.2.11.4. Protease Inhibition

Low contents of cystatins in natural resources may limit their applications (Nakai 2000). However, genetic modification and expression of cystatin has been examined, not only providing a source for increased quantities of cystatin but also resulting in the production of recombinant cystatin with enhanced proteinase inhibitory activity (Ogawa et al. 2002).

4.2.12. Avidin

Hen egg avidin, a trace component (0.05%) of egg white, was discovered in the 1920s around the time of the discovery of the vitamin biotin (Boas 1924; Eakin et al. 1941; Mine 2000). Avidin is a basic tetrameric glycoprotein, composed of four identical subunits, each containing 128 amino acids and with a molecular weight reported between 15 and 15.8 kDa, giving a total molecular weight of 66–69 kDa (Korpela 1984). The complete amino acid sequence of the protein subunit was elucidated from studies on cyanogen bromide peptides and tryptic peptides obtained from the protein (DeLange and Huang 1971). Each subunit has a intrachain disulfide bond between Cys4 and Cys83. The carbohydrate moiety, which constitutes about 10% of the molecular weight of avidin, consists of a single oligosaccharide chain with four or five mannose and three *N*-acetylglucosamine residues per subunit linked to Asp17 of the polypeptide chain (Bruch and White 1982).

Functional peptides from avidin have not been reported. However, avidin, which possesses the unique ability to specifically bind the water-soluble vitamin biotin, has been found to inhibit the growth of biotin-requiring bacteria and yeasts (Banks et al. 1986; Green 1975). Avidin binds four biotin molecules, one per subunit (Green 1964). The avidin–biotin interaction is one of the strongest noncovalent protein–ligand interactions found in nature, with a dissociation constant K_a of about $10^{15} M^{-1}$ (Green 1963a, 1963b). The avidin–biotin complex is resistant to denaturation and proteolytic enzymes. The structure of the binding site residues is highly complementary to that of the incoming biotin molecule, accounting for prompt and specific recognition (Pugliese et al. 1994). The crystal structure of the holoavidin complex has shown that biotin sits in a deep protein core pocket displaying polar residues at its deadend, recognizing the ureido group of the biotin, while the remaining part of the pocket is essentially hydrophobic (Pugliese et al. 1993). The high specific affinity of avidin for biotin has many uses in applications such as affinity chromatography, histochemistry, pathological probe, diagnostics, signal amplification, immunoassay, hybridoma technology and blotting technology (Wilchek and Bayer 1990). The efficient expression of biologically active recombinant avidin in *Baculovirus*-infected Lepidopteran insect cells offers a tool for the production of avidin fusion proteins, as well as avidin in studies of the structure–function relationship through site-directed mutagenesis (Airenne et al. 1997).

4.2.12.1. Anticancer Activity

Avidin has been used in cancer treatment, to localize and image cancer cells and to pre-target drugs to tumors. Because of its tetrameric structure, tight biotin binding, and signal amplification, it leads to the accumulation of higher effective doses and increased persistence of biotinylated anticancer drugs, as compared to other immunotherapeutic procedures. Tumor pretargeting with avidin has also been found effective in increasing the uptake of TNF-α conjugated to biotin *in vitro*, thus improving the antitumor activity of TNF (Corti et al. 1998). Pretargeting with avidin could be a realistic possibility to improve the therapeutic index of TNF (Gasparri et al. 1999; Hytonen et al. 2003; Moro et al. 1997). Yao et al. (1998) found that radio labeled avidin also bound to lectins expressed on the surface of tumor cells, and localized highly and rapidly in various types of tumors in mice, thereby reducing radioactivity accumulation in other organs. It has also been found to be essential for the activity of adoptively transferred T cells at tumor sites (Guttinger et al. 2000). The conjugation of avidin to polyethylene glycol chains ($n = 7$) originates a compound with a suitable blood clearance, low immunogenicity, and concurrent low cross-reactivity with avidin (Chinol et al. 1998).

4.2.12.2. Antimicrobial Activity

The antimicrobial activity of avidin has also been attributed to its ability to bind to various Gram-negative and Gram-positive bacteria, including *E. coli* K-12, *Klebsiella pneumoniae, Serratia marcescens, Pseudomonas aeruginosa, Staphylococcus aureus*, and *S. epidermidis* (Korpela et al. 1984). It inhibits the growth of biotin-requiring microbes (Eakin et al. 1941; Green 1975).

Like ovalbumin, avidin inhibits ConA-stimulated and pokeweed mitogen-stimulated mouse spleen T-lymphocyte proliferative responses, while ovomucin enhances LPS-stimulated B lymphocytes, suggesting that avidin, ovalbumin, and ovomucin enhance antibody production at the cell level, and that these proteins are related to the potential allergenicity of hen eggs. The inhibitory activity of avidin on ConA-stimulated proliferative responses is due not only to direct binding and deactivativation of ConA but also to interaction with accessory cells, and the inhibition is due to both the carbohydrate and polypeptide portions of avidin (Otani and Maenishi 1994) through either the inhibition of binding of IL-2 to its receptors on ConA-activated spleen lymphocytes or the inhibition of IL-2 receptor expression on the spleen lymphocytes (Otani and Maenishi 1995). Ingestion of high-affinity protein binding sites can establish an absorptive barrier at the gastrointestinal mucosa to prevent the uptake of unwanted low-molecular-weight chemicals (Rasmussen et al. 2001).

The unique feature of this binding is the strength and specificity of the formation of the avidin–biotin complex, formed when avidin binds four molecules of biotin. In addition, like streptavidin, avidin can be derivatized with various molecules and probes (Wilchek and Bayer 1990; Livnah et al. 1993).

4.2.12.3. Biospecific Ligand Activity

The utilization of avidin as a biospecific ligand in drug delivery through the blood–brain barrier has been demonstrated, facilitating delivery of therapeutics to the brain (Bickel et al. 2001).

4.2.12.4. Immunomodulating Activity

The induction of avidin in inflammation and cellular damage and its secretion by macrophages has suggested a role for avidin as a possible host-defense factor (Korpela 1984).

4.2.13. Ovofactor 1

A proteinaceous component with 88.2% homology to pleiotrophin (neurotrophic factors that promote the development of the nervous system) of basement membrane of chicken embryos and with more than 80% homology with pleiotrophins from humans and cattle, was identified and found to stimulate DNA synthesis and proliferation of cultured cells from chicken embryos (Nakamura et al. 1995).

4.2.14. Extracellular Fatty Acid–Binding Protein

Extracellular fatty acid–binding protein (ex-FABP), also referred to as Ch21 or *quiescence–specific protein*, belongs to the subfamily of chondrogenesis-related lipocalins (lipophilic molecule carrier proteins), and has a molecular weight of 21 kDa (Larsen et al. 1999), which is much higher than its theoretical molecular weight of 18 kDa deduced from the genomic sequence (Descalzi-Cancedda et al. 1990), suggesting possible glycosylation (Guérin-Dubiard et al. 2006). Its apparent pI (5.6) is consistent with its theoretical pI calculated from the amino acid sequence (Desert et al. 2001). This protein was initially identified in developing chick embryonic bone and cartilage, and is suspected to be a stress protein expressed during cartilage formation, inflammatory responses, and chondrocyte and myoblast differentiation (Descalzi-Cancedda et al. 2000, 2001). Ex-FABP is involved in heart development, fatty acid transport, and lipid metabolism, and may act as a cell survival protein by protecting cells from the toxic effect of fatty acid accumulation (Gentili et al. 2005). Such functions of lipocalins make them good candidates for the development of ligand-binding proteins, because of the lipocalin scaffolds (anticalins), which offer therapeutic applications as antidotes antagonists and tissue-targeting vehicles (Schlehuber and Skerra 2005).

4.2.15. Chondrogenesis-Associated Lipocalins β and γ

The chondrogenesis-associated lipocalins β (CAL-β) and γ (CAL-γ) have been detected in hen egg white (Guérin-Dubiard et al. 2006). Pagano et al. (2002, 2003, 2004) isolated and characterized chicken CAL-β. It is reported that CAL-β, CAL-γ, ex-FABP, and Ggal-C8CC (an ortholog of the mammalian complement factor 8γ chain) form a genomic cluster resulting in the four genes being coordinately regulated, the orthology suggesting that Ggal-C8CC was present in the ancestor common to reptiles and mammals. CAL-β, CAL-γ, and ex-FABP are located in sequence from the 5′ end to the 3′ end, respectively, within the genomic locus, have similar spatial and temporal expression patterns during embryonic development, and are functionally related.

CAL-γ is known to act synergistically with ex-FABP in bone formation in chicken embryos (Pagano et al. 2003). The gene has been sequenced (Pagano et al. 2003). The protein also occurs in unfertilized egg white (Guérin-Dubiard et al. 2006).

4.2.16. Tenp

A gene transiently expressed in neural precursors (*tenp*) was identified for the first time in unfertilized hen egg white by Guérin-Dubiard et al. (2006). The gene was reported earlier to be present specifically and briefly in developing neural tissues, including the brain and retina (Wang and Alder 1994; Yan and Wang 1998). Tenp belongs to the bacterial permeability increasing (BPI) protein family, and, on the basis of homology searches, is assumed to have antibacterial activity by binding to lipid A of Gram-negative bacterial lipopolysaccharide (Bingle and Craven 2004). The toxic action of BPI occurs by immediate bacterial growth arrest due to alterations in the outer membrane followed by damage to the inner membrane (Elsbach and Weiss 1998).

4.2.17. Clusterin

Clusterin has been identified in hen egg white (Guérin-Dubiard et al. 2006), as was immunodetected earlier in several chicken tissues (Mann et al. 2003) and also found in biological fluids of humans (Mahon et al. 1999). Clusterin has been classified as a α/β heterodimer with a molecular mass of 35 kDa. Mammalian clusterin is a chaperone protein known to bind and stabilize unfolded or partially folded proteins to prevent their aggregation or precipitation, and has been reported to interact with slowly aggregating proteins associated with Alzheimer's, Creutzfeldt–Jakob, and Parkinson's diseases (Poon et al. 2002). Unlike mammalian clusterin, clusterin in chickens occurs as a single-chain polypeptide, and is absent from circulation, but is present in the somatic granulosa cells of the ovary during the developmental stages of individual follicles, and could serve as a marker for follicular atresia and resorption, as well as a carrier for receptor-mediated endocytosis into oocytes during embryonic development (Mahon et al. 1999). Clusterin is suggested to function in the uterine fluid toward the prevention of premature aggregation and precipitation of eggshell matrix components before and during their assembly into the rigid protein scaffold necessary for ordered mineralization (Mann et al. 2003).

4.2.18. Hep21

Hep21 is a glycoprotein expressed predominantly in hen egg white and the magnum (Nau et al. 2003) as two isoforms with significantly different molecular masses. It is a member of the multifunctional uPAR/CD59/Ly6/snake neurotoxin superfamily. However, its biological function has yet to be determined.

4.2.19. VMO-1 Protein

VMO-1 (VMO-1 family of proteins) is an alkaline protein with a molecular mass of 17 kDa, and was detected in the egg white by Guérin-Dubiard et al. (2006). VMO-1 and VMO-2 were first identified in the outer layer of the egg vitelline membranes by Back et al. (1982) and Kido et al. (1992). Under denaturing conditions, VMO-2 appeared as two bands of 5 and 8 kDa (Schäfer et al. 1998). The proteins along with lysozyme bind tightly to ovomucin in the vitelline membrane. They are released from the membrane, leading to membrane deterioration under nonrefrigerated conditions of egg storage (Schäfer et al. 1998).

4.2.20. Thiamin-Binding Protein

Thiamin-binding protein isolated from hen egg white has a molecular mass of 38 kDa (Muniyappa and Adiga 1979). It is not a glycoprotein and binds thiamin with a molar ratio of 1.0 and a dissociation constant of 0.3 μM. It has a high specificity for riboflavin-binding protein, binding with a molar ratio of 1:1. This property led to the development of a purification procedure for thiamin binding protein that used riboflavin-binding protein immobilized on CNBr-activated Sepharose.

4.2.21. Minor Glycoproteins

Two minor glycoproteins with apparent molecular weight of 52 kDa and globulin-like properties have been identified (Itoh et al. 1993).

REFERENCES

Abdallah FB, Chahine JM (1999). Transferrins, the mechanism of iron release by ovotransferrin. Eur J Biochem 263:912–920.

Abrahamson M, Alvarez-Fernandez M, Nathanson CM (2003). Cystatins. Biochem Soc Symp 70:179–199.

Agarwal V, Nazzal S, Reddy IK, Khan MA (2001a). Transport studies of insulin across rat jejunum in the presence of chicken and duck ovomucoids. J Pharm Pharmacol 53:1131–1138.

Agarwal V, Reddy IK, Khan MA (2001b). Polymethylacrylate based microparticulates of insulin for oral delivery: Preparation and *in vitro* dissolution stability in the presence of enzyme inhibitors. Int J Pharm 225:31–39.

Aguilera O, Quiros LM, Fierro JF (2003). Transferrins selectively cause ion efflux through bacterial and artificial membranes. FEBS Lett 548:5–10.

Ahlborn GJ, Clare DA, Sheldon BW, Kelly RW (2006). Identification of egg shell membrane proteins and purification of ovotransferrin and beta-NAGase from hen egg white. Protein J 25:71–81.

Airenne KJ, Oker-Blom C, Marjomäki VS, Bayer EA, Wilchek M, Kulomaa MS (1997). Production of biologically active recombinant avidin in *Baculovirus*-infected insect cells. Protein Exper Purif 9:100–108.

Alderton G, Ward WH, Fevold HL (1945). Isolation of lysozyme from egg white. J Biol Chem 157:43–58.

Ames BN, Shigenaga MK, Hagen TM (1993). Oxidants, antioxidants, and the degenerative diseases of aging. Proc Natl Acad Sci USA 90:7915–7922.

Amoresano A, Brancaccio, A, Andolfo A, Perduca M, Monaco HL, Marino G (1999). The carbohydrates of the isoforms of three avian riboflavin-binding proteins. Eur J Biochem 263:849–858.

Arima H, Ibrahim HR, Kinoshita T, Kato A (1997). Bactericidal action of lysozymes attached with various sizes of hydrophobic peptides to the C-terminal using genetic modification. FEBS Lett 415:114–118.

Asakura K, Kojima T, Shirasaki H, Kataura A (1990). Evaluation of the effects of antigen specific immunotherapy on chronic sinusitis in children with allergy. Auris Nasus Larynx 17:33–38.

Auerswald EA, Nägler DK, Assfalg-Machleidt I, Stubbs MT, Machleidt W, Fritz H (1995). Harpin loop mutations of chicken cystatin have different effects on the inhibition of cathepsin B, cathepsin L and papain. FEBS Lett 361:179–184.

Babini GS, Livermore DM (2000). Effect of conalbumin on the activity of Syn 2190, a 1, 5 dihyroxy-4-pyridon monobactam inhibitor of AmpC beta-lactamases. J Antimicrob Chemother 45:105–109.

Back JF, Bain JM, Vadehra DV, Burley RW (1982). Proteins of the outer layer of the vitelline membrane of hen's eggs. Biochim Biophys Acta 705:12–19.

Banks JG, Board RG, Sparks NHC (1986). Natural antimicrobial systems and their potential in food preservation of the future. Biotechnol Appl Biochem 8:103–147.

Barrett AJ (1986). The cystatins: A diverse superfamily of cysteine peptidase inhibitors. Biomed Biochim Acta 45:1363–1374.

Begum S, Saito A, Xu X, Kato A (2003). Improved functional properties of the ovoinhibitor by conjugating with galactomannan. Biosci Biotechnol Biochem 67:1897–1902.

Bickel U, Yoshikawa T, Pardridge WM (2001). Delivery of peptides and proteins through the blood-brain barrier. Adv Drug Deliv Rev 46:247–279.

Bingle CD, Craven CJ (2004). Meet the relatives: A family of BPI- and LBP-related proteins. Trends Immunol 25:53–55.

Biziulevičius GA, Kislukhina OV, Kazlauskaite J, Žukaite V (2006). Food protein enzymatic hydrolysates possess both antimicrobial and immunostimulatory activities: A "cause and effect" theory of bifunctionality. FEMS Immunol Med Microbiol 46:131–138.

Bjorck L (1990). Proteinase inhibition, immunoglobulin-binding proteins and a novel antimicrobial principle. Mol Microbiol 4:1439–1442.

Blankenvoorde MF, Henskens YM, van't Hof W, Veerman E, Nieuw Amerongen AV (1996). Inhibition of the growth and cysteine proteinase activity of *Porphyromonas gingivalis* by human salivary cystatin S and chicken cystatin. Biol Chem 377: 847–850.

Blankenvoorde MF, van't Hof W, Walgreen-Weterings E, van Steenbergen TJ, Brand HS, Veerman EC, Nieuw Amerongen AV (1998). Cystatin and cystatin-derived peptides have antibacterial activity against the pathogen *Porphyromonas gingivalis*. Biol Chem 379:1371–1375.

Boas MA (1924). An observation on the value of egg white as the sole source of nitrogen for young growing rats. Biochem J 18:422–424.

Bode W, Engh RA, Musil D, Thiele U, Huber R, Karshikov A, Brzin J, Kos J, Turk V (1988). The 20 A X-ray crystal structure of chicken egg white cystatin and its possible mode of interaction with cysteine proteinases. EMBO J 7:2593–2599.

Brissette R, Prendergast JK, Goldstein NI (2006). Identification of cancer targets and therapeutics using phage display. Curr Opin Drug Discov Devel 9:363–369.

Bruch RC, White HB (1982). Compositional and structural heterogeneity of avidin glycopeptides. Biochemistry 21:5334–5341.

Cegnar M, Premzl A, Zavasnik-Bergant V, Kristl J, Kos J (2004). Poly(lactide-co-glycolide) nanoparticles as a carrier system for delivering cysteine protease inhibitor cystatin into tumor cells. Exp Cell Res 301:223–231.

Chen TL, Lo YC, Hu WT, Wu MC, Chen ST, Chang HM (2003). Microencapsulation and modification of synthetic peptides of food proteins reduces the blood pressure of spontaneously hypertensive rats. J Agric Food Chem 51:1671–1675.

Chinol M, Casalini, P, Maggiolo M, Canevari S, Omodeo ES, Caliceti P, Veronese FM, Cremonesi M, Chiolerio F, Nardone E, Siccardi AG, Paganelli G (1998). Biochemical modifications of avidin improve pharmacokinetics and biodistribution, and reduce immunogenicity. Br J Cancer 78:189–197.

Corti A, Gasparri A, Sacchi A, Curnis F, Sangregorio R, Columbo B, Siccardi AG, Magni F (1998). Tumor targeting with biotinylated tumor necrosis factor alpha:

Structure-activity relationships and mechanism of action on avidin pretargeted tumor cells. Cancer Res 58:3866–3872.

Das S, Banerjee S, Gupta JD (1992). Experimental evaluation of preventative and therapeutic potential of lysozyme. Chemotherapy 38:350–357.

Das L, Datta N, Bandyopadhyay S, Das PK (2001). Successful therapy of lethal murine visceral leishmaniasis with cystatin involves up-regulation of nitric oxide and a favourable T cell response. J Immunol 166:4020–4028.

Davalos A, Miguel M, Bartolome B, Lopez-Fandino R (2004). Antioxidant activity of peptides derived from egg white proteins by enzymatic hydrolysis. J Food Protect 67:1939–1944.

DeLange RJ, Huang T-S (1971). Egg white avidin. III. Sequence of the 78-residue middle cyanogen bromide peptide. Complete amino acid sequence of the protein subunit. J Biol Chem 246:698–709.

Descalzi-Cancedda F, Dozin B, Rossi F, Molina F, Cancedda R, Negri A, Ronchi S (1990). The Ch21 protein, developmentally regulated in chick embryo, belongs to the superfamily of lipophilic molecule carrier proteins. J Biol Chem 265: 19060–19064.

Descalzi-Cancedda F, Dozin B, Zerega B, Cermelli S, Cancedda R (2000). Ex-FABP: A fatty acid binding lipocalin developmentally regulated in chicken endochondral bone formation and myogenesis. Biochim Biophys Acta 1482:127–135.

Descalzi-Cancedda F, Dozin B, Zerega B, Cermelli S, Cancedda R (2001). Extracellular fatty acid binding protein (Ex-FABP) is a stress protein expressed during chondrocyte and myoblast differentiation. Osteoarthr Cartil 9:118–122.

Desert C, Guérin-Dubiard C, Nau F, Jan G, Val F, Mallard J (2001). Comparison of different electrophoretic separations of hen egg white proteins. J Agric Food Chem 49:4553–4561.

Donovan JW, Mapes CJ, Davis JG, Hamburg RD (1969). Dissociation of chicken egg-white macroglobulin into subunite in acid. Hydrodynamic, spectrophotometric, and optical rotatory measurements. Biochemistry 8:4190–4199.

During K, Porsch P, Mahn A, Brinkmann O, Gieffers W (1999). The non-enzymatic microbicidal activity of lysozymes. FEBS Lett 449:93–100.

Eakin RE, Snell EE, Williams RJ (1941). The concentration and assay of avidin, the injury-producing protein in raw egg white. J Biol Chem 140:535–543.

Ebina T, Tsukada K (1991). Protease inhibitors prevent the development of human rotavirus-induced diarrhea in suckling mice. Microbiol Immunol 35:583–588.

Elsbach P, Weiss J (1998). Role of the bactericidal/permeability-increasing protein in host defence. Curr Opin Immunol 10:45–49.

Fan X, Subramaniam R, Weiss MF, Monnier VM (2003). Methylglyoxal-bovine serum albumin stimulates tumor necrosis factor alpha secretion in RAW 264.7 cells through activation of mitogen-activating protein kinase, nuclear factor kappaB and intracellular reactive oxygen species formation. Arch Biochem Biophys 409:274–286.

Farrell HM Jr, Mallette MF, Buss EG, Clagett CO (1969). The nature of the biochemical lesion in avian renal riboflavinuria. 3. The isolation and characterization of the riboflavin-binding protein from egg albumen. Biochim Biophys Acta 194:433–442.

Feeney RE, Abplanalp H, Clary JJ, Edwards DL, Clark JR (1963). A genetically varying minor protein constituent of chicken egg white. J Biol Chem 238:1732–1736.

Feldman SR, Pizzo SV (1984). Comparison of the binding of chicken alpha macroglobulin and ovomacroglobulin to the mammalian alpha 2-macroglobulin receptor. Arch Biochem Biophys 235:267–275.

Ferreira SH, Greene LJ, Alabaster VA, Bakhle YS, Vane JR (1970). Activity of various fractions of bradykinin potentiating factor against angiotensin 1 converting enzyme. Nature 225:379–380.

Fleming A (1922). On a remarkable bacteriolytic element found in tissue and secretions. Proc Roy Soc Lond B 39:306–317.

Frenkel K, Chrzan K, Ryan CA, Wiesner R, Troll W (1987). Chymotrypsin-specific protease inhibitors decrease H_2O_2 formation by activated human polymorphonuclear leukocytes. Carcinogenesis 8:1207–1212.

Fujita H, Usui H, Kusahahi K, Yoshikawa M (1995a). Isolation and characterization of ovokinin, a bradykinin B1 agonist peptide derived from ovalbumin. Peptides 16:785–790.

Fujita H, Sasaki R, Yoshikawa M (1995b). Potentiation of the antihypertensive activity of orally administered ovokinin, a vasorelaxing peptide derived from ovalbumin, by emulsification in egg phosphatidylcholine. Biosci Biotechnol Biochem 59:2344–2345.

Fujita H, Yokoyama K, Yoshikawa M (2000). Classification and antihypersensitive activity of angiotensin 1-converting enzyme inhibitory peptides derived from food proteins. J Food Sci 65:564–569.

Gasparri A, Moro M, Curnis F, Sacchi A, Pagano S, Veglia F, Casorati G, Siccardi AG, Dellabona P, Corti A (1999). Tumor pretargeting with avidin improves the therapeutic index of biotinylated tumor necrosis factor alpha in mouse models. Cancer Res 59:2917–2923.

Gentili C, Tutolo G, Zerega B, Di Marco E, Cancedda R, Descalzi Cancedda F (2005). Acute phase lipocalin Ex-FABP is involved in heart development and cell survival. J Cell Physiol 202:683–689.

Giansanti F, Rossi P, Massucci MT, Botti D, Antonini G, Valenti P, Seganti L (2002). Antiviral activity of ovotransferrin discloses an evolutionary strategy for the defensive activities of lactoferrin. Biochem Cell Biol 80:125–130.

Gill AO, Holley RA (2000). Surface application of lysozyme, nisin, and EDTA to inhibit spoilage and pathogenic bacteria on ham and bologna. J Food Protect 63:1338–1436.

Goldberg J, Shrikant P, Mescher MF (2003). *In vivo* augmentation of tumor-specific CTL responses by class I/peptide antigen complexes on microspheres (large multivalent immunogen). J Immunol 170:228–235.

Green NM (1963a). Avidin 3. The nature of the biotin-binding site. Biochem J 89:599–609.

Green NM (1963b). Avidin: The use of (^{14}C) biotin for kinetic studies and for assay. Biochem J 89:585–591.

Green NM (1964). The molecular weight of avidin. Biochem J 92:16–17.

Green NM (1975). Avidin. Adv Protein Chem 29:85–133.

Guérin-Dubiard C, Pasco M, Hietanen A, Quiros del Bosque A, Nau F, Croguennec T (2005). Hen egg white fractionation by ion-exchange chromatography. J Chromatogr A 1090:58–67.

Guérin-Dubiard C, Pasco M, Mollé D, Désert C, Croguennec T, Nau F (2006). Proteomic analysis of hen egg white. J Agric Food Chem 54:3901–3910.

Guttinger M, Guidi F, Chinol M, Reali E, Veglia F, Viale G, Paganelli G, Corti A, Siccardi AG (2000). Adoptive immunotherapy by avidin-driven cytotoxic T lymphocyte-tumor bridging. Cancer Res 60:4211–4215.

He X, Tsang TC, Luo P, Zhang T, Harris DT (2003). Enhanced tumor immunogenicity through coupling cytokine expression with antigen presentation. Cancer Gene Ther 10:669–677.

Henskens YMC, Veerman ECI, Nieuw AAV (1996). Cystatins in health and disease. Biol Chem 377:71–86.

Herzog J, Maekawa Y, Cirrito TP, Illian BS, Unanue ER (2005). Activated antigen-presenting cells select and present chemically modified peptides recognized by unique CD4 T cells. Proc Natl Acad Sci USA 102:7928–7933.

Hilpert K, Wessner H, Schneider-Mergener J, Welfle K, Misselwitz R, Welfle H, Hocke AC, Hippenstiel S, Hohne W (2003). Design and characterization of a hybrid miniprotein that specifically inhibits porcine pancreatic elastase. J Biol Chem 278:24986–24993.

Hirota Y, Yang MP, Araki S, Yoshihara K, Furusawa S, Yasuda M, Mohamed A, Matsumoto Y, Onodera T (1995). Enhancing effects of chicken egg white derivatives on the phagocytic response in the dog. J Vet Med Sci 57:825–829.

Holen E, Bolann B, Elsayed S (2001). Novel B and T cell epitopes of chicken ovomucoid (Gal d 1) induce T cell secretion of IL-6, IL-13, and IFN-gamma. Clin Exp Allergy 31:952–964.

Hosono A, Shashikanth KN, Otani H (1988). Antimutagenic activity of whole casein on the pepper-induced mutagenicity to streptomycin-dependent strain SD 510 of *Salmonella typhimurium* TA98. J Dairy Res 55:435–442.

Hunter HN, Jing W, Schibli DJ, Trinh T, Park IY, Kim SC, Vogel HJ (2005). The interactions of antimicrobial peptides derived from lysozyme with model membrane systems. Biochim Biophys Acta 1668:175–189.

Huntington JA, Stein PE (2001). Structure and properties of ovalbumin. J Chromatogr B Biomed Sci Appl 756:189–198.

Hytonen VP, Laitinen OH, Grapputo A, Kettunen A, Savolainen J, Kalkkinen N, Martilla AT, Norlund HR, Nyholm TKM, Paganelli G, Kulomaa MK (2003). Characterisation of poultry egg-white avidins and their potential as tools in pretargeting cancer treatment. Biochem J 372:219–225.

Ibrahim HR (1997). Insights into the structure-function relationships of ovalbumin, ovotransferrin, and lysozyme. In: Yamamoto T, Juneja LR, Hatta H, Kim M, eds. Hen Eggs, Their Basic and Applied Science. New York: CRC Press; pp 37–56.

Ibrahim HR (2000). Ovotransferrin. In: Naidu AS, ed. Natural Food Antimicrobial Systems. New York: CRC Press; pp 211–226.

Ibrahim HR, Kato A, Kobayashi K (1991). Antimicrobial effects of lysozyme against Gram-negative bacteria due to covalent binding of palmitic acid. J Agric Food Chem 39:2077–2082.

Ibrahim HR, Yamada M, Kobayashi K, Kato A (1992). Bactericidal action of lysozyme against Gram-negative bacteria due to insertion of a hydrophobic pentapeptide into its C-terminus. Biosci Biotechnol Biochem 56:1361–1363.

Ibrahim HR, Kobayashi K, Kato A (1993). Length of hydrocarbon chain and antimicrobial action to Gram-negative bacteria of fatty acylated lysozyme. J Agric Food Chem 41:1164–1168.

Ibrahim HR, Yamada M, Matsushita K, Kobayashi K, Kato A (1994a). Enhanced bactericidal action of lysozyme to *Escherichia coli* by inserting a hydrophobic pentapeptide into its C terminus. J Biol Chem 18:5059–5063.

Ibrahim HR, Hatta H, Fujiki M, Kim M, Yamamoto T (1994b). Enhanced antimicrobial action of lysozyme against Gram-negative and Gram-positive bacteria due to modification with perillaldehyde. J Agric Food Chem 42:1813–1817.

Ibrahim HR, Sugimoto Y, Aoki T (2000). Ovotransferrin antimicrobial peptide (OTAP-92) kills bacteria through a membrane damage mechanism. Biochim Biophys Acta 1523:196–205.

Ibrahim HR, Matsuzaki T, Aoki T (2001a). Genetic evidence that antibacterial activity of lysozyme is independent of its catalytic function. FEBS Lett 506:27–32.

Ibrahim HR, Thomas U, Pellegrini A (2001b). A helix-loop peptide at the upper lip of the active site cleft of lysozyme confers potent antimicrobial activity with membrane permeabilization action. J Biol Chem 276:43767–43774.

Ibrahim HR, Aoki T, Pellegrini A (2002). Strategies for new antimicrobial proteins and peptides: Lysozyme and aprotinin as model molecules. Curr Pharm Des 8:671–693.

Ijiri Y, Yamamoto T, Kamata R, Aoki H, Matsumoto K, Okamura R, Kambara T (1993). The role of *Pseudomonas aeruginosa* elastase in corneal ring abscess formation in pseudomonal keratitis. Graefes Arch Clin Exp Ophthalmol 231: 521–528.

Ikai A, Nakashima M, Aoki Y (1989). Inhibition of inflammatory proteinases, medullasin, by alpha 2-macroglobulin and ovomacrogloulin. Biochem Biophys Res Commun 158:831–836.

International Chicken Gene Sequencing Consortium (2004). Sequence and comparative analysis of the chicken genome provide unique perspectives on vertebrate evolution. Nature 432:695–716.

Itoh T, Takeuchi S, Saito T (1993). New minor glycoproteins isolated from hen's egg white by heparin-affinity chromatography. Biosci Biotechnol Biochem 57: 1018–1019.

Jiang ST, Chen GH, Tang SJ, Chen C-S (2002). Effect of glycosylation modification (N-Q-108K–N-Q-108T) on the freezing stability of recombinant chicken cystatin overexpressed in *Pichia pastoris* X-33. J Agric Food Chem 50:5313–5317.

Kato I, Schrode J, Kohr WJ, Laskowski M Jr (1987). Chicken ovomucoid: determination of its amino acid sequence, determination of the trypsin reactive site, and preparation of all three of its domains. Biochemistry 26:193–201.

Kato T, Imatani T, Miura T, Minaguchi K, Saitoh E, Okuda K (2000). Cytokine-inducing activity of family 2 cystatins. Biol Chem 381:1143–1147.

Kennedy AR (1993). Anticarcinogenic activity of protease inhibitors. In: Troll W, Kennedy AR, eds. Protease Inhibitors as Cancer Chemopreventative Agents. New York: Plenum Press; pp 9–64.

Ketterer B (1965). Ovoglycoprotein, a protein of hen's-egg white. Biochem J 96:372–376.

Kido S, Morimoto A, Kim F, Doi YK (1992). Isolation of a novel protein from the outer layer of the vitelline membrane. Biochem J 286:17–22.

Kinoshita K, Shimogiri T, Okamoto S, Yoshizawa K, Mannen H, Ibrahim HR, Cheng HH, Maeda Y (2004). Linkage mapping of chicken ovoinhibitor and ovomucoid genes to chromosome 13. Anim Genet 35:356–358.

Kitamoto T, Nakashima M, Ikai A (1982). Hen egg white ovomacroglobulin has a protease inhibitory activity. J Biochem 92:1679–1682.

Kitts DD, Weiler K (2003). Bioactive proteins and peptides from food sources. Applications of bioprocesses used in isolation and recovery. Curr Pharm Des 9:1309–1323.

Kobayashi K, Hattori M, Hara-Kudo Y, Okubo T, Yamamoto S, Takita T, Sugita-Konishi Y (2004). Glycopeptide derived from hen egg ovomucin has the ability to bind enterohemorrhagic *Escherichia coli* O157:H7. J Agric Food Chem 52:5740–5746.

Kojima S, Takagi N, Minagawa T, Fushimi N, Miura K (1999). Effects of amino acid replacements around the reactive site of chicken ovomucoid domain 3 on the inhibitory activity toward bovine a-chymotrypsin, bovine trypsin and porcine pancreatic elastase. Protein Eng 12:857–862.

Konduri SD, Yanamandra N, Siddique K, Joseph A, Dinh DH, Olivero WC, Gujrati M, Kouraklis G, Swaroop A, Kyritsis AP, Rao JS (2002). Modulation of cystatin C expression impairs the invasive and tumorigenic potential of human glioblastoma cells. Oncogene 21:8705–8712.

Korant BD, Brzin J, Turk V (1995). Cystatin, a protein inhibitor of cysteine proteinases alters viral protein cleavages in infected human cells. Biochem Biophys Res Commun 127:1072–1076.

Korant BD, Towatari T, Ivanoff L (1986). Viral therapy: Prospects for protease inhibitors. J Cell Biochem 32:91–95.

Korhonen H, Pihlanto A (2003). Food-derived bioactive peptides—opportunities for designing future foods. Curr Pharm Des 9:1297–1308.

Korpela J (1984). Avidin, a high affinity biotin-binding protein as a tool and subject of biological research. Med Biol 62:5–26.

Korpela J, Salonen E-M, Kuusela P, Sarvas M, Vaheri A (1984). Binding of avidin to bacteria and to the outer membrane porin of *Escherichia coli*. FEMS Microbiol Lett 22:3–10.

Kovacs-Nolan J, Phillips M, Mine Y (2005). Advances in the value of eggs and egg components for human health. J Agric Food Chem 53:8421–8431.

Krol J, Sato S, Rettenberger P, Assfalg-Machleidt I, Schmitt M, Magdolen V, Magdolen U (2003a). Novel bi- and trifunctional inhibitors of tumor-associated proteolytic systems. Biol Chem 384:1085–1096.

Krol J, Kopitz C, Kirschenhofer A, Schmitt M, Magdolen U, Kruger A, Magdolen V (2003b). Inhibition of intraperitoneal tumor growth of human ovarian cancer cells by bi- and trifunctional inhibitors of tumor-associated proteolytic systems. Biol Chem 384:1097–1102.

Krowarsch D, Cierpicki T, Jelen F, Otlweski J (2003). Canonical protein inhibitors of serine proteases. Cell Mol Life Sci 60:2427–2444.

Laber B, Krieglstein K, Henschen A, Kos J, Turk V, Huber R, Bode W (1989). The cysteine proteinase inhibitor chicken cystatin is a phosphoprotein. FEBS Lett 248:162–168.

Lang T, Hansson GC, Samuelsson T (2006). An inventory of mucin genes in the chicken genome shows that the mucin domain of Muc13 is encoded by multiple exons and that ovomucin is part of a locus of related gel-forming mucins. BMC Genom 7:197–206.

Lanni F, Sharp DG, Eckert E, Dillon E, Beard O, Beard JW (1949). Egg white inhibitor of influenza virus hemagglutination. I. Preparation and properties of semi purified inhibitor. J Biol Chem 179:1275–1287.

Larsen LB, Hammershoj M, Rasmussen JT (1999). Identification of Ch21, a developmentally regulated chicken embryo protein in egg albumen. Proc 8th Eur Symp Quality of Eggs and Egg Products, vol II Sept 19–23, Bologna, Italy; pp 61–67.

Laskowski M Jr, Kato I (1980). Protein inhibitors of proteinases. Annu Rev Biochem 49:593–626.

Lee MY, Shin SH, Yang NU, Lee SE, Rhee JH, Park Y, Kim IS (1996). Effect of limitation on the production of siderophore and hemolysin in *Stapylococcus aureus*. J Kor Soc Microbiol 31:331–337.

Lee-Huang S, Huang PL, Sun Y, Huang PL, Kung HF, Blithe DL, Chen HC (1999). Lysozyme and RNases as anti-HIV components in beta-core preparations of human chorionic gonadotropin. Proc Natl Acad Sci USA 96:2678–2681.

Li-Chan E, Nakai S (1989). Biochemical basis for the properties of egg white. Crit Rev Poultry Biol 2:21–58.

Li-Chan ECY, Powrie WD, Nakai S (1995). The chemistry of eggs and egg products. In: Stadelman WJ, Cotterill OJ, eds. Egg Science and Technology, 4th ed. New York: Haworth Press; pp 105–175.

Liang AH, Xue BY, Liang RX, Wang JH, Wang D (2003). Inhibitory effect of egg white lysozyme on ceftazidime-induced release of endotoxin from *Pseudomonas aeruginosa*. Yao Xue Xue Bao 38:801–804.

Lim Y, Shin SH, Lee SI, Kim IJ, Rhee JH (1998). Iron repressibility of siderophore and transferrin-binding protein in *Staphylococcus aureus*. FEMS Microbiol Lett 163:19–24.

Liu WH, Means GE, Feeney RE (1971). The inhibitory properties of avian ovoinhibitor against proteolytic enzymes. Bochim Biophys Acta 229:176–185.

Liu H, Zheng F, Cao Q, Ren B, Zhu L, Striker G, Vlassara H (2006). Amelioration of oxidant stress by the defensin lysozyme. Am J Physiol Endocrinol Metab 290:E824.

Livnah O, Bayer EA, Wilchek M, Sussman JL (1993). Three-dimensional structures of avidin and the avidin-biotin complex. Proc Natl Acad Sci USA 90:5076–5080.

Longsworth LG, Cannan RK, MacInnes DA (1940). An electrophoretic study of the proteins of egg white. J Am Chem Soc 62:2580–2590.

Losso JN, Nakai S, Charter EA (2000). Lysozyme. In: Naidu AS, ed. Natural Food Antimicrobial Systems. New York: CRC Press; pp 185–210.

Lovitch SB, Unanue ER (2005). Conformational isomers of a peptide-class II major histocompatibility complex. Immunol Rev 207:293–313.

Machleidt W, Thiele U, Laber B, Assfalg-Machleidt I, Esterl A, Wiegandn G, Kosn J, Turk V, Bode W (1989). Mechanism of inhibition of papain by chicken egg white cystatin. Inhibition constants of N-terminally truncated forms and cyanogen bromide fragments of the inhibitor. FEBS Lett 243:234–238.

Maeda H, Akaike T, Sakata Y, Maruo K (1993). Role of bradykinin in microbial infection: Enhancement of septicemia by microbial proteases and kinin. Agents Actions Suppl 42:159–165.

Maehashi K, Matano M, Kondo A, Yamamoto Y, Udaka S (2007). Riboflavin-binding protein exhibits selective sweet suppression toward protein sweeteners. Chem Senses 32:183–190.

Mahon MG, Lindsted KA, Hermann M, Nimpf J, Schneider WJ (1999). Multiple involvement of clusterin in chicken ovarian follicle development. Binding to two oocyte-specific members of the low density lipoprotein receptor gene family. J Biol Chem 274:4036–4044.

Maliar T, Balaz S, Tandlich R, Sturdik E (2002). Viral proteinases—possible targets of antiviral drugs. Acta Virol 46:131–140.

Mann K, Gautron J, Nys Y, McKee MD, Badjari T, Schneider WJ, Hincke MT (2003). Disulfide-linked heterodimeric clusterin is a component of the chicken eggshell matrix and egg white. Matrix Biol 22:397–407.

Maruo K, Akaike T, Ono T, Maeda HH (1998). Involvement of bradykinin generation in intravascular dissemination of *Vibrio vulnificus* and prevention of invasion by a bradykinin antagonist. Infect Immun 66:866–869.

Masschalck B, Van Houdt R, Van Haver EGR, Michiels CW (2001). Inactivation of Gram-negative bacteria by lysozyme, denatured lysozyme, and lysozyme-derived peptides under high hydrostatic pressure. Appl Environ Microbiol 67: 339–344.

Matoba N, Usui H, Fujita H, Yoshikawa M (1999). A novel anti-hypertensive peptide derived from ovalbumin induces nitric oxide-mediated vasorelaxation in an isolated SHR mesenteric artery. FEBS Lett 452:181–184.

Matoba N, Doyama N, Yamada Y, Maruyama N, Utsumi S, Yoshikawa M (2001a). Design and production of genetically modified soybean protein with anti-hypertensive activity by incorporating potent analogue of ovokinin(2–7). FEBS Lett 497:50–54.

Matoba N, Yamada Y, Usui H, Nakagiri R, Yoshikawa M (2001b). Designing potent derivatives of ovokinin(2–7), an anti-hypertensive peptide derived from ovalbumin. Biosci Biotechnol Biochem 65:736–739.

Matoba N, Yamada Y, Yoshikawa M (2003). Design of a genetically modified soybean protein preventing hypertension based on an anti-hypertensive peptide derived from ovalbumin. Curr Med Chem Cardiovasc Hematol Agents 1:197–202.

Matsushima K (1958). An undescribed trypsin inhibitor in egg white. Science 127: 1178–1179.

Miguel M, Aleixandre A (2006). Antihypertensive peptides derived from egg proteins. J Nutr 136:1457–1460.

Miguel M, Recio I, Gómez-Ruiz JA, Ramos M, López-Fandiño R (2004). Angiotensin I-converting enzyme inhibitory activity of peptides derived from egg white proteins by enzymatic hydrolysis. J Food Protect 67:1914–1920.

Miguel M, López-Fandiño R, Ramos M, Aleixandre A (2005). Short-term effect of egg white hydrolysate products on the arterial blood pressure of hypertensive rats. Br J Nutr 94:731–737.

Miguel M, Aleixandre A, Ramos M, López-Fandiño R (2006a). Effect of simulated gastrointestinal digestion on the antihypertensive properties of ACE-inhibitory peptides derived from ovalbumin. J Agric Food Chem 54:726–731.

Miguel M, López-Fandiño R, Ramos M, Aleixandre A (2006b). Long-term intake of egg white hydrolysate attenuates the development of hypertension in spontaneously hypertensive rats. Life Sci 78:2960–2966.

Miguel M, Alvarez Y, López-Fandiño R, Alonso MJ, Salaices M (2007). Vasodilator effects of peptides derived from egg white proteins. *Regul Peptides* 140:131–135.

Mine Y (2000). Avidin. In: Naidu N, ed. Boca Raton, FL: CRC Press; pp 228–253.

Mine Y, Kovacs-Nolan J (2004). Biologically active hen egg components in human health and disease. J Poultry Sci 41:1–29.

Mine Y, Zhang JW (2002a). Comparative studies on antigenicity and allergenicity of native and denatured egg white proteins. J Agric Food Chem 50:2679–2683.

Mine Y, Zhang JW (2002b). Identification and fine mapping of IgG and IgE epitopes in ovomucoid. Biochem Biophys Res Commun 292:1070–1074.

Mine Y, Ma F, Lauriau S (2004). Antimicrobial peptides released by enzymatic hydrolysis of hen egg white lysozyme. J Agric Food Chem 52:1088–1094.

Mire-Sluis AR (1999). Analytical characterization of cytokines and growth factors. In: Brown F, Mire-Sluis AR, eds. Biological Characterization and Assay of Cytokines and Growth Factors. Switzerland: Karger; vol. 97, pp 3–9.

Miyagawa S, Kamata R, Matsumoto K, Okamura R, Maeda H (1991a). Inhibitory effects of ovomacroglobulin on bacterial keratitis in rabbits. Graefes Arch Clin Exp Ophthalmol 229:281–286.

Miyagawa S, Matsumoto K, Kamata R, Okamura R, Maeda H (1991b). Spreading of *Serratia marcescens* in experimental keratitis and growth suppression by chicken egg white ovomacroglobulin. Jpn J Ophthalmol 35:402–410.

Miyagawa S, Nishino N, Kamata R, Okamura R, Maeda H (1991c). Effects of protease inhibitors on growth of *Serratia marcescens* and *Pseudomonas aeruginosa*. Microb Pathog 11:137–141.

Miyagawa S, Kamata R, Matsumoto K, Okamura R, Maeda H (1994). Therapeutic intervention with chicken egg white ovomacroglobulin and a new quinolone on experimental *Pseudomonas keratitis*. Graefes Arch Clin Exp Ophthalmol 232:488–493.

Modun B, Kendall D, Williams P (1994). *Staphylococci* express a receptor for human transferrin: Identification of a 42-kilodalton cell wall transferrin-binding protein. Infect Immun 62:3850–3858.

Molla A, Matsumura Y, Yamamoto T, Okamura R, Maeda H (1987). Pathogenic capacity of proteases from *Serratia marcescens* and *Pseudomonas aeruginosa* and their suppression by chicken egg white ovomacroglobulin. Infect Immun 55:2509–2517.

Molla A, Yamamoto T, Akaike T, Miyoshi S, Maeda H (1989). Activation of Hageman factor and prekallikrein and generation of kinin by various microbial proteases. J Biol Chem 264:10589–10594.

Moro M, Pelagi M, Fulci G, Paganelli G, Dellabona P, Casorati G, Siccardi AG, Corti A (1997). Tumor cell targeting with antibody-avidin complexes and biotinylated tumor necrosis factor alpha. Cancer Res 57:1922–1928.

Muehlenweg B, Assfalg-Machleidt I, Parrado SG, Burgle M, Creutzburg S, Schmitt M, Auerswald EA, Machleidt W, Magdolen V (2000). A novel type of bifunctional inhibitor directed against proteolytic activity and receptor/ligand interaction. Cystatin with a urokinase receptor binding site. J Biol Chem 275:33562–33566.

Muniyappa K, Adiga PR (1979). Isolation and characterization of thiamin-binding protein from chicken egg white. Biochem J 177:887–894.

Nagai A, Terashima M, Harada T, Shimode K, Takeuchi H, Murakawa Y, Nagasaki M, Nakano A, Kobayashi S (2003). Cathepsin B and H activities and cystatin C concentrations in cerebrospinal fluid from patients with leptomeningeal metastasis. Clin Chim Acta 329:53–60.

Nagaoka S, Masaoka M, Zhang Q, Hasegawa M, Watanabe K (2002). Egg ovomucin attenuates hypercholesterolemia in rats and inhibits cholesterol absorption in Caco-2 cells. Lipids 37:267–272.

Nagase H, Harris ED Jr (1983). Ovostatin: A novel proteinase inhibitor from chicken egg white. II. Mechanism of inhibition studied with collagenase and thermolysin. J Biol Chem 258:7490–7498.

Nagase H, Harris ED Jr, Woessner JF Jr, Brew K (1983). Ovostatin: A novel proteinase inhibitor from chicken egg white. I. Purification, physicochemical properties, and tissue distribution of ovostatin. J Biol Chem 258:7481–7489.

Nakai S (2000). Molecular modifications of egg proteins for functional improvement. In: Sim JS, Nakai S, Guenter W, eds. Egg Nutrition and Biotechnology. UK: CAB Int; pp 205–217.

Nakamura S, Kato A (2000). Multi-functional biopolymer prepared by covalent attachment of galactomannan to egg-white proteins through naturally occurring Maillard reaction. Nahrung 44:201–206.

Nakamura R, Matsuda T (1984). Heat inactivation of ovoinhibitor in the alkaline pH region. J Agric Food Chem 32:483–486.

Nakamura S, Kato A, Kobayashi K (1991). New antimicrobial characteristics of lysozyme-dextran conjugate. J Agric Food Chem 39:647–650.

Nakamura S, Gohya Y, Losso JN, Nakai S, Kato A (1996). Protective effect of lysozyme-galactomannan or lysozyme-palmitic acid conjugates against *Edwardsiella tarda* infection in carp, *Cyprinus carpio* L. FEBS Lett 383:251–254.

Nakamura R, Takayama M, Nakamura K, Umemura O (1980). Constituent proteins of globulin fraction obtained from egg white. Agric Biol Chem 44:2357–2361.

Nakamura T, Saito T, Kitazawa H, Takeuchi S, Itoh T (1995). Isolation of a new minor protein (ovofactor-1), which has a cell growth promoting activity, from hen's egg white by heparin affinity chromatography. Biosci Biotech Biochem 59:1946–1948.

Nau F, Guérin-Dubiard C, Desert C, Gautron J, Bouton S, Gribonval J, Lagarrigue S (2003). Cloning and characterization of HEP21, a new member of the uPAR/Ly6 protein superfamily predominantly expressed in hen egg white. Poultry Sci 82:242–250.

Norioka N, Okada T, Hamazume Y, Mega T, Ikenaka J (1985). Comparison of the amino acid sequences of hen plasma, yolk, and white-riboflavin binding proteins. J Biochem (Tokyo) 97:19–28.

Odashima M, Takano F, Otani H (2000). Inhibitory properties of Japanese quail ovoinhibitor toward lymphocyte proliferations. Anim Sci J 71:92–102.

Ofuji Y, Suzuki T, Yoshie H, Hara K, Adachi M (1992). The effects of ovomacroglobulin on gingival wound healing in rats. Periodon Clin Invest 14:13–22.

Ogawa M, Nakamura S, Scaman CH, Jing H, Kitts DD, Dou J, Nakai S (2002). Enhancement of proteinase inhibitory activity of recombinant human cystatin C using random-centroid optimization. Biochim Biophys Acta 1599:115–124.

Oguro T, Ohaki Y, Asano G, Ebina T, Watanabe K (2001). Ultrastructural and immunohistochemical characterization on the effect of ovomucin in tumor angiogenesis. Jpn J Clin Electron Microsc 33:89–99.

Onishi K, Matoba N, Yamada Y, Doyama N, Maruyama N, Utsumi S, Yoshikawa M (2004). Optimal designing of beta-conglycinin to genetically incorporate RPLKPW, a potent antihypertensive peptide. Peptides 25:37–43.

Otani H, Maenishi K (1994). Effect of hen egg proteins on proliferative responses of mouse spleen lymphocytes. Lebens Wissen Technol 27:42–47.

Otani H, Maenishi K (1995). Inhibition of concanavalin A-induced proliferative responses of mouse spleen cells and rabbit Peyer's patch cells by hen egg-white avidin. Food Agric Immunol 7:43–53.

Otani H, Odashima M (1997). Inhibition of proliferative responses of mouse spleen lymphocytes by lacto- and ovotransferrins. Food Agric Immunol 9:193–202.

Pacor S, Giacomello E, Bergamo A, Clerici K, Zacchigna M, Boccu E, Sava G (1996). Antimetastatic action and lymphocyte activation by the modified lysozyme mPEG-Lyso in mice with MCa mammary carcinoma. Anticancer Res 16:2559–2564.

Pacor S, Gagliardi R, Di Daniel E, Vadori M, Sava G (1999). *In vitro* down regulation of ICAM-1 and E-cadherin and *in vitro* reduction of lung metastases of TS/A adenocarcinoma by a lysozyme derivative. Int J Mol Med 4:369–375.

Pagano A, Giannoni P, Zambotti A, Randazzo N, Zerega B, Cancedda R, Dozin B (2002). CALbeta, a novel lipocalin associated with chondrogenesis and inflammation. Eur J Cell Biol 81:264–272.

Pagano A, Crooijmans R, Groenen M, Randazzo N, Zerega B, Cancedda R, Dozin B (2003). A chondrogenesis-related lipocalin cluster includes a third new gene, CALgamma. Gene 305:185–194.

Pagano A, Giannoni P, Zambotti A, Sánchez D, Ganfornina MD, Gutiérrez G, Randazzo N, Cancedda R, Dozin B (2004). Phylogeny and regulation of four lipocalin genes clustered in the chicken genome: Evidence of a functional diversification after gene duplication. Gene 331:95–106.

Pellegrini A (2003). Antimicrobial peptides from food proteins. Curr Pharm Des 9:1225–1238.

Pellegrini A (2004). Proteolytic fragments of ovalbumin display antimicrobial activity. Biochim Biophys Acta 1672:76–85.

Pellegrini A, Thomas U, Bramaz N, Klauser S, Hunziker P, von Fellenberg R (1997). Identification and isolation of a bactericidal domain in chicken egg white lysozyme. J Appl Microbiol 82:372–378.

Pellegrini A, Thomas U, Wild P, Schraner E, von Fellenberg R (2000). Effect of lysozyme or modified lysozyme fragments on DNA and RNA synthesis and membrane permeability of *Escherichia coli*. Microbiol Res 155:69–77.

Plate NA, Valuev IL, Sytov GA, Valuev LI (2002). Mucoadhesive polymers with immobilized proteinase inhibitors for oral administration of protein drugs. Biomaterials 23:1673–1677.

Poon S, Treweek TM, Wilson MR, Easterbrook-Smith SB, Carver JA (2002). Clusterin is an extracellular chaperone that specifically interacts with slowly aggregating proteins on their off-folding pathway. FEBS Lett 513:259–266.

Premzl A, Puizdar V, Zavasnik-Bergant V, Kopitar-Jerala N, Lah TT, Katunuma N, Sloane BF, Turk V, Kos J (2001). Invasion of ras-transformed breast epithelial cells depends on the proteolytic activity of cysteine and aspartic proteinases. Biol Chem 382:853–857.

Pugliese L, Coda A, Malcovait M, Bolognesi M (1993). Three dimensional structure of the tetragonal crystal form of egg white avidin in its functional complex with biotin at 2.7 A° resolution. J Mol Biol 231:698–710.

Pugliese L, Malcovati M, Coda A, Bolognesi M (1994). Crystal structure of apo-avidin from hen egg-white. J Mol Biol 235:42–46.

Qasim MA, Van Etten RL, Yeh T, Saunders C, Ganz PJ, Qasim S, Wang L, Laskowski M Jr (2006). Despite having a common P_1 Leu, eglin C inhibits α-lytic proteinase a million-fold more strongly than does turkey ovomucoid third domain. Biochemistry 45:11342–11348.

Rada JA, Huang Y, Rada KG (2001). Identification of choroidal ovotransferrin as a potential ocular growth regulator. Curr Eye Res 22:121–132.

Raikos V, Hansen R, Campbell L, Euston SR (2006). Separation and identification of hen egg protein isoforms using SDS-PAGE and 2D gel electrophoresis with MALDI-TOF mass spectrometry. Food Chem 99:702–710.

Rasmussen MV, Barker TT, Silbart LK (2001). High affinity binding site-mediated prevention of chemical absorption across the gastrointestinal tract. Toxicol Lett 125:51–59.

Rezansoff AJ, Hunter HN, Jing W, Park IY, Kim SC, Vogel HJ (2005). Interactions of the antimicrobial peptide AC-FRWWHR-NH(2) with model membrane system and bacterial cells. J Peptide Res 65:491–501.

Rohrer JS, White HB III (1992). Separation and characterization of the two Asn-linked glycosylation sites of chicken serum riboflavin-binding protein. Glycosylation differences despite similarity of primary structure. Biochem J 285: 275–280.

Rudra JS, Dave K, Haynie DT (2006). Antimicrobial polypeptide multilayer nanocoatings. J Biomater Sci Polym Ed 17:1301–1315.

Rutherfurd-Markwick KJ, Moughan PJ (2005). Bioactive peptides derived from food. J AOAC Int 88:955–966.

Sadakane Y, Matsunaga H, Nakagomi K, Hatanak Y, Haginaka J (2002). Protein domain of chicken alpha(1)—acid glycoprotein is responsible for chiral recognition. Biochem Biophys Res Commun 295:587–590.

Saleh Y, Siewinski M, Kielan W, Ziolkowski P, Grybos M, Rybka J (2003). Regulation of cathepsin B and L expression *in vitro* in gastric cancer tissues by egg cystatin. J Exp Ther Oncol 3:319–324.

Salton MJR (1957). The properties of lysozyme and its action on microorganisms. Bacteriol Rev 21:82–98.

Sampson HA, Ho DG (1997). Relationship between food-specific concentrations and risk of positive food challenges in children and adolescents. J Allergy Clin Immunol 100:444–457.

Sava G (1989). Reduction of B16 melanoma metastases by oral administration of egg-white lysozyme. Cancer Chemother Pharmacol 25:221–222.

Sava G (1996). Pharmacological aspects and therapeutic applications of lysozymes. EXS 75:433–449.

Sava G, Benetti A, Ceschia V, Pacor S (1989). Lysozyme and cancer: Role of exogenous lysozyme as anticancer agent. Anticancer Res 9:583–591.

Sava G, Ceschia V, Pacor S, Zabucchi G (1991). Observations on the antimetastatic action of lysozyme in mice bearing Lewis lung carcinoma. Anticancer Res 11:1109–1113.

Sava G, Pacor S, Dasic G, Bergamo A (1995). Lysozyme stimulates lymphocyte response to ConA and IL-2 and potentiates 5-fluorouracil action on advanced carcinomas. Anticancer Res 15:1883–1888.

Saxena I, Tayyab S (1997). Protein proteinase inhibitors from avian egg whites. Cell Mol Life Sci 53:13–23.

Schäfer A, Drewes W, Schwägele F (1998). Analysis of vitelline membrane proteins of fresh and stored eggs via HPLC. Z. Lebens Untersforsch A 206:329–333.

Schimmoller F, Higaki JN, Cordell B (2002). Amyloid forming proteases: Therapeutic targets for Alzheimer's disease. Curr Pharm Des 28:2521–2531.

Schlehuber S, Skerra A (2005). Anticalins as an alternative to antibody technology. Exp Opin Biol Ther 5:1453–1462.

Scott MJ, Huckaby CS, Kato I, Kohr WJ, Laskowski M Jr, Tsai MJ, O'Malley BW (1987). Ovoinhibitor introns specify functional domains as in the related and linked ovomucoid gene. J Biol Chem 262:5899–5907.

Scruggs P, Filipeanu CM, Yang J, Chang JK, Dun NJ (2004). Interaction of ovokinin(2–7) with vascular bradykinin 2 receptors. Regul Peptides 120:85–91.

Shah RB, Khan MA (2004). Protection of salmon calcitonin breakdown with serine proteases by various ovomucoid species for oral drug delivery. J Pharm Sci 93:392–406.

Sharon N, Ofek I (2002). Fighting infectious diseases with inhibitors of microbial adhesion to host tissues. Crit Rev Food Sci Nutr 42:267–272.

Shcherbakova EG, Bukhman VM, Isakova EB, Bodiagin DA, Arkhipova NA, Rastunova GA, Vorob'eva LS, Lipatov NN (2002). Effect of lysozyme on the growth of murine lymphoma and antineoplastic activity of cyclophosphamide. Antibiot Khimioter 47:3–8.

Smith SR, Pala I, Benore-Parsons M (2006). Riboflavin binding protein contains a type II copper binding site. J Inorg Biochem 100:1730–1733.

Stadelman WJ (1999). The incredibly functional egg. Poultry Sci 78:807–811.

Stubbs MT, Laber B, Bode W, Huber R, Jerala R, Lenarcic B, Turk V (1990). The refined 2.4A° X-ray crystal structure of recombinant human stefin B in complex with the cysteine proteinase papain: A novel type of proteinase inhibitor intearaction. EMBO J 9:1939–1947.

Sugahara T, Murakami F, Yamada Y, Sasaki T (2000). The mode of actions of lysozyme as an immunoglobulin production stimulating factor. Biochim Biophys Acta 1475:27–34.

Sugino H, Nitoda T, Juneja LR (1997). General chemical composition of hen eggs. In: Yamamoto T, Juneja L, Hatta H, Kim M, eds. New York: CRC Press; pp 13–24.

Tan F, Gippner C, Fink E (1988). On the genomic organization of the human pancreatic secretory trypsin inhibitor. Biol Chem Hoppe-Seyler 365:51–54.

Tanizaki H, Tanaka H, Iwata H, Kato A (1997). Activation of macrophages by sulfated glycopeptides in ovomucin, yolk membrane, and chalazae in chicken eggs. Biosci Biotechnol Biochem 61:1883–1889.

Tenuovo J (2002). Clinical applications of antimicrobial host proteins lactoperoxidase, lysozyme and lactoferrin in xerostomia: Efficacy and safety. Oral Dis 8:23–29.

Terato K, Ye XJ, Miyahara H, Cremer MA, Griffiths MM (1996). Induction by chronic autoimmune arthritis in DBA/1 mice by oral administration of type II collagen and *Escherichia coli* lipopolysaccharide. Br J Rheumatol 35:828–838.

Tezuka H, Yoshikawa M (1995). Abstract Annual Mtg Jpn Soc Bioscience, Biotechnology, and Agrochemistry, Tokyo; p 163.

Tomimatsu Y, Clary JJ, Bartulovich JJ (1966). Physical characterization of ovoinhibitor, a trypsin and chymotrypsin inhibitor from chicken egg white. Arch Biochem Biophys 115:536–544.

Tranter HS, Board RG (1984). The influence of incubation temperature and pH on the antimicrobial properties of hen egg albumen. J Appl Bacteriol 56:53–61.

Tsuge Y, Shimoyamada M, Watanabe K (1996a). Binding of egg white proteins to viruses. Biosci Biotechnol Biochem 60:1503–1504.

Tsuge Y, Shimoyamada M, Watanabe K (1996b). Differences in hemagglutination inhibition activity against bovine rotavirus and hen newcastle disease virus based on the subunits in hen egg white ovomucin. Biosci Biotechnol Biochem 60:1505–1506.

Tsuge Y, Shimoyamada M, Watanabe K (1997a). Structural features of newcastle disease virus- and anti-ovomucin antibody-binding glycopeptides from pronase-treated ovomucin. J Agric Food Chem 45:2393–2398.

Tsuge Y, Shimoyamada M, Watanabe K (1997b). Binding of ovomucin to newcastle disease virus and anti-ovomucin antibodies and its heat stability based on binding abilities. J Agric Food Chem 45:4629–4634.

Valenti P, Antonini G, Von Hundstein C, Visca P, Orsi N, Antonini E (1983). Studies of the antimicrobial activity of ovotransferrin. Int J Tissue React 5:97–105.

Valenti P, Visca P, Antonini G, Orsi N (1985). Antifungal activity of ovotransferrin towards genus *Candida*. Mycopathologia 89:169–175.

Valueva TA, Matveev NL, Mosolov VV, Penin VA (1992). Protein proteinase inhibitor therapy in experimental pancreatitis: pharmacological characterization of the inhibitor. Agents Actions Suppl 38:203–210.

Verdot L, Lalmanach G, Vercruysse V, Hartmann S, Lucius R, Hoebeke J, Gauthier F, Vray B (1996). Cystatins up-regulate nitric oxide release from interferon-gamma-activated mouse peritoneal macrophages. J Biol Chem 271:28077–28081.

Verdot L, Lalmanach G, Vercruysse V, Hoebeke J, Gauthier F, Vray B (1999). Chicken cystatin stimulates nitric oxide release from interferon-gamma-activated mouse peritoneal macrophages via cytokine synthesis. Eur J Biochem 266:1111–1117.

Vered M, Gertler A, Berstein Y (1981). Inhibition of porcine elastase II by chicken ovoinhibitor. Int J Peptide Protein Res 18:169–179.

Vidovic D, Graddis T, Chen F, Slagle P, Diegel M, Stepan L, Laus R (2002). Antitumor vaccination with HER-2-derived recombinant antigens. Int J Cancer 102:660–664.

Vis EH, Plinck AF, Alink GM, van Boekel MAJS (1998). Antimutagenicity of heat-denatured ovalbumin, before and after digestion, as compared to caseinate, BSA, and soy protein. J Agric Food Chem 46:3713–3718.

Wahn V (2003). Primary immunodeficiencies caused by defects of cytokines and cytokine receptors. In: Korholz D, Kiess W, eds. Methods in Molecular Biology: Cytokines and Colony Stimulating Factors: Methods and Protocols. New Jersey: Humana Press; p 215.

Wang SZ, Alder RA (1994). A developmentally regulated basic-leucine zipper-like gene and its expression in embryonic retina and lens. Proc Natl Acad Sci USA 91:1351–1355.

Watanabe K, Tsuge Y, Shimoyamada M, Ogama N, Ebina T (1998a). Antitumor effects of pronase-treated fragments, glycopeptides, from ovomucin in hen egg white in a double grafted tumor system. J Agric Food Chem 46:3033–3038.

Watanabe K, Tsuge Y, Shimoyamada M (1998b). Binding activities of pronase treated fragments from egg white ovomucin with anti-ovomucin antibodies and newcastle disease virus. J Agric Food Chem 46:4501–4506.

Wilchek EA, Bayer M (1990). Introduction to avidin-biotintechnology. Meth Enzymol 184:5–13.

Wu J, Akaike T, Hayashida K, Okamoto T, Okuyama A, Maeda H (2001). Enhanced vascular permeability in solid tumor involving peroxynitrite and matrix metalloproteinases. Jpn J Cancer Res 92:439–451.

Xie H, Huff GR, Huff WE, Balog JM, Rath NC (2002). Effects of ovotransferrin on chicken macrophages and heterophil-granulocytes. Devel Compar Immunol 26:805–815.

Yamada Y (2002). Design of a highly potent anti-hypertensive peptide based on ovokinin(2–7). Biosci Biotechnol Biochem 66:1213–1217.

Yan RT, Wang SZ (1998). Identification and characterization of *tenp*, a gene transiently expressed before overt cell differentiation during neurogenesis. J Neurobiol 34:319–328.

Yao Z, Zhang M, Sakahara H, Saga T, Arano Y, Konishi J (1998). Avidin targeting of intraperitoneal tumor xenografts. J Natl Cancer Inst 90:25–29.

Yolken RH, Willoughby R, Wee SB, Miskuff R, Vonderfecht S (1987). Sialic acid glycoproteins inhibit *in vitro* and *in vivo* replication of rotaviruses. J Clin Invest 79:148–154.

Yoshikawa M, Fujita H (1994). Studies on the optimum conditions to utilize biologically active peptides derived from food proteins. In: Yano T, Matsuno R, Nakamura K, eds. Developments in Food Engineering. New York: Blackie Academic and Professional; pp 1053–1055.

Younvanich SS, Britt BM (2006). The stability curve of hen egg white lysozyme. Protein Peptide Lett 13:769–772.

Yuan W, Zhao GF, Dong XY, Sun Y (2006). High-capacity purification of hen egg-white proteins by ion-exchange electrochromatography with an oscillatory transverse electric field. J Sep Sci 29:2383–2389.

5

BIOACTIVE COMPONENTS IN EGG YOLK

HAJIME HATTA
Department of Food and Nutrition, Kyoto Women's University, Kyoto, Japan
MAHENDRA P. KAPOOR AND LEKH RAJ JUNEJA
Taiyo Kagaku Co. Ltd., Yokkaichi, Mie, Japan

5.1. INTRODUCTION

Egg and milk have both been considered as excellent sources of nutrition; however, their biological roles are decisively different. When milk is warmed up for 21 days, it turns sour, whereas the egg hatches and produces a chick. Milk is the primarily food for newborn mammals and contains the necessary nutrient compositions for their growth. In addition to being highly nutritious, the egg contains a complete set of biological substances, which allow it to hatch 21 days after fertilization. The egg can be considered as a primary capsule of life, in which all ingredients are incubated in the precise amounts necessary to sustain development of an embryo and give rise to birth of a chick. Thus, the egg contains a number of materials that are necessary to generate new life, particularly entire cell components including energy supply substances, which are necessary for development and growth of an embryo. Therefore, an egg-laying hen transfers these substances (lipids, proteins, etc.) into the egg white and egg yolk on a daily basis.

Egg white consists of a solution of proteins, containing more than 40 kinds of proteins. In addition to 10% proteins, it also contains nearly 0.5% carbohydrate and 0.15% sodium and potassium ions. (Burley and Vadehra 1989; Li-Chan et al. 1995). The main biological role of egg white is to protect egg yolk within the shell of the egg, wherein both thick and thin whites surround the yolk, safely maintaining it in the center of the egg. It is also well known

Egg Bioscience and Biotechnology Edited by Yoshinori Mine

that egg white contains *lysozyme*, a *lytic enzyme* and antibacterial protein, which helps protect the ovum against bacteria that could potentially invade it from the eggshell surface. Interestingly, all protease inhibitors in egg white, proteins for binding vitamins, and metal ions are arranged naturally in such a way as to prevent the egg from being decomposed by any microorganisms. Nearly 60% of the egg proteins are included in egg white, and its nutritional value can be considered as the principal source of amino acid for the chick.

Egg yolk consists of nearly 48% water, 34% lipids, and 17% proteins as illustrated in Figure 5.1. The yolk lipids exist mainly as lipoproteins in combination with other proteins (Parkinson 1966; Sugino et al. 1997). Physiologically active substances needed for cell growth along with other nutrients are available in sufficient amounts in egg yolk for hatching and growth of a chick.

Egg yolk lipids also exist in sufficient quantities, particularly as duplexes of protein, while egg yolk lipids are the main active ingredient in the hatching process. Egg yolk lipids are usually responsible for building the cell membrane, the atomic structure that is necessary to sustain life, and provide the main source of energy for development of the cerebral component, which is an aggregate of nerve cells. Egg yolk is also considered as a food because it contains an abundant amount of cholesterol, which, in turn, is stored in sufficient quantities in the egg as an indispensable ingredient in the chick birthing process.

Among the various physiological functions of egg yolk components (proteins, lipids, carbohydrates, etc.), the most widely considered are their antibac-

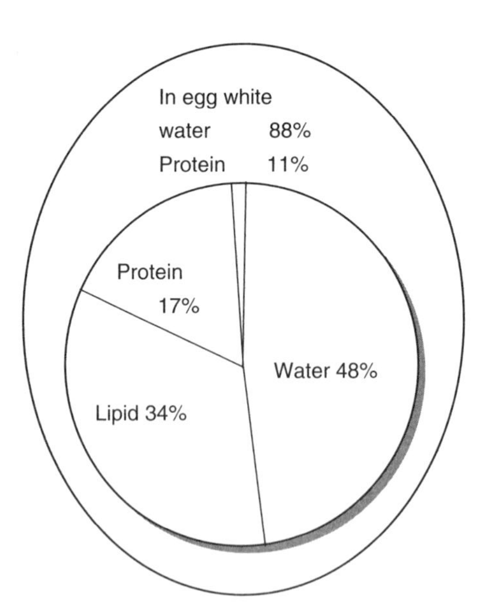

Figure 5.1. Composition of an egg.

terial, antiviral, anticancer, antioxidant, anti-inflammatory, and immunological enhancement activities, and will be discussed in this chapter. The bioactive components that have been used in the most recent practical applications are summarized in this chapter. We have focused in particular on the use of egg yolk lipids in human infant formula, sialyloligosaccharides as the basic, initial ingredients in anti-inflammation or antineoplastic medicine, and egg yolk antibody (IgY) for the prevention of infectious diseases.

5.2. EGG YOLK PROTEINS AND THEIR BIOLOGICAL ACTIVITIES

Egg yolk consists of nearly 48% water, 32.0–35.0% lipids, 15.7–16.6% proteins, 0.2–1.0% carbohydrates, and 1.1% ash content (Sugino et al. 1997). Usually, once egg yolk protein is synthesized as a precursor protein by the hen's liver, it is secreted in the bloodstream and subsequently transferred into the oocyte by the ovary. The various types of egg yolk proteins and their characteristics are listed in Table 5.1.

TABLE 5.1. Properties of Egg Yolk Proteins

Name of Protein	(%)	Localization	Molecular Weight	Characteristics
Low-density lipoprotein (LDL)	65	Plasma and granules	10,300 kDa (LDL_1) 3,300 kDa (LDL_2)	Lipid content is ~90%, Apovitellenins I–VI are known as a *lipoproteins*
Lipovitellin; high-density lipoprotein (HDL)	16	Granules	400 kDa (α, β-lipovitellin, complex)	Lipid content is ~25%; molecular weights of apoprotein are 125, 80, 40, 30 kDa
Livetin	10	Plasma	80 kDa α-livetin 40, 42 kDa β-livetin 180 kDa γ-livetin	Serum albumin Fragment of *C*-terminus in vitellogenin IgY (hen's serum IgG)
Phosvitin	4	Granules	33, 45 kDa	Most phosphorylated protein in nature
Egg yolk riboflavin-binding protein	0.4	Plasma	36 kDa	Similar to flavoprotein in egg white and to serum riboflavin binding protein
Others	4.6	Mainly plasma		Binding proteins against either biotin, thiamine, vitamin B_{12}, retinol, egg yolk transferrin, or other compounds

Egg yolk can be divided into supernatant (plasma) and precipitate (granules) using ultracentrifugation in 4:1 proportion, respectively (Sugino et al. 1997). The major components of plasma are low-density lipoproteins and livetin, which normally occupies nearly 62% and 10% of overall egg yolk solid contents, respectively. The main components of the yolk granules are high-density lipoproteins (usually known as α- and β-lipovitellins) and phosvitin, including low-density lipoprotein, which occupy nearly 16%, 4%, and 3% of overall egg yolk solid fractions, respectively.

5.2.1. Yolk Plasma Proteins

5.2.1.1. Low-Density Lipoprotein (LDL)

The LDL of yolk is generally subdivided into the super-low-density lipoprotein [very low-density lipoprotein (VLDL)] and is equivalent to a serum VLDL because of its extremely high lipid content (85–89%) and low specific gravity (0.98) (Cook and Martin 1969). In most egg yolk lipoproteins, nearly 75% are the neutral lipids and the remainder are the phospholipids. Phosphatidylcholine (71–76%), phosphatidylethanolamine (16–20%), and sphingomyelin (8%) are the main phospholipids included in egg yolk. Yolk LDL can be further divided into two components differentiated according to buoyancy (i.e., density) and size, defined as LDL-1, with a molecular weight (MW) of 10 $\times$ 10^6 Da; and LDL-2, with MW 3 $\times$ 10^6 Da (Martin et al. 1964). In particular, yolk LDLs are generated by serum VLDL, which is usually absorbed in the egg yolk via the LDL receptors on the vitellin membrane in cases of rapid growth of the oocyte prior to ovulation (Fig. 5.2) (Yamamura et al. 1995).

Serum VLDL, which is normally a precursor to yolk LDL, usually consists of apoprotein II (apolipoprotein II; apoVLDL) and the apoprotein B (apolipoprotein B; apoB) as a major apoprotein. A gene expression of apoVLDL is controlled by estrogen and is a component peculiar to egg yolk LDL. Although apoB in yolk LDL is a common apoprotein that is similar to serum VLDL, when the apoB in serum VLDL passes through the vitellin membrane into the egg yolk, it receives limited digestion to be an apoB of yolk LDL (Evans and Burley 1987).

A yolk LDL apoprotein consists of apovitellenin I–IV fractions. Apovitellenin I (MW 9.4 kDa), a main apoprotein, is identical to a serum apoVLDL, where, in the case of a chicken, a dimer is formed by an S—S bond. Apovitellenin II (MW 20 kDa) is a glycoprotein, which tends to dissolve in salt solution; however, it does not seem to be a component indispensable to egg yolk LDL (Inglis et al. 1982). Apovitellenins III and IV (MW 65 and 170 kDa, respectively) are water-insoluble, and in the delipidated state (wherein lipids are completely removed) cannot easily be rendered soluble by using a denaturant such as urea. Apovitellenins III and IV both originate from apoB (Evans and Burley 1987).

As for the physiological function of egg yolk LDL, the production efficiency of an IgM antibody is increased in a human–human hybridoma cell. The

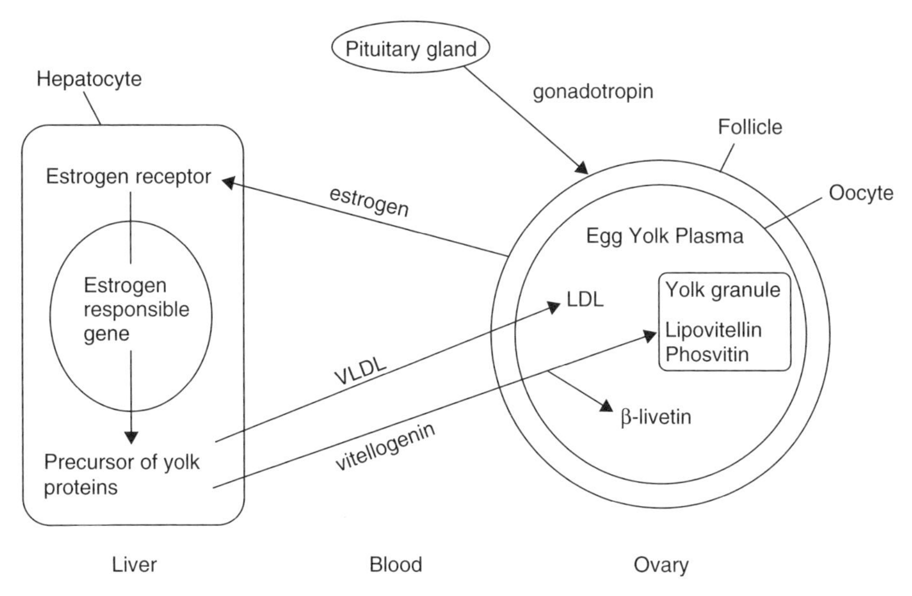

Figure 5.2. Synthesis of yolk proteins and its transfer into the egg yolk. [Reproduced from Matsuda T (1998), in Tamago no Kagaku, Nakamura R, ed., Asakura Shoten, p. 33.]

mechanism that promotes human monocyte formation and multiplication of a leukocyte cell has also been identified.

5.2.1.2. Livetins

Livetins are major components of water-soluble egg yolk protein. There are three types: α-livetin (MW 80 kDa), β-livetin (MW 40–42 kDa), and γ-livetin (MW 180 kDa). Livetins can also be transferred from blood serum and are identified as serum albumin (α-livetin) and immunoglobulin G (IgG) (γ-livetin), respcctively (Williams 1962). In particular, γ-livetins are denoted as IgY because they are structurally different from the IgG of a mammal (MW 150 kDa) as well as in nature. More recently, the applicability of IgY in preventing infectious disease has received much attention as the specific antibody obtained from egg-based foods. β-Livetin is a glycoprotein possessing a higher cysteine content, and can be classified into two types in terms of molecular weight: 40 kDa and 42 kDa. These liveins are usually transferred by enzyme degradation of serum vitellogenin, which occurs during passage through the vitellin membrane. Serum vitellogenin is a known precursor, whereas the lipovitellin and the phosvitin are the granule proteins (Yamamura et al. 1995) (Fig. 5.2).

5.2.1.3. Egg Yolk Riboflavin-Binding Protein

Yolk riboflavin-binding protein combines glycoprotein (MW 36 kDa) and riboflavin in 1:1 ratio and is phosphorylated at several serine residues ranging from Ser 185 to 147. Riboflavin-binding protein (RBP) is present in both egg yolk and egg white; however, the egg yolk RBP differs from the egg white RBP by a deletion of 11–13 amino acids at the C terminus due to limited proteolytic cleavage during oocyte uptake (Vasudevan et al. 2001). In order to differentiate them, the egg white RBP was termed *ovoflavoprotein*. Both RBP are glycosylated at sites Asn36 and Asn147, but differ in their carbohydrate composition (Amoresanno et al. 1999).

5.2.1.4. *Other Proteins*

The biotin-binding proteins combine a biotin molecule in protein (MW 68 kDa) that consists of four 17-kDa subunits. Such proteins are similar to an avidin of egg white; however, because of the other proteins present, the biotin-binding protein in serum is shifted to egg yolk (Niskanen et al. 2005). The thiamine-binding protein is the simple protein of MW 38 kDa that combines a thiamine molecule ($K_d = 0.41\,\mu M$), which acts on a mutual specificity with flavoprotein (Muniyappa and Adiga 1981). Similar binding proteins also exist in egg white as well as in serum. The combination ($K_d = 0.40\,\mu$M) of vitamin B_{12} molecule with serum glycoprotein (MW 37 kDa) creates a vitamin B_{12}-binding protein that is identical to the vitamin B_{12}-binding protein of egg white (Levine and Doscherholmen 1983). Retinol-binding protein (MW 21 kDa) also seems identical to the retinol in the serum-binding protein (Vieira et al. 1995). Also, the egg yolk *transferrin* protein is identical to that of egg white (*ovotransferrin*) or serum but with different glucose organization. Proteins with cholinesterase activity (MW 100, 440 kDa) also exist in egg yolk.

5.2.2. Proteins of Yolk Granules

5.2.2.1. *Lipovitellin*

Lipovitellin, which is a globular lipoprotein, is also called *high-density lipoprotein* (HDL), which can be further subdivided into two components: α-lipovitellin and β-lipovitellin. Although the phosphorus and sugar content of their amino acid compositions are different, both lipovitellins include zinc. The abundance ratio of α:β is 2:1 (Kurisaki et al. 1964).

In both lipovitellins, the protein content is nearly 75%, while lipid consists mainly of phospholipids (15–17%) and triglycerides (7–8%), and usually exist as dimers of MW 400 kDa. Lipovitellins dissociate easily under ionic conditions depending on the pH of the solution. Also, lipovitellins have structural features similar to those of lipoprotein but somewhat different from those of serum lipoproteins. They form a complex when binding with a phosvitin in an egg yolk granule; however, the resulting complex can be easily dissociated by changing the ionic strength. Similarly, an apoprotein of α-lipovitellin consists of four subunits of different molecular weights (e.g., 125, 80, 40, 30 kDa), whereas the two subunits exist in the case of β-lipovitellin (e.g., 125, 30 kDa) (Groche et al. 2000).

5.2.2.2. Phosvitin

Phosvitin is a glycoprotein containing a higher phosphorus content, (nearly 10%) and is considered one of the most highly phosphorylated natural (i.e., nonsynthetic) proteins (Ito and Fujii 1962). The special feature in its amino acid composition is that 30–50% of amino acids are serine, wherein most serine residues are phosphorylated to form phosphoserines. Phosvitin also combines with many bivalent metal ions; for instance, 95% of the Fe ions in an egg are present in the egg yolk, which helps the phosvitin combine with other species via its strong binding ability, and this eventually inhibits the reverse distribution of Fe ions in the living body. The strong binding ability of Fe with phosvitin also exhibits an antibacterial effect in Fe-deficient bacteria (Sattar Khan et al. 2000; Choi et al. 2004) and imparts an antioxidation function to the phospholipids (Lu and Baker 1986). Phosvitin and its enzymatic digests provide protection against iron-catalyzed hydroxyl radical formation and protect DNA against oxidative damage induced by Fe(II) and peroxide (Ishikawa et al. 2004). Therefore, phosvitin may be useful for the prevention of iron-mediated oxidative stress-related diseases, such as colorectal cancer (Ishikawa et al. 2004).

Several molecular species of phosvitin of different molecular weights are reported in the literature on their phosphoric acid and sugar contents. There are five known types of phosvitin: B, C, E1, E2, and F with molecular weights of 40, 33, 18, 15, and 13 kDa, respectively; phosvitins B and C are extensively studied major components (Wallance and Morgan 1986). Phosvitin is usually derived from serum vitellogenin (types I, II and III), whereas phosvitins C and F are derived from vitellogenin I. While phosvitin B is derived from vitellogenin II, Wallance and Morgan (1986) reported that phosvitins E1 and E2 could be formed from vitellogenin III. Phosphorylation of phosvitin occured earlier, prior to the secretion of vitellogenin from the liver. A calcium absorption facilitatory effect of phosvitin-origin phosphopeptide was also described. Phosvitin phosphopeptides derived via tryptic hydrolysis followed by partial alkaline dephosphorylation have been reported to enhance calcium-binding capacity and inhibit insoluble calcium phosphate formation (Jiang and Mine 2000, 2001).

5.2.2.3. LDL in Granules

Although LDL occurs in a limited quantity, it also exists in granules. Structurally, it seems to be identical to those of plasma LDLs derived from the protein and similar in constitution to the lipids.

5.3. YOLK LIPIDS

The lipid content of the egg is mainly present in the egg yolk. The white consists of 88% water, 11% protein, and a small amount of carbohydrate and mineral. On the other hand, the weight proportions for egg yolk are about one-half water (48%), one-third lipids (34%), and one-sixth proteins (17%), as shown in Figure 5.1 (Li-Chan et al. 1995). All portions of the egg yolk lipid exist as a lipoprotein combined with a protein. The nutritients provided by egg yolk are sufficient to facilitate the birth and growth of a chick, and the lipids also exist in duplex quantities of the protein. The egg yolk lipid facilitates the hatching process of the egg and provides important ingredients for the cell membrane, which is a basic life-sustaining atomic structure and also the main energy source for the chick. Yolk lipids in precise quantities, are also crucial in development of the chick's cerebral component, which is an aggregate of nerve cells.

In this section, we briefly introduce the components of egg yolk lipid and their characteristics and potential health benefits. The latest information on a cholesterol study is also discussed, revealing that egg cholesterol is indeed not a risk factor for heart disease.

5.3.1. Components of Egg Yolk Lipid

In general, the lipid content total egg yolk weight is nearly 30%; thus the 20 g of yolk in the average egg contains approximately 6 g of lipids. Neutral lipid (65%), the phospholipids (30%), and cholesterol (4%) are the major components of egg yolk lipids (Figs. 5.3 and 5.4) (Li-Chan et al. 1995). The role of neutral lipid in the hatching process is mainly as an energy-supplying source, while the phospholipids and cholesterol are important ingredients promoting the formation of somatic cell structures and the cell membrane of cranial nerve cells (phospholipid bilayer) of the chick. Thus, the most important function of the egg yolk lipid is to provide a high content of phospholipids and cholesterol as the constituent materials in the cell membrane (Juneja 1997).

Egg yolk phospholipids, also known collectively as *yolk lecithin*, consist of 84% phosphatidylcholine (PC), 12% phosphatidylethanolamine (PE), 2% sphingomyelin, and 2% lysophosphatidylcholine (LPC) and other minor components. A systematic comparison of soybean-derived phosphatides (soy leci-

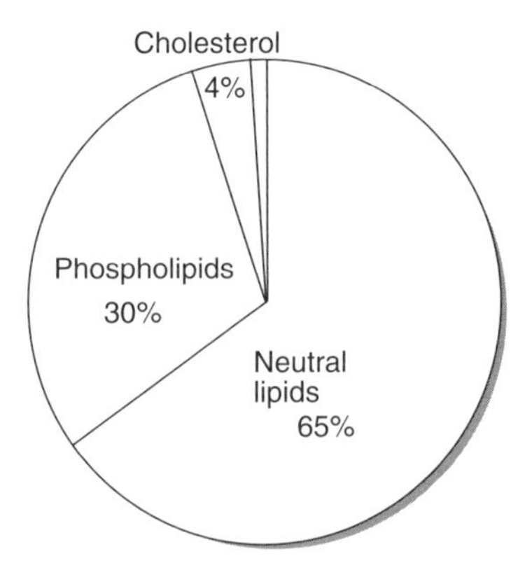

Figure 5.3. Composition of egg yolk lipids.

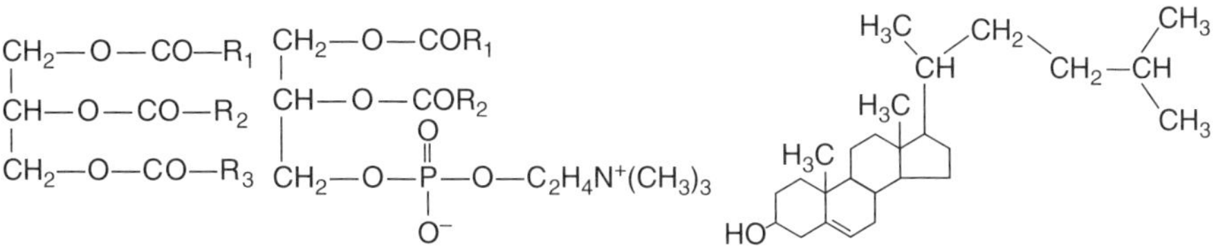

$CH_2—O—CO—R_1$
$CH—O—CO—R_2$
$CH_2—O—CO—R_3$

Neutral lipids

$CH_2—O—COR_1$
$CH—O—COR_2$
$CH_2—O—P(=O)(O^-)—O—C_2H_4N^+(CH_3)_3$

Phospholipids:
Phosphatidylcholine (PC)

Cholesterol

Figure 5.4. Basic structure of egg yolk lipids.

TABLE 5.2. Comparison of Phospholipids

Name of Phospholopid	Abbreviation	Egg Yolk (%)	Soybean (%)
Phosphatidylcholine	PC	84.30	33.00
Phosphatidylethanolamine	PE	11.90	14.10
Phosphatidylinositol	PI		16.80
Phosphatidic acid	PA		6.40
Sphingomyelin	SM	1.90	
Lysophosphatidylcholine	LPC	1.90	0.90
Others			28.80

thin) is presented in Table 5.2 (Weiner et al. 1989). Because of their high phosphatidylcholine content, egg yolk lipids are exceptionally promising for applications in biomedical as well as cosmetics industries.

Typically, in a 100-g egg yolk fraction, lipid constitutes ~30 g, which consists of ~25.4 g of fatty acids. The typical fatty acid composition of egg yolk lipid is described in Table 5.3. The major fatty acids are oleic acid (OA; 43.6%), palmitic acid (PA; 25.1%), linoleic acid (LA; 13.4%), stearic acid (SA; 8.6%), palmitoleic acid (PcA; 3.6%), docosahexaenoic acid (DHA; 1.8%), and arachidonic acid (AA; 1.7%). In addition, either α-linolenic acid (α-LA) or eicosapentaenoic acid (EPA) exists in fatty acid (Tesedo et al. 2006).

Regarding the composition of egg yolk fatty acid, the most of the oleic acid content is reported to reduce serum cholesterol values. Also, while the DHA and AA found in sufficient quantities in egg yolk lipids are indispensable for development of the human newborn's brain and retina, the mother's milk is an additional source of such nutrients (Makrides et al. 2002).

5.3.2. Qualitative Value of the Yolk Lipid

Fatty acids can be classified in terms of qualitative values. The total volume of egg yolk lipids (25.4 g), containing saturated fatty acid (8.7 g; 34 vol%), monounsaturated fatty acid (12.2 g; 48 vol%), and polyunsaturated fatty acid (4.5 g; 18 vol%) have been estimated. A qualitative estimation of these values in egg yolk lipid revealed a polyunsaturated fatty acid : saturated fatty acid (P/S) ratio

TABLE 5.3. Fatty Acid Composition and Structure of Eggyolk Lipids

Fatty Acid	C Chain	Clasification	*n* Series	Composition
Oleic acid	C18:1	Monounsaturated	*n*-9	$CH_3(CH_2)7CH{=}CH(CH_2)7COOH$
Palmitic acid	C16:0	Saturated	—	$CH_3(CH_2)14COOH$
Linoleic acid	C18:2	Polyunsaturated	*n*-6	$CH_3(CH_2)4CH{=}CHCH_2{-}CH{=}CH(CH_2)7COOH$
Stearic acid	C18:0	Saturated	—	$CH_3(CH_2)16COOH$
Palmitoleic acid	C16:1	Monounsaturated	*n*-9	$CH_3(CH_2)7CH{=}CH(CH_2)5COOH$
Docosahexaenoic acid	C22:6	Polyunsaturated	*n*-3	$CH_3CH_2CH{=}CH{-}CH{=}CHCH_2CH{=}CHCH_2CH{=}CHCH_2CH{=}CHCH_2CH{=}CH{-}(CH_2)3COOH$
Arachidonic acid	C20:4	Polyunsaturated	*n*-6	$CH_3(CH_2)4CH{=}CHCH_2CH{=}CHCH_2CH{=}CHCH_2CH{=}CH{-}(CH_2)3COOH$
α-Linolenic acid	C18:3	Polyunsaturated	*n*-3	$CH_3CH_2CH{=}CHCH_2CH{=}CHCH_2CH{=}CH(CH_2)7COOH$
Eicosapentaenoic acid	C20:5	Polyunsaturated	*n*-3	$CH_3CH_2CH{=}CHCH_2CH{=}CHCH_2CH{=}CHCH_2CH{=}CHCH_2CH{=}CH{-}(CH_2)3COOH$

of 0.5. This implies that the content of saturated fatty acid (which raises the serum cholesterol level) more than doubled that of polyunsaturated fatty acid (which lowers total serum cholesterol). The ratio P/S = 1 is considered as the standard value needed to prevent a rise in serum cholesterol. In a comparative study of the fat content of the Japanese diet, the P/S value has somewhat fallen (P/S < 1.0) since the mid-1980s because of westernization of dietary habits, in some countries the P/S value is nearly 0.5, which is claimed to be a cause of heart-related diseases.

Regarding the ratio of *n*-6 fatty acids (linoleic acid and arachidonic acid) and *n*-3 fatty acids (α-linolenic acid, DHA, and EPA), the *n*-6/*n*-3 ratio in yolk lipids is approximately 7. Individuals unable to produce *n*-6 and *n*-3 fatty acids must obtain these substances from other sources. Currently, the Japanese *n*-6/*n*-3 ratio is nearly 4 and has dramatically changed since the mid-1980s. However, in Europe and North America the *n*-6/*n*-3 ratio has already exceeded 10 because of the lack of seafood in their diet. Because of the relatively low *n*-6/*n*-3 ratio in Japan, the average Japanese life expectancy is longer, with a lower incidence of heart-related diseases. Generally, according to the established dietary guideline, the *n*-6/*n*-3 ratio should be under 4 (Sugano and Hirahara 2000).

5.3.3. Health-Promoting Effects of Egg Yolk Lipids

Other characteristic features of egg yolk lipids are their emulsification and moisture-retaining properties. Therefore, egg yolk lipids can be employed as emulsifiers in food, cosmetics, and other moisturizing applications. Since egg yolk lipids play a decisive biological role in the chick hatching process their potential role in promoting building blocks and restoration of human somatic cells and brain nerve cells is highly anticipated.

In addition, because egg oil has been used as a health food for several decades and although its mechanisms are not yet clear, it has been used as a traditional dietary supplement, which has been very popular among consumers in Japan. In one (non-Japanese) study, volunteering senior citizens each received 10 g of egg yolk lipid preparation (active lipid; AL721) daily for 3 consecutive weeks, and the results revealed a conspicuous rise in immunoreactivity of leukocytes. It was suggested that absorption of AL721 could restore or enhance the fluidity of cell membranes aquiring lower fluidity on aging (Rabinowich et al. 1987).

Further, polyunsaturated fatty acid, particularly the DHA of *n*-3 and an AA of *n*-6 found in egg yolk lipids, is being considered as a lipid-rich source for infant formula milk (modified milk powder). Both DHA and AA fatty acids are necessary for the development of brain cells as well as the ocular retina in newborn infants, and the proportions should be well balanced (AA = 0.28–0.60%; DHA = 0.22–1.0%) in the mother's milk (Makrides et al. 2002). A finding released by American and Singaporean researchers claimed that breastmilk reduces the risk of myopia (nearsightedness) developing later in

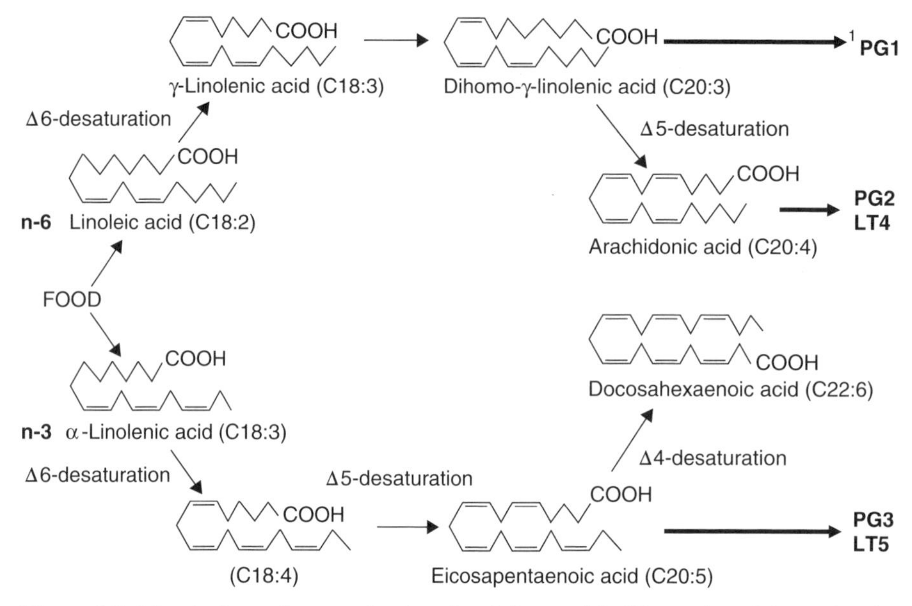

Figure 5.5. Metabolic pathways of *n*-6 and *n*-3 fatty acids. [Reproduced from Juneja LR (1997). Egg yolk lipids. In: Yamamoto T, Juneja LR, Hatta H, Kim M, eds. Hen Eggs, Their Basic and Applied Science. New York: CRC Press; p. 89.]

the child's life. It was revealed that DHA in the mother's milk contributes to development of a photoreceptor cell of the infant's retina (Chong et al. 2005).

The content of formula milk has consistently improved, more closely approaching the mother's milk, which contains the correctly balanced protein and nutrient composition such as carbohydrates and fat, which is extremely important. Since soybean oil (linoleic acid, 53%; an α-linolenic acid, 7%) is used as a fat source in infant formula, the AA/DHA fatty acid composition is insufficient. Generally, ingested linoleic acid and α-linolenic acid are enzymatically converted into AA, EPA, and a DHA (Fig. 5.5) but this conversion is insufficient for infants because of a lack of desaturating enzyme activity. Thus, it is important to combine a DHA and AA in infant formula milk, to simulate the mother's milk as closely as possible. Therefore, egg yolk lipids can be regarded as a new fat source for infant formula milk (Makrides et al. 2002).

5.3.4. Egg Yolk–Derived Phospholipids and Their Uses

Phospholipids are natural biosurfactants as well as important material for many applications in the food, cosmetic, and pharmaceutical industries (Prosise 1985; Ostro 1987; Baker 1988, 1989). They also are becoming important, often crucial, components in numerous medicinal, biotechnological, and nanotech applications. Phospholipids have been widely used in the manufacture of pharmaceutical emulsions and liposome applications. Choline is definitely a nutrient necessary in ample supply for good health and is a key component of many fat-containing structures in cell membranes, whose flexibility and integrity depend on adequate supplies of choline. Two fat-like molecules in the brain, *phosphatidylcholine* and *sphingomyelin*, account for an unusually high percentage of the brain's total mass, so choline is particularly important for brain function and health. Choline is also a key component of *acetylcholine*. A neurotransmitter that carries messages from and to nerves, acetylcholine is the body's primary chemical means of sending messages between nerves and muscles. Usually, a large hen egg provides nearly 300 μg of choline, and also contains about 300 mg of *phosphatidylcholine*. Although most sources report free choline content as only 300 μg, it is the phosphatidylcholine that is the most common form in which choline is incorporated into cell membrane phospholipids. Nakane et al. (2001) have shown that hen egg yolk contain a high amount of lysophosphatidic acids (acyl form) in addition to a small amount of lysoplasmanic acid (alkyl form). Thus, the phospholipids constitute nearly 25% of the total egg yolk, wherein the major component is phosphatidylcholine (around 80% of the total phospholipids). It also revealed that egg yolk consists of more than threefold the amount of phospholipids compared to natural soy lecithin (Table 5.2).

It is widely recognized that mammals derive a major fraction of choline from the dietary sources. In contrast very minor amount (<1%) of choline is usually present in the diets in the form of free -base. Hirsch and Wurtman (2002) have shown that consumption of a single meal containing lecithin, the major source of choline present naturally in the diet, increases the concentration of choline and acetylcholine in rat brain and adrenal glands. Eventually, the consumption of phosphatidylcholine increases plasma and brain choline levels and promotes neuronal acetylcholine synthesis. Hence, it was noted that the concentration of acetylcholine in the tissue might normally be under direct, short-term nutritional control. Consumption for 3–11 days of a diet supplemented with choline sequentially increases the concentration of serum choline, brain choline, and brain acetylcholine synthesis in rats (Hirsch and Wurtman 2002). Such precursor-induced changes in brain and adrenomedullary acetylcholine concentration are probably associated with parallel alterations in neurotransmitter release (Zeisel 1981).

Currently, it is recommended to reduce the fat in daily diets. Western countries are reducing the fat in their diets at an alarming rate, which leads to a low intake of phospholipids and choline. This cannot be described as a healthy trend, because choline is an essential nutrient for everyone (Zeisel et al. 1991). It has been shown that even a few weeks on a choline-deficient diet can cause abnormal hepatic (liver) function in adults. The role of phosphatidylcholine is to protect against fibrosis in alcohol-fed baboons (Lieber et al. 1994). Also, the high content of phosphatidylcholine in the egg yolk is useful for the infant's brain function. Choline is routinely added to commercial infant formula as an essential nutrient. In other words, an inadequate choline intake might lower folic acid level's in women. Thus, excess choline can endanger a current or future pregnancy, and phosphatidylcholine deficiency in animals has been associated with infertility, growth retardation, and bone abnormalities (Zeisel 1981; Best and Huntsman 1932; Michael et al. 1975; Griffith and Wade 1939; Patterson and McHenry 1944; Chang and Jensen 1975; Jukes 1940; Pawelczyk and Lowenstein 1993). Choline has been reported to enhance learning in young rats when administered to mother rats before birth and to young rats after birth (Zeisel 1981).

The presence of choline has also proved critical in intercellular communication, a virtual process in which cells work together. Defects in cell communication can cause hazardous health effects such as cancer and other tumors. Jenike et al. (1986) have reported that disruption of the cholinergic system is a major factor in producing the cognitive impairment that occurs in patients with Alzheimer's disease, which is the most common form of dementia. Davies and coworkers proposed an administration of acetylcholine precursors as one treatment approach and suggested that treatment with egg yolk lipids, a useful source of choline as precursors of acetylcholine, would appreciably alleviate the symptoms of Alzheimer's disease (Davies and Maloney 1976).

The conventional methods used for the extraction of phospholipids from egg yolk are modifications of the original Pangborn method (Pangborn 1951). The preferred solvent is acetone for the egg yolk extraction, which allows facile and clear separation of a complex mixture of natural lipids (Lungberg 1973; Yano et al. 1979). However, the modern technology frequently reported in the more recent literature for phospholipids extraction favors treatment of lipid mixtures with supercritical gases. In addition to the fundamental drawbacks of postsynthesis treatment, the process is not very economical because of the high equipment and infrastructure cost. However, Juneja et al. (1994) have established a large-scale preparation method of chromatographically homogeneous and high-purity phospholipids synthesis in which the ion exchange column was employed.

In addition to egg yolk, phospholipid molecules are present in many other types of cells, where they constitute the major components of membrane structure. Vegetable oils, such as corn, soybean, cottonseed, linseed, peanut, safflower, sunflower oils and rapeseed, are commonly used as lecithin sources. Among these, soybean oil is a main source of commercial lecithin (Scholfield 1985; Van Nieuwenhuyzen 1976). Several phospholipids, which are not found in egg yolk, can be prepared by enzymatic transphosphatidylation of phosphatidylcholine in presence of respective acceptors (Fig. 5.6). The transphosphatidylation reaction mentioned above could be performed (with 100% selectivity) for phosphatidylethanolamine, phosphatidylglycerol, or phosphatidylserine preparation from phosphatidylcholine in the biphasic reaction mixture in the presence of phospholipase D extracted from streptomyces (Juneja et al. 1989).

As we have mentioned earlier, egg yolk is a natural surfactant (surface-active agent) and has diverse applications in the cosmetic, pharmaceutical, food, and related industries. Phosphatidylcholine-rich phospholipids are currently in high demand, and are also known as *modified lecithin*. Pure phospholipids or their lyso- derivatives alter the hydrophilic–lipophilic balance as well

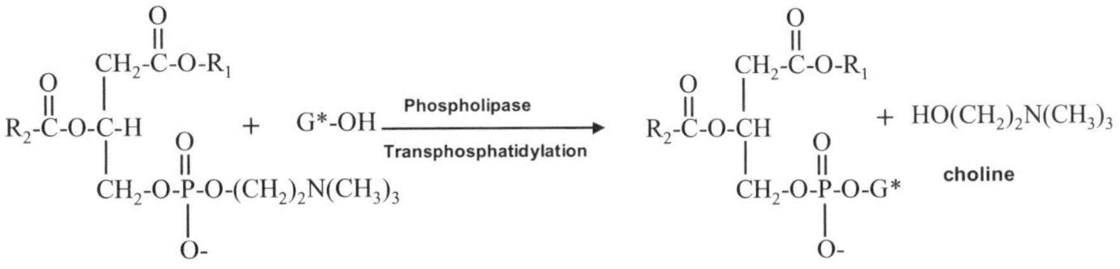

$$\begin{array}{l} CH_2\text{-}C(=O)\text{-}O\text{-}R_1 \\ R_2\text{-}C(=O)\text{-}O\text{-}C\text{-}H \\ CH_2\text{-}O\text{-}P(=O)(O^-)\text{-}O\text{-}(CH_2)_2N(CH_3)_3 \end{array} + G^*\text{-}OH \xrightarrow[\text{Transphosphatidylation}]{\text{Phospholipase}} \begin{array}{l} CH_2\text{-}C(=O)\text{-}O\text{-}R_1 \\ R_2\text{-}C(=O)\text{-}O\text{-}CH \\ CH_2\text{-}O\text{-}P(=O)(O^-)\text{-}O\text{-}G^* \end{array} + HO(CH_2)_2N(CH_3)_3$$

Phosphatidylcholine

choline

Figure 5.6. Enzymatic transformation of polar headgroup of phospholipids, where, G* is glycerol, ethanolamine or serine, and so on for the respective final products.

as enhance oil–water (O/W)-type emulsification properties compared to commercial lecithin (Prosise 1985). Generally, the phospholipids are used in confectioneries, bakery products, noodles, snack foods, fast foods, margarines, dairy products, ice cream, milk, milk drinks and other products, and meat and poultry processing. In the abovementioned applications, the phospholipids are usually applied as emulsifying, wetting, dispersing, or releasing agents. Phospholipids can also be used as sealants and lubricants. Apart from these applications, phospholipids are used in soaps, detergents, dyes, fertilizers, leather, paints, papers, and textiles, which are the increasingly common non-food-related applications of phospholipids.

Some phospholipids are attached to unique fatty acids like arachidonic acid (AA) or docosahexaenoic acid (DHA), which are now regarded as essential in the diet for brain function and visual acuity in humans. It was shown that content of these acids increases according to the of phospholipids content.

Antioxidant properties of phospholipids have been well demonstrated in fish oil, vegetable oil, and animal oil. Lecithin has usually been considered as an antioxidative synergist that helps improve or prolong the initial action of primary antioxidants such as tocopherols. The unique ability of phospholipids to enhance the antioxidative effect of the abovementioned primary oxidants appears to vary with the type of oil being stabilized (Dziedzic and Hudson 1984). It was shown that phospholipids or their components regenerate primary antioxidants via hydrogen radicals or proton donation pathways. The antioxidative properties of egg yolk lipids were widely studied in docosahexaenoic acid oil, and it was revealed that the order of antioxidative activity increases with increase in phospholipid content. Curiously, the antioxidative activity of egg yolk phospholipids was superior compared to that of soy phospholipids. This also confirmed that antioxidative ability is dependent on phospholipid concentration. However, the antioxidative activity of phospholipids was found to decrease with hydrogenation (Dziedzic 1986).

Egg yolk phospholipids are also highly desirable raw materials for the liposome and liposome-based drug synthesis in pharmaceutical industries. Furthermore, partial and complete hydrogenation of egg yolk phospholipids provides additional opportunities for their diverse uses. Egg yolk phospholipids have also frequently been used in the manufacture of pharmaceutical emulsions (Baker 1988). It has been established that liposomes of required specification can be prepared on an industrial scale for different purposes by mixing the phospholipids in their appropriate ratios. Liposomes of phospholipids have been used as carriers and sustained-release vehicles for enzymes, vaccines, and drugs, with particular emphasis on antitumor drugs, cell-modifying molecules, hormones, and so on to be administered orally or intravenously (Tyrrell et al. 1976). Liposomes of phospholipids are also shown to deliver medicine to affected tissues in a novel way wherein instead of being diluted in the blood, drug entrapped in the vesicles would reach the target site in concentrated doses (Ostro 1987). The molecular shapes of phospholipids are also important for liposome modeling. Hydrated phosphatidylcholine, with its approximate cylindrical shape, forms lamellar structures, and lysophosphatidylcholine forms spherical micelles when dispersed in water. It was also shown that lysophosphatidylcholine acts as a natural modulator of biological membranes because of its inverted-cone shape. Furthermore, hydrogenated phosphatidylcholine liposomes are turbid, but liposomes contain nearly 30% molar ratios of lysophosphatidylcholine, which are transparent because of their small (~50 nm) size. Encapsulation efficiency can also be controlled without affecting the transparency properties.

Finally, egg yolk phospholipids are also found in applications in the cosmetics industry. Phospholipids are preferable because of their "skin feel," skin absorption, and dispersing, moisturizing, spreading, wetting, emulsifying, and several other useful properties. Overall, because of their multifunctioanlity, egg yolk phospholipids are considered an important material in cosmetics, as they reduce the oily and greasy feeling on the skin and provide an elegant feel by replenishing the lipid deficiency in skin cells and providing a texture that closely structurally resembles that provided by the skin's own lipids. In addition, egg yolk phospholipids also increase the skin absorptivity of the active ingredients of cosmetic formulations, and owing to improved film adhesion, the resulting products have reduced transfer to clothing (Sagarin 1957). Currently, egg yolk phospholipids are the potential base materials for cleansing creams, moisturizing creams, hair products, aftershave lotions, bath lotions, body lotions, shampoos, rinses, skin creams, soaps and lotions, makeup, lipsticks, sun protective creams and related products, gel and cream for the eyes, and several other products used every day. Apart from the abovementioned uses of egg yolk phospholipids, the hydrated form of phospholipids has been applied in liposome synthesis, which is used in many other cosmetic applications to increase the value of cosmetic products by encapsulating and stabilizing a variety of ingredients such as moisturizers, skincare agents, vitamins, coenzymes, tanning agents, and sunscreen agents.

5.4. SIALIC ACID

Sialic acid, which has a simple chemical structure, is a component of a number of the more complex chemical structures in the human body. Sialic acid functions as a carbohydrate molecule and is known to exist in all kinds of animal and human cells (Warren 1959; Sillanaukee et al. 1999; Huttunen 1966). It is found in several body fluids, including cerebrospinal fluid, saliva, urine, serum, amniotic fluid, mother's milk, and certain microbes; it occurs most abundantly in glycoproteins and glycolipids. In mammals, it is found in high levels in the brain, adrenal glands, and the heart. In humans, high concentrations are found in the brain and kidney as well as many other tissues. Typically, sialic acid is an unusual nine-carbon monosaccharide ubiquitously displayed on the surfaces of mammalian cells; in humans the more common form of this sugar is *N*-acetylneuraminic acid. An acidic aminosugar was first isolated and named *sialic acid*, while other independent research groups isolated a similar crystallized form and called it *neuraminic acid*. Later, the correct structure was proposed and it was agreed to use sialic acid as the generic name for more than 30 derivatives of neuraminic acid, with *N*-acetylneuraminic acid and *N*-glycolylneuraminic acid forming the core structures. One major type of sialic acid is *N*-acetylneuraminic acid (abbreviated Neu5Ac), which is the biosynthetic precursor for most of the other types (Varki 1997; Varki et al. 1999; Kelm and Schauer 1997). The addition of a single oxygen atom to this sialic acid gives rise to a very common variation, *N*-glycolylneuraminic acid (Neu5Gc). This irreversible conversion is catalyzed by a specific hydroxylase that converts CMP-Neu5Ac to CMP-Neu5Gc (Kozutsumi et al. 1990; Muchmore et al. 1989; Kawano et al. 1995; Shaw and Schauer 1988). The most common form of sialic acid found in humans is *N*-acetylneuraminic acid. In most other animals (except chickens) Neu5Ac peacefully coexists with *N*-glycolylneuraminic acid (Neu5Gc), a slightly modified form of sialic acid bearing a hydroxyl group at the *N*-acyl position (Fig. 5.7).

These sugar nucleotides are the high-energy donors necessary for the addition of sialic acid to glycoconjugates. Thus, the sialic acids are a family of acidic sugars found in all vertebrates and are frequently located at the outer end of glycoconjugates on cell surfaces or on secreted glycoconjugates (Varki 1997; Varki et al. 1999; Kelm and Schauer 1997; Schauer 1982a, 1982b). A disturbance in a gene responsible for sialic acid metabolism may lead to an abnormality reflected in sialic acid concentration in blood, urine, and solid tissue.

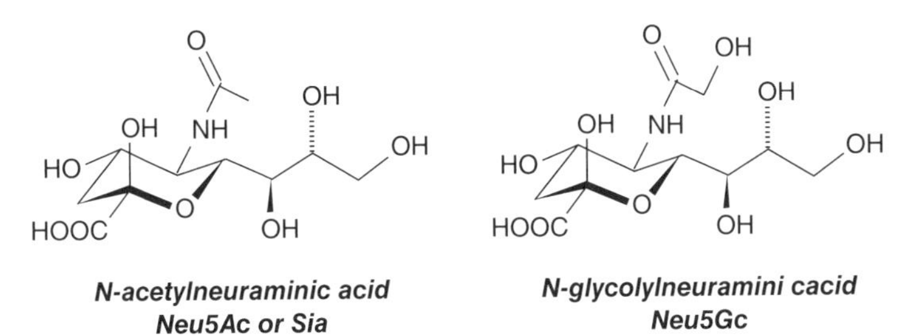

Figure 5.7. Illustration showing the structures of *N*-acetylneuraminic acid (Neu5Ac) and *N*-glycolylneuraminic acid (Neu5Gc).

The abnormalities of sialic acid metabolism almost invariably lead to both physical and mental deterioration. Sialic acid can be analyzed in cells cultured from solid tissue such as skin, or can be measured in blood or urine. Abnormalities of sialic acid metabolism may be suspected in infants who fail to grow, regress in development, or have hepatosplenomegaly coarsening of facial features, or failure of pigmentation of hair and skin. Therefore, sialic acid is important in all of our lives. For example, binding sugar is one of the determinants for influenza virus, and an interesting hypothesis is that the lack of Neu5Gc expression due to inactivation of an enzymatic step in human sialic acid pathway may provide a molecular explanation for the differences between humans and other primates (and the rest of the animal kingdom) as far as their development of an increased level of intelligence is concerned (Chou et al. 2002).

5.4.1. Sialic Acid from Hen Eggs

Hen eggs are recommended dietary sources of sialic acids. Hen eggs contain sialic acid in such amounts that they are thought to be a potential source for the industrial-scale isolation of sialic acid (Koketsu et al. 1992). Juneja et al. (1991) have disclosed the large-scale preparation of sialic acid from chalazas and egg yolk membrane. The sugar chains of the glycoproteins play several important physiological and physiochemical roles in the development of cells of newborn chicks and in eggs (Sharon 1975). Thus, sialic acids in hen eggs are small aminosaccharides consisting of a neuraminic acid backbone with one or multiple *O*- or *N*-linked sidechains. Most hen egg proteins are glycol derivatives wherein the *N*-linked type is formed through the β-amino group of the asparagine residue and the *O*-linked type is formed through the hydroxyl group of serine or threonine residue. Hence the free sugars are major components in hen eggs and their concentrations in whole egg, albumen, and yolk are 0.7%, 0.8%, and 0.7%, respectively.

More than 30 types of sialic acid derivatives have thus far been isolated from numerous sources (Schauer 1982a, 1982b), and approximately 14 varieties have been found in the bovine submandibular gland. Interestingly, the only sialic acid found in hen eggs is *N*-acetylneuraminic acid (Koketsu et al. 1992; Li et al. 1978), suggesting that hen eggs are a potential source of sialic acid and can be isolated in very pure form. Juneja et al. (1991) have reported that sialic acid is quantitatively present in several fractions of hen eggs, where the egg yolk constitutes the highest percentage among the various structural fractions. The content of sialic acid was estimated to be approximate 0.95 g/kg of fresh hen egg yolk.

Hen egg white proteins have been extensively utilized as ingredients in food processing because of their unique functional properties (e.g. gelling and foaming). Because egg white proteins possess multiple functional properties (foaming, emulsification, heat setting, binding adhesion, etc.), these proteins are desirable ingredients in many foods such as bakery products, meringues, meat products, and cookies. Investigations into the physicochemical characteristics of these proteins have aided the elucidation of their structure–function relationships for the benefit of food technologists. Several methods have achieved efficient isolation of sialic acid from hen egg yolk. A simple (as yet unannounced) method is via ethanol extraction of dry egg yolk powder followed by filtration and reslurry in acidified water (pH 1.4 with 3 *M* sulfuric acid) and subsequent heat treatment at 80°C, followed by neutralization with saturated barium hydroxide (pH 6.0), filtration, and finally electrodialysis. The method details are described elsewhere showing 97% purity of sialic acid, which can be further purified to 100% pure *N*-acetylneuraminic acid (Koketsu et al. 1992).

5.4.2. Hen Egg Glycoproteins and Glycolipids

Ovomucoid, ovalbumin, ovotransferrin, ovomucin, phosvitin, egg yolk immunoglobulin, and yolk riboflavin-binding protein are the major glycoproteins in hen eggs. Sialic acid is a glycoconjugate (consisting mainly of glycoproteins and glycolipids), existing in many biological materials, in which it is usually bound to the carbohydrate chains of glycoproteins and glycolipids. Also, the sialic acid residues are located in the terminal ends of carbohydrate chains of many glycoproteins (e.g., immunoglobulins and peptide hormones), and when the terminal sialic acid is removed by neuraminidase of the vascular endothelium, a galactose molecule is revealed. Specific receptors on hepatocytes recognize these asialoglycoproteins and remove them from the circulation (Stryer 1988).

Oligosaccharides with a terminal sugar sialic acid have been shown to have unique functionality. Ovomucoid constitutes about 11% of the total egg white protein (Lin and Feeney 1972), having four major binding sites with sugar chains (Kato et al. 1987; Beeley 1976), while ovalbumin makes up about 54% of the total white protein and is known as a major phosphoglycoprotein of egg white with 3.2% sugars, which form an *N* -linked complex with mannose-rich structure (Nisbet et al. 1981; McReynolds et al. 1978). ovotransferrin constitutes about 12% of total egg white protein, while ovomucin makes up about 3.5%. The oligosaccharides of ovomucin are linked to serine and threonine through their hydroxyl groups, which gives egg white its thick-fluid consistency. Phosvitin and egg yolk immunoglobulin are phosphoglycoproteins, which make up approximately 11% and 5% of total egg yolk protein, respectively. The major structure of the oligosaccharides of phosvitin is triantennary, and two of its sugar chains are terminated with sialic acid residues (Brockbank and Vogel 1990). Egg yolk immunoglobulin (IgY), formerly known as *gamma-livetin*, has four major sites to bind oligosaccharides, while yolk riboflavin-binding protein has a more complex structure, with two sites bound with sugar chains, in which all the oligosaccharides have sialic acid at their nonreducing ends. The structural details of all types of glycoproteins are comprehensively described elsewhere (Koketsu 1997).

Egg yolks are also rich in glycolipids. Li et al. (1978) have isolated galactosylceramide, *N*-acetylneuraminosylgalactosylceramide (GM4), *N*-acetylneuraminosyllactosyl ceramide (GM3), and di-4*N*-acetylneuraminosyllactosylceramide (GD3) from egg yolks in reasonable yields and studied their structures.

5.4.3. Sialyloligosaccharides

In hen egg yolks the majority of carbohydrates (70%) are oligosaccharides, mainly mannose and glucosamine, bound to protein, while remaining 30%, the free carbohydrates are present in the form of glucose (Sugino et al. 1997; Koketsu et al. 1996). Several sialyloligosaccharides in egg yolk have also been isolated and their structures determined (Holmgren et al. 1980). The major sialyloligosaccharides isolated from egg yolk were of the *N*-acetyllactosamine type, and their structures are in Figure 5.8.

Sialylglycoconjugates, such as gangliosides, sailyloligosaccharides, and sailylglycoproteins, have been reported to play various important roles in animal and human tissue cells. The terminal glycosylation sequences, in particular sialyloligosaccharides, are responsible for a variety of complex biological events such as cell adhesion, viral infections, and toxin neutralization. For example, they act as receptors of viruses such as Sendai virus (Holmgren et al. 1980), influenza virus (von Itzstein et al. 1993), and corona virus (Schultze and Herrler 1994). Other studies also revealed that sialylglycoconjugates on the cell surface are known to serve as receptors for microorganisms, viruses, and toxins (Fishman and Brady 1976; Paulson et al. 1984; Morschhäuser et al. 1990; Smith et al. 1984). Sialyloligosaccharides such as sialyl-LeX and sialyl-LeA have been characterized as carbohydrate ligands of the inflammatory response (Phillips et al. 1990), and cancer metastasis (Kannagi et al. 1988). Sialyloligosaccharides have also been applied to rotaviral infections. Rotavirus is a well-known major pathogen of infectious gastroenteritis, diarrhea, and vomiting in infants (Tabassum et al. 1994). The clinical studies of vaccines for

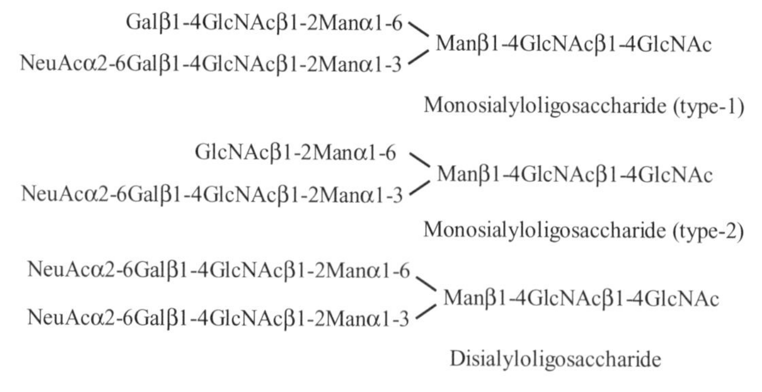

Figure 5.8. Chemical formula of *N*-acetyllactosamine-type sialyloligosaccharides.

the prevention of rotavirus infections have been unsuccessful because of the difficulty in inducing a specific antibody in the intestinal tract of infants who have not yet developed general immunity (DeMol et al. 1986). Therefore, the prevention of rotaviral infections has been the long-term goal for the researchers. Fukudome et al. (1999) have proposed a method in which a suspension of delipidated egg yolk was incubated with proteinase to release the peptides bearing the oligosaccharides from the glycoproteins. These peptidyloligosaccharides appeared in the supernatant of the suspension, and its concentrates were able to inhibit rotaviral infection *in vitro*. The chromatographic separation of sialyloligosaccharides (which are acidic in nature) and neutral oligosaccharides was achieved using an anion exchanger. The sialyloligosaccharide fraction was found to be positive in inhibition rotaviral replication, whereas the neutral oligosaccharide fraction was totally inert. Also, the peptidyloligosaccharide fraction and sialyloligosaccharides fraction were incubated with neuraminidase while the asialo products, from which sialic acid has been chemically removed, showed no inhibitory effect on rotavirus propagation. Those results led them to conclude that sialic acid derivatives having peptide linked oligosaccharides and sialyloligosaccharides are potential candidates for the rotaviral inhibition.

Koketsu et al. (1995b) have investigated the inhibitory effect of the sialyloligosaccharide fraction on rotavirus *in vivo* using suckling mice, which were infected with rotavirus (SA-11) for 3 h before administration of sialyloligosaccharide. The group administered the sialyloligosaccharide fraction showed a significantly lower incidence of rotaviral diarrhea. The incidence of diarrhea was inspected 1, 3, and 5 days after oral administration of 2.5 mg of sialyloligosaccharides per mouse, and was lowered by 24% and 43% on days 3 and 5, respectively, compared to the control group. This result indicates that egg yolk–derived sailyloligosaccharides are useful in the inhibition of rotavirus and can be orally administrated in infected individuals.

Sialyloligosaccharides are likely to play their most important role as defense mechanisms against diseases caused by pathogenic microorganisms, including pneumonia, diarrhea, gastritis, and ulcers. It has also been reported that the level of sialyloligosaccharides in mother's milk is very high at the time of parturition (Juneja 1999). Sialic acid derivatives (gangliosides) are also known to be involved in brain function and are also important in protecting infants from various diseases. More recently, sialic acid and sialyloligosaccharides have attracted attention from the pharmacological and food chemical industries because of their potential biological functions. The development of various industrial preparations from egg yolk has resulted in the production of various grades of products for use in infant formula, healthfoods, and nutritional supplements (Koketsu et al. 1995b, 1995c, 1996). The incorporation of hen egg sialylglycoconjugates into functional foods or pharmaceuticals has also been investigated (Seko et al. 1997).

Isolation of sialyloligosaccharides from egg yolk has been extensively studied, while in other reports the efficient preparation process and characterization of sialylglycopeptides from protease-treated egg yolk was revealed (Seko et al. 1997; Koketsu et al. 1993). In a very simple commercial process, supernatant of the water extract of delipidated egg yolk was dialyzed using a membrane for isolating the molecular species of molecular weight <1000 Da. In a subsequent step, asparagine-linked oligosaccharides in the concentrate were liberated by hydrazinolysis and *N*-acetylation, and the reducing ends of the resulting oligosaccharides were labeled with *p*-amino benzoic ethyl ester. This resulted in a UV-absorbing compound that could be fractionated by anion exchange and reverse-phase HPLC (Matsuura et al. 1992; Matsurra and Imaoka 1988; Ohta et al. 1990). The process provides reasonable yields of monosialyloligosaccharides (47.7%) and disialyloligosaccharides (50.6%). The structures confirmed by HPLC and NMR analyses suggest that major sialyloligosaccharides isolated from egg yolk belong to the *N*-acetyllactosamine family.

Koketsu et al. (1995a) have also studied the effect of egg yolk–derived sialyloligosaccharides on the learning performance using the maze test. Egg yolk sialyloligosaccharides were administrated to infant rats aged 14–21 achieved days, and the target timeframe and success ratio were achieved after observing the rats until they reached 42–49 days of age, by applying identical measures in the maze test. Interestingly, the target timeframe was significant shorter and the target success ratio was much higher in case of sialyloligosaccharide-doped rats. The results also suggested that sialyloligosaccharide intake during the lactation period improve learning performance and also protect the rats from various infectious diseases. Further details on the impact of sialyloligosaccharides on learning can be found elsewhere (Koketsu et al. 1995a).

In summary, sialyloligosaccharides are likely to play their most prominent role in the protection mechanisms against diseases due to pathogenic microorganisms such as gastritis, pneumonia, ulcers, and diarrhea. Sialic acid and its derivatives are also known to be involved in brain function and are critical in protecting infants from various diseases. In general, sialic acid and sialyloligosaccharides have attracted much attention from pharmacological and food chemical industries because of their useful biological functions. Moreover, the development of a variety of industrial preparations from egg yolk has resulted in increased production of products with varying grades of use in infant formula, healthcare products, nutritional supplements, and food products.

5.5. EGG YOLK ANTIBODY (IgY)

The IgG protein is generally found in hen blood serum is known to accumulate in egg yolk to provide an acquired immunity to offspring. This biological immune transfer of hens was been reported over 100 years ago (Klemperer 1893). An antibody in egg yolk has been identified as IgY. Currently, a numerous number of hens are subjected to systematic immunization processes and are injected with numerous antigens (in the form of vaccination) to protect them from infectious diseases, and to help them lay eggs as scheduled for commercial transactions. Thus, hen eggs are presently considered as a potential source of a large-scale production of antibody (IgY) (Mine and Kovacs-Nolan 2002).

One useful application of IgY is as an immunologic tool in the fields of medical diagnosis as well as in pure research, where its unique binding specificity to certain antigens that have not yet been identified using other methods is currently being studied. Another important use of IgY is for passive immunization therapy wherein its specific binding ability to the antigens (*pathogens, venoms*, etc.) serves to neutralize the pathogenicities of antigens. Passive immunization has been recognized as one of the most valuable of antibody applications, in which pathogen-specific IgY is administered to prevent the development of infectious diseases (Larsson and Carlander 2003; Hatta et al. 1997).

In the following section, a systematic comparison between egg yolks–derived IgY and serum mammalian IgG is summarized, followed by application of IgY as an immunologic tool in the fields of diagnosis as well as pure and applied research. In addition, clinical use of IgY in passive immunization is discussed as a preventive measure against the severe diarrhea caused by rotavirus infection, dental caries (by *S. mutans*), stomach ulcer (by *H. pylori*), fish disease, and other disorders.

5.5.1. Egg Yolk Antibody Versus Serum Antibody

In case of invading bacteria, viruses' and non-self-components of heteroprotein, generally known as *pathogenic antigens* invading the human body, it is necessary to produce the immune protein (i.e., specific antibody) in the blood to protect the living body from these antigens. Typically, an antibody possesses the unique function of eliminating the infectivity and toxicity of antigens by recognizing and binding with the antigen in a specific manner. Such antibody functions have been widely studied in reference to a humoral immunity governed by one of the most important biophylaxis functions possessed by a living body. In general, an antibody is a glycoprotein of the group called *immunoglobulins* (Igs), and exists in living body fluid (blood, saliva, nasal cavity liquid, lactation, milk, etc.) as well as in an egg. Basically, an antibody of the mammals is classified according to its structure and function into five subclasses, abbreviated IgG, IgM, IgA, IgD, and IgE, and the basic structures of which are presented in Figure. 5.9 (Maddison and Reimer 1976).

Nearly 75% of the antibody present in the blood is IgG. An antibody equivalent to IgG, IgM and IgA in mammals (chicken) exists in 5.0, 1.25, and 1.25 mg per 1 mL of blood serum, respectively. These antibodies can also be found in the egg, where IgM and IgA include nearly 0.2 and 0.7 mg per 1 mL of egg white, respectively (Leslie and Martin 1973), while IgG exists only in egg yolk, where its concentration ranges between 5 and 25 mg per 1 mL egg yolk (Fig. 5.10) (Rose et al. 1974; Schade et al. 1991; Li et al. 1997). A receptor specific to IgG translocation is known to exist on the surface of the yolk membrane (Loeken and Roth 1983; Morrison et al. 2002). An antibody in the egg

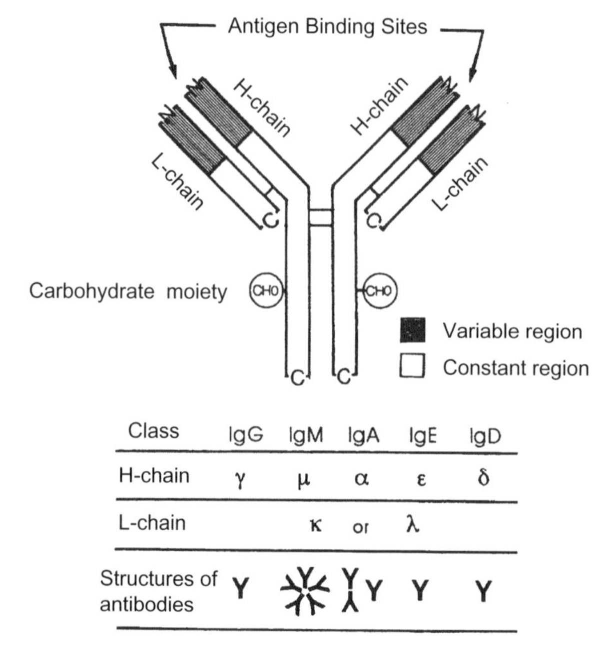

Figure 5.9. Basic structure of antibodies.

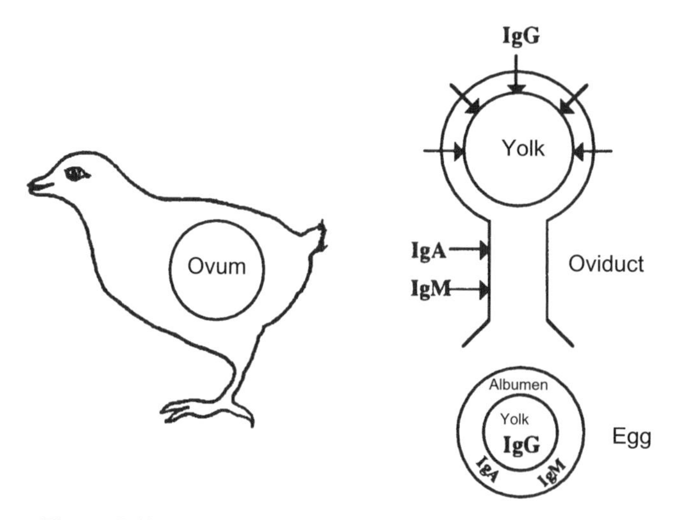

Figure 5.10. Distribution of immunoglobulins in hen egg.

can be defined as the maternal antibody for a parent chicken to transfer an acquired immunity to her progeny. After the egg hatches, the IgG of the egg yolk is transferred into the blood serum of a chick, while IgA and IgM of the egg white goes into the intestinal tract, which plays an important role in pre-

TABLE 5.4. Comparison of a Yolk Antibody (IgY) and a Serum Antibody in Mammals (IgG)

Molecular weight: IgY ~180,000, IgG ~150,000, the H chain is larger; the constant H-chain region in IgY contains 4 domains (IgG, 3 domains)
Isoelectric point: IgY ~6.0, ~1 pH unit lower than IgG
Thermal denaturation temperature: IgY, 73.9°C, rabbit IgG, 77.0°C
Structure of oligosaccharides in IgY differ from those in manmmlian IgG, containing unusual monoglucosylated oligomannose-type oligosaccharides
IgY does not activate complement of mammals
IgY does not combine with proteins A and G (IgG binding proteins)
IgY does not combine with a rheumatic factor (autoantibody to an Fc radical of a IgG).
IgY does not combine with Fc receptor of mammalian a cell

Source: Hatta et al. (1997).

vention of infection until the chick can produce antibody by itself. Regarding the egg of a fowl, that's the main characteristic of the mother→child immunity function, where the female mammals of a viviparous species transfer antibodies to their offspring via the placenta as well as their own milk. Similarly, an acquired immunity of a chicken is handed down to its offspring via the egg (Mine and Kovacs-Nolan 2002).

An egg yolk antibody is equivalent to the IgG class of the mammalian antibody, but is somewhat different from IgG antibodies as defined in protein chemistry and in immunochemistry, which are blood serum antibodies (Table 5.4). Since the egg yolk antibody is not a blood antibody, it is called an *egg yolk antibody* (abbreviated as IgY) in the field of comparative immunology (Leslie and Clem 1969).

5.5.2. Comparison of Specific Antibody Preparations

Antibody-producing cells, called *B lymphocytes* circulate in body fluids and represent a barrier against potential invading pathogens. In event of an antigen invasion into the body, an immunologic mechanism is stimulated, and an antibody having a specific ability to bind to the particular antigen is produced in specific quantities required to neutralize the toxicity of the invading antigen. To utilize such antibody production for medicinal purposes, a selected antigen is inoculated into the animal, producing specific antibody and acquiring specific antigen recognition in the blood. Therefore, it's possible to mass-produce a specific antibody by using animals.

Usually a small mammal such as a rabbit, goat, and guinea pig is used to prepare a specific antibody IgG (polyclonal antibody) from their blood serum, whereas, using the maternal antibody of a chicken, only the specific antibody IgY would be obtained from the egg yolk (Fig. 5.11) (Hatta et al. 1997). Table 5.5 lists the typical preparations of specific antibodies along with significance of each method used. In case of hen immunization method, blood collection is not required and mass rearing is easier. Immunization can also be performed

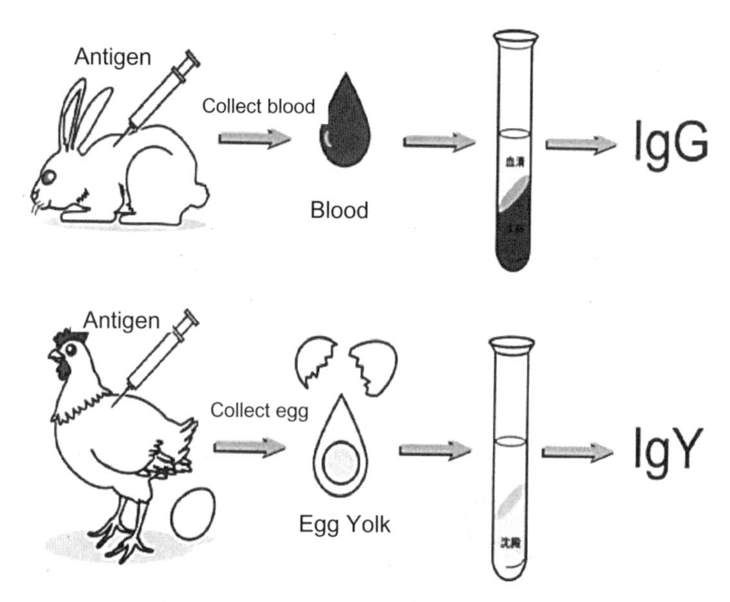

Figure 5.11. Preparation of specific antibody. (See insert for color representation.)

easily using a continuous-pistol-type syringe equipped with a needle; immunization of ~10,000 hens in one day by one worker is possible.

A hen lays approximately 300 eggs per year, and yolk separation from the egg can be automated.

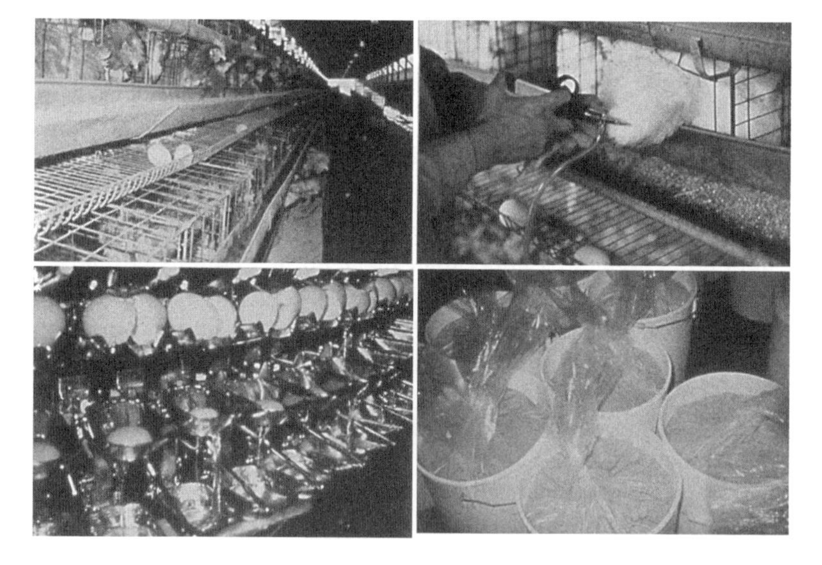

Therefore, it is wise to implement a method for egg immunization that is economically feasible for large-scale production of a specific antibody compared to the conventional animal immunization method (Hatta et al. 1997).

TABLE 5.5. Comparison of Specific Antibody Preparations

Method	Rabbit Immunization	Egg Immunization Method
Extraction source of an antibody	Rabbit blood	Hen egg yolk
A specific antibody preparation	Immunization of rabbit	Immunization of chicken
	Exsanguination	Egg collection and yolk separation
	Serum separation	Water-soluble protein separation
	Purification of IgG	Purification of IgY
Class of antibody	An IgG, in addition to a IgA and a IgM, is included in the serum	Yolk includes only IgY and purification is easy
Animal feeding	Mass rearing is difficult	Large-scale poultry farming is possible
Immunization	Rabbit restrained and injection performed	An immunization is systematized to prevent disease in chickens (vaccination)
Antibody manufacturing scale	Laboratory level	Industrial-scale mass production is possible

TABLE 5.6. Comparison in Productivity of Specific Antibodies

Factor	Rabbit Immunization	Egg-Laying Hen Immunization
Antibody source	Blood serum	Hen egg yolk
Antibody type	Polyclonal IgG	Polyclonal IgY
Antibody protein amount	1400 mg (one rabbit)	~40,000 mg (one hen)
Anti-HRV (MO strain) antibody	6×10^6 neutralizing antibody titer	600×10^6 neutralizing antibody titer
Anti-HRV (Wa strain) antibody	38×10^6 neutralizing antibody titer	520×10^6 neutralizing antibody titer
Anti-mouse IgG antibody	700 mg (50%)[a]	11,200 mg (28%)[a]
Anti-insulin antibody	0 mg (0%)[a]	2000 mg (5%)[a]

[a]Percentage of the specific antibody that accounts for the adhesion and eluted polyclonal antibody for the immunosorbent in which the antigen binds as a ligand.
Source: Hatta et al. (1997).

5.5.3. Comparison in Productivity of Specific Antibodies

A productivity comparison between specific antibodies production using hen immunization and the rabbit immunization method is summarized in Table 5.6. Even considering a repeated immunity injection, an egg-laying hen could bear nearly 250 eggs a year and still yield up to 40 g of purified IgY from the yolks of all of its eggs. In contrast, the exsanguination method produces an antiserum of nearly 40–50 mL per rabbit, where only 1400 mg of purified IgG could be extracted (Hatta et al. 1997).

After purifying both polyclonal antibodies, a titer to the viral antigen was measured and the corresponding antibody amounts were compared. In case of the IgY to a human rotavirus, the Wa and Mo strains were observed, while both of the rabbit IgG values indicated a generalized neutralizing antibody titer 14 and 100 times lower than that for IgY counterparts, respectively (Hatta et al. 1993). Further, the amount of antibody needed against specific protein antigens was estimated using immunoaffinity chromatography, which entailed coupling of the antigen on resin to absorb specific IgY. The binding of the specific IgY antibody to a mouse IgG (antigen) was 16 times more than that of the rabbit IgG antibody, revealing that specific antibody was not significant with the rabbit immunization method, whereas with hen immunization method provided the production of a sufficient quantity of specific antibody. Gottstein and Hemmeler (1985) reported that the specific IgY antibody amount produced per month was 18 times that of the rabbit IgG antibody. Jensenius et al. (1981) reported the production of 500 mL per month of antibody equivalent antiserum using the hen immunization method. Therefore, in terms of antibody production, the hen immunization method seems to be an excellent procedure, compared to the conventional rabbit immunization method.

5.5.4. IgY Purification Method

Egg yolk is a so-called liquid emulsion of proteins and lipids, consisting of 48% water, 34% of lipids, and 17% of protein. Therefore, it would be very tedious to purify IgY efficiently from the yolk, because it is a water-soluble protein where the lipids exist as lipoproteins, which are combined with a protein. For this reason, the yolk water-soluble protein and the yolk lipoprotein (the yolk fat) must be separated for the purification of IgY. A separation method using ultracentrifugation of lipoprotein has been reported (Akita and Nakai 1992; Kim and Nakai 1998), and organic solvent delipidation (Svendsen et al. 1995; McLaren et al. 1994) followed by IgY extraction have been suggested. Lipoprotein-coagulating agents, including poly(ethylene glycol) (Svendsen et al. 1995; Polson et al. 1980; Akita and Nakai 1993), dextran sulfate sodium (Svendsen et al. 1995, Akita and Nakai 1993), and polyacrylic resin (Hamada et al. 1991) were used in the refining process of water-soluble proteins containing IgY. However, large-scale preparation was rather difficult for the conventional production of IgY because of problems related to food safety and the cost constraints.

Earlier it was revealed that alginate sodium (Hatta et al. 1988), a food additive, could coagulate the yolk lipoprotein. Later, a more suitable and effective yolk lipoprotein-coagulating agent such as natural polysaccharide, λ-carrageenan (Hatta et al., 1990), often used as a food-grade thickening stabilizer, was found to possess a strong yolk lipoprotein cohesion property. The water-soluble protein was first separated from egg yolk by anion exchange chromatography followed by a sodium sulfate desalting procedure that yielded IgY >95% purity. Later, higher purity IgY was prepared using the λ-carrageenan method (Hatta et al. 1990). In another report, Hassl and Aspock (1988) compared the various IgY purifying methods and reported poly(ethylene glycol) as a yolk lipoprotein-coagulating agent. In their IgY-refining process nearly 40 mg of IgY per egg could be obtained with 70% purity. It was also possible to prepare the IgY with a high degree of purity (>98%) and a higher yield (70–100 mg) per egg using the carrageenan method. Carrageenan is a food-grade thickening stabilizer generally used for cost-effective synthesis of high-grade IgY to prepare ice cream and other products.

5.5.5. Application of IgY as an Immunologic Tool

Antibody IgG isolated from serum of superimmunized mammals such as rabbits, cows, and goats has been widely applied as an immunologic tool for medical diagnosis as well as pure applied research (Larsson et al. 1993). While antibody IgY is useful because of its ability to bind to specific antigens, either IgG or IgY can be used to detect antigens with a higher specificity than that ever achieved by other methods.

5.5.5.1. Medical Diagnosis

Altschuh et al. (1984) reported that IgY specific to human antibody (IgG and IgM) could be employed to determine the concentrations of these agents in biological fluid using the rocket–immunoelectrophoresis method. In the case of rabbit IgG, carbamylation of the IgG is generally needed to change its isoelectric point from that of the human antibody. However, in the case of IgY, carbamylation was not required since the isoelectric point of IgY is different from that of the human antibody. Fertel et al. (1981) demonstrated the application of IgY in determining prostaglandin levels in serum using radioimmunoassay, in which prostaglandin conjugated with hemocyanin (*hapten*) was used as an antigen for immunization of hens. Gardner and Kaye (1982) prepared IgY specific to rotavirus, adenovirus, and influenza virus, and used the IgY as the primary antibody and FITC-conjugated rabbit IgG specific to IgY as the secondary antibody for detection of these viruses. They suggested that the preparation of IgY specific to these viruses was achieved with much more convenience in comparison to the conventional rabbit IgG method because it was not necessary to purify the virus as antigen. Since these viruses can be cultivated using the fertilized egg, the virus culture must be free of any contaminants from the egg components and must not be immunogenic to hens, as the hens are immunized using the virus culture as antigen. They also suggested that IgY is a suitable antibody for detecting pathogens in stool samples, because it does not bind the protein A derived from *Staphylococcus aureus* usually found in stool. Owing to this property of IgY, false-positive detection of pathogens in stool could be avoided.

The use of IgY in immunologic assays for clinical testing can also eliminate interference and false positives normally experienced when using IgG. Freshly obtained human serum samples often contain an active complement system, which is frequently activated by mammalian antibodies. IgY, which does not activate the human complement system, helps eliminate the interference that would otherwise be caused by IgG (Larsson et al. 1991). Serum samples may also contain rheumatoid factor (RF) and human anti–mouse IgG antibodies (HAMA), which are well-known causes of false-positive reactions in immunologic assays (Carlander et al. 1999a, 1999b). The RRF factor reacts with the Fc portion of IgG, and HAMA, occasionally found occurring naturally in human serum, will bind to any mouse antibodies being used in an immunoassay, producing false-positive results (Carlander et al. 1999a). IgY does not react with RF (Larsson et al. 1991) or HAMA (Larsson and Mellstedt 1992), and their use has been suggested in lieu of IgG for immunologic assays dealing with human serum. The Fc portion of mammalian antibodies may also interact with the Fc receptor, found on many types of blood cells and bacteria. IgY does not interact with Fc receptors, and can therefore be used to avoid interference due to Fc binding (Carlander and Larsson 2001).

Many researchers have also demonstrated the application of IgY for determination of various important but very minor biological substances such as plasma kallikrein (Burger et al. 1990), 1,25-dihydroxyvitamin D (Bauwens et al. 1988), hematoside (NeuGc) (Hirabayashi et al. 1983), and human transferrin (Ntakarutimana et al. 1992). IgY has been applied in many diagnostic and other medical applications, including the diagnosis of gastric cancer (Noack et al. 1999), the detection of breast and ovarian cancer markers (Grebenschikov et al. 1997; Lemamy et al. 1999; Al-Haddad et al. 1999), the detection of African horsesickness virus (Du Plessis et al. 1999), *Campylobacter* fetus diagnosis (Cipolla et al. 2001), and human hemoclassification of blood group antigens (Gutierrez Calzado et al. 2001). IgY has been produced against the human thymidine kinase 1 (TK1) enzyme, and suggested for the early prognosis of cancer and for monitoring patients undergoing cancer treatment (Wu et al. 2003). Other applications of IgY include screening for human papilloma virus for the early detection of cervical cancer (Di Lonardo et al. 2001), and detection of the protein YRL40 as a marker for disease models of arthritis, cancer, atherosclerosis, and liver fibrosis (De Ceuninck et al. 2001).

Another advantage in the use of IgY as an immunological tool over using rabbit IgG is the sensitivity of hens against antigens originating from mammalians. A number of proteins exist whose amino acid sequence are well preserved among mammals, and many of these proteins have no or only minimal antigenicity toward mammals. Evolutionary differences allow the production of IgY against conserved mammalian proteins, resulting in an enhanced immune response not possible in mammals, and can minimize the cross-reactivity normally observed among mammalian IgG (Carlander et al. 1999a).

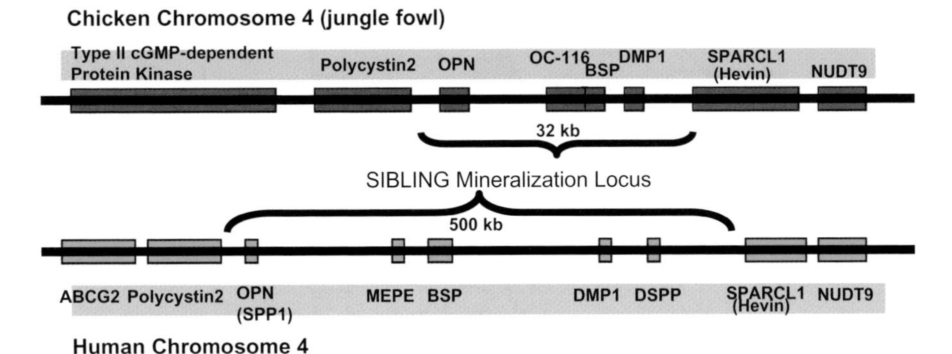

Figure 2.3. Comparison of the SIBLING (*s*mall *i*ntegrin-*b*inding *li*gand, *n*-linked *g*lycoprotein) mineralization loci in chicken and human.

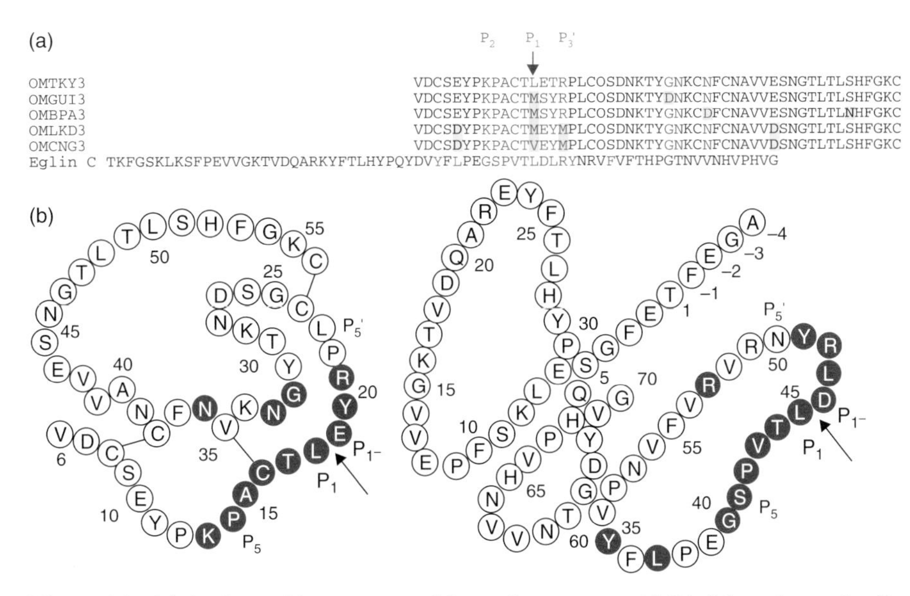

Figure 4.1. (a) Amino acid sequences of five avian ovomucoid third domains and eglin C; (b) covalent structures of turkey ovomucoid third domain, OMTKY3, and eglin C.

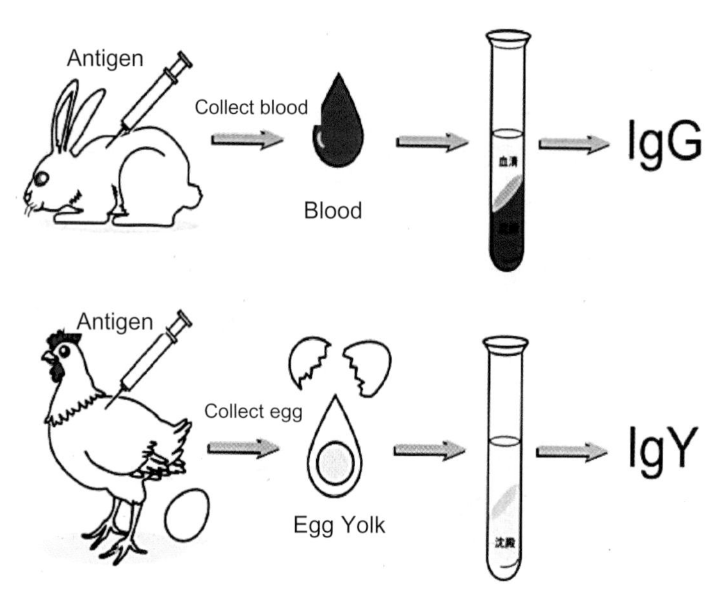

Figure 5.11. Preparation of specific antibody.

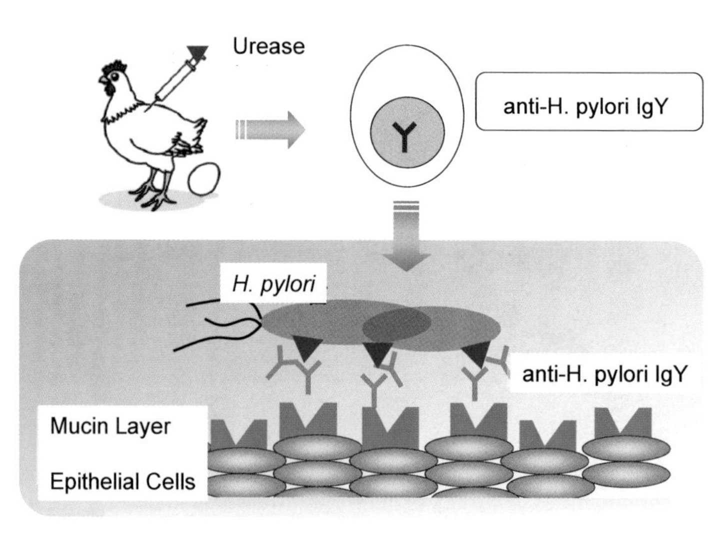

Figure 5.13. Competitive inhibition of *Helicobacter pylori* adherence by anti–*H. pylori* IgY.

BeanStalk
ハキラ

光泉
身体秩序
雙效優酪乳
Piro
Piro

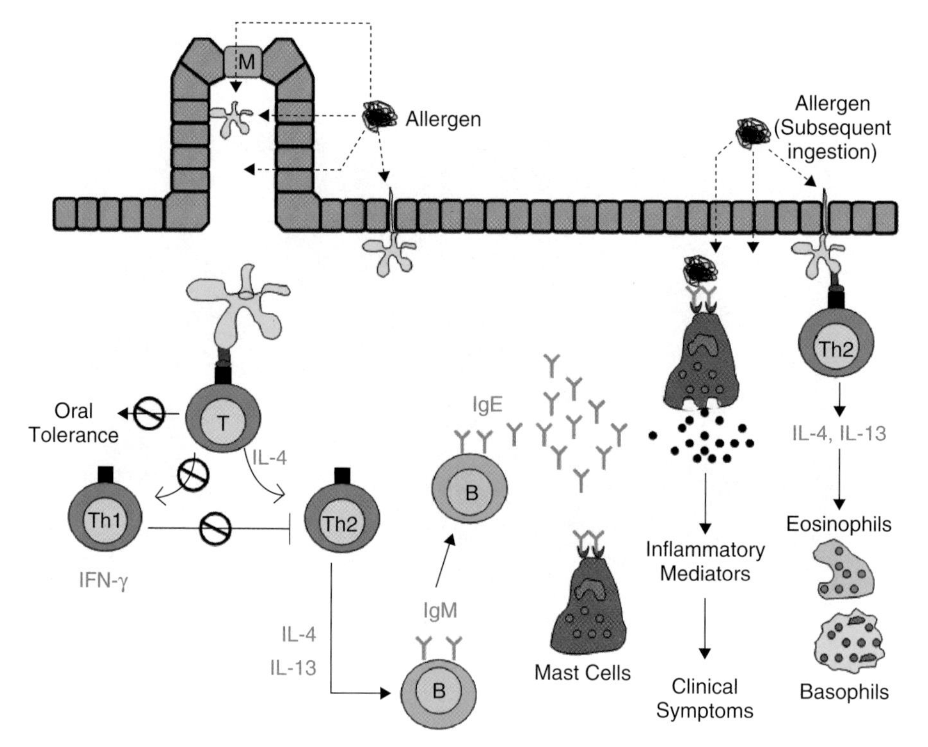

Figure 6.3. Cellular and molecular mechanisms of food allergy.

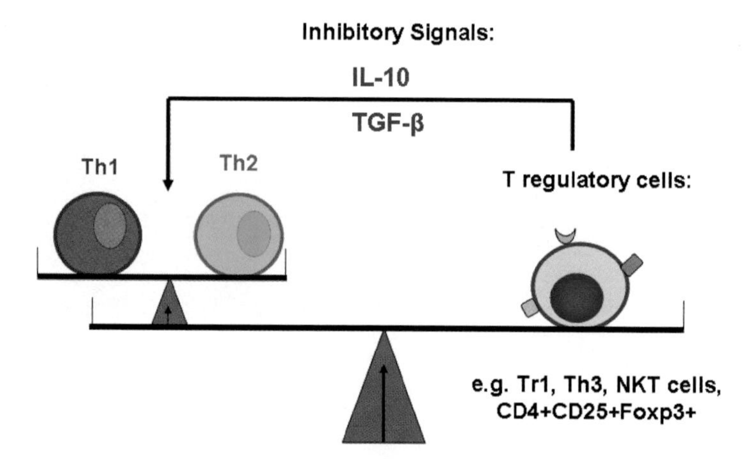

Figure 6.4. Currently accepted view on T-cell regulation.

Therefore, for preparing antibodies against those proteinous antigens, the hen seems highly promising as an alternative animal, because of the wide immunologic distinction between chickens and mammals. In fact, Carroll and Stollar (1983) succeeded in preparing IgY against RNA polymelase 2, which has failed to generate specific antibodies in mammals. Preparation of IgY against glutathion peroxidase (Yoshimura et al. 1991) and human insulin (Lee et al. 1991) was also successful, as it has a significantly lower antigenicity among mammals.

5.5.5.2. IgY as an Immunosorbent Ligand

Immunoaffinity chromatography has been widely applied as a useful tool for purification of proteins (antigens). Rabbit IgG has been traditionally used as a ligand for attachment to an immunoadsorbent such as cellulose or agarose. However, several disadvantages of this affinity chromatography have been pointed out when rabbit IgG was used as a ligand, because acidic solution of pH values ($pH < 2$) is necesary for the dissociation of protein attached to the rabbit IgG on the immunoadsorbent. Therefore, the dissociated protein is often denatured depending on its nature. Moreover, production of rabbit IgG in large amounts is generally highly expensive. Immunoaffinity chromatography has thus been applied for isolation of only certain specific proteins.

Specific IgY, which can easily be produced on a large scale suitable for industrial applications, would provide an ideal replacement for other polyclonal or monoclonal antibodies currently used in immunoaffinity chromatography (Li-Chan 2000). More recently, it has been demonstrated that IgY is an effective alternative antibody as a ligand for an immunoadsorbent. In the experiment, IgY and rabbit IgG specific to mouse IgG were immobilized on Sepharose 4B, respectively, in order to compare its dissociation efficiency for purification of mouse IgG. Mouse serum was applied on immunoadsorbents, and the adsorbent was eluted stepwise with a buffer solution of pH 4.0 and 2.0. The mouse IgG dissociated at pH 4.0 was only a half of that applied, and the remaining IgG was eluted with pH 2.0 buffer solutions in an immunoadsrbent using rabbit IgG as a ligand. However, 97% of the mouse IgG was dissociated even at pH 4.0 on the immunoadsrbent using IgY as a ligand (Mine and Kovacs-Nolan 2002). These results indicate that using IgY as an immunoaffinity ligand may permit the use of less harsh elution conditions. Although IgY is more sensitive to low pH than is IgG, it was found that an IgY immunoaffinity column was capable of retaining stability when subjected to standard affinity chromatography conditions, and could be reused over 50 times without a significant decrease in binding capacity (Akita and Li-Chan 1998). To extend the use of IgY immunoaffinity columns, Kim et al. (1999) examined the reusability of avidin-biotinylated IgY columns, in which biotinylated IgY is held by the strong noncovalent interaction on columns containing immobilized avidin. It was also found that when the antibody-binding activity had been reduced by prolonged use, the column could be regenerated by dissociating the avidin-biotinylated IgY complex and applying new biotinylated IgY, thereby restoring the binding activity of the column. Immobilized yolk antibodies have been used for the isolation of value-added proteins from dairy products, including the purification of lactoferrin (Li-Chan et al. 1998), and the isolation and separation of IgG subclasses from colostrum, milk, and cheese whey (Akita and Li-Chan 1998), as well as for the purification of biological molecules from human serum.

5.5.6. Passive Immunization by Use of IgY

5.5.6.1. Active versus Passive Immunity

The concept of active immunity versus passive immunity is understood by studying the infection defense for which the immunity function of the animal is used (Fig. 5.12). An active immunization is called *vaccine therapy*; vaccination against measles and Japanese encephalitis are typical examples. The antigen (virus and bacteria) that lost the infectivity is inoculated to a person or a cow to activate the immune system, which helps the body to produce a specific antibody such as that used for biophylaxis. Alternatively, the concept of passive immunization is based on use of a pathogen-specific antibody obtained from other animals immunized with that pathogen. The pathogen-specific antibody administered to individuals protects them against to that pathogen disease related prevents (Mine and Kovacs-Nolan 2002). Serum therapy consists in administering antiserum, containing specific antitoxin or antibacterial antibodies in the patient. Adoptively-transferred immunity therapy, in which the T cells cultivated in the existence of cytokines, are inoculated to an immunodeficient patient or a cancer patient is an exemplary application of passive immunity.

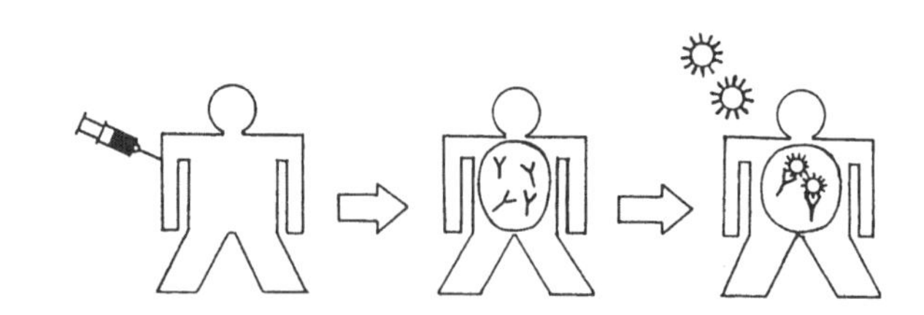

Active Immunization (Vaccination)

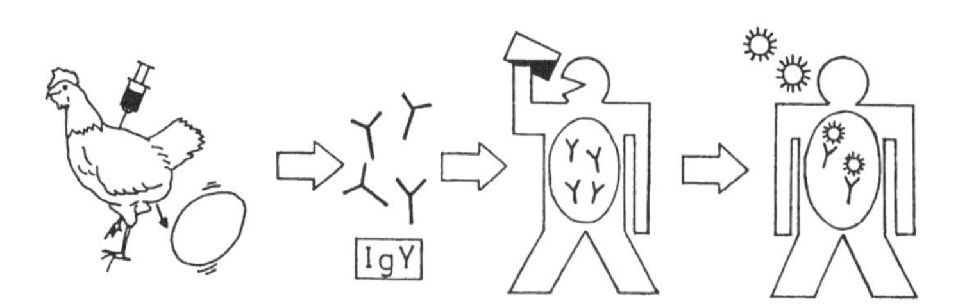

Passive Immunization (Orally)

Figure 5.12. Differences between active immunization and passive immunization.

5.5.6.2. Yolk Antibody and Passive Immunization

Passive immunization, in which a specific antibody is used, can be administered in several ways to prevent development of infection. Oral administration of a specific antibody (oral passive immunization) is suitable for preventing adherent infection from a pathogen in the mouth or the digestive tract. Another way to administer a specific antibody is via intramuscular or intravenous injection. For example, to neutralize the effects of snake venom or a bacterial toxin using specific antibodies against these venom or toxin is in practice a well krown form of antiserum therapy. Considering the oral passive immunization of IgY, many specific antibodies can be obtained from egg yolk, since the more recent use of egg-containing functional foodstuff. IgY also replaces horse serum used in the past for antiserum treatment.

5.5.7. Systemic Administration of IgY

IgY is an excellent antibody capable of neutralizing the venom of a viper or a scorpion, and staphylococcal enterotoxin and also prevents the development of Newcastle virus infection and rabies, in which cases direct injection of specific IgY antibody is recommended. Moreover, more recent study has revealed that IgY may be used to reduce the risk of hyperacute rejection in transplantation of pig organs into humans.

5.5.7.1. Neutralizing of Snake Venom

Approximately 1.7 million people worldwide are bitten each year by a snake, a scorpion, a spider, or a jellyfish; nearly 40,000–50,000 of these incidents are fatal. An antiserum therapy is now available to such patients. Previously, horse antiserum was often used to neutralize the toxicity of a toxin. However, "serum sickness" often resulted from impurities in the serum, and in some cases IgG antibodies in horse antiserum combined with the infected person's alexin to produce inflammation as well as side effects. Currently, direct injection of specific IgY antibody is in practice. Antivenom IgY has been produced and found to have a bioactivity higher than that of antiserum traditionally prepared in horses (Thalley and Carroll 1990; Almeida et al. 1998). Since IgY differs from the IgG of the mammals and does not combine with a person's alexin, IgY antibody can be regarded as a safe alternative to the conventional horse serum. Since a commercial preparation of high-purity IgY antibody from egg yolk is already established, use of IgY as an antibody as a safer means of detoxication is highly anticipated.

5.5.7.2. Neutralization of Staphylococcal Enterotoxins

Staphylococcal enterotoxins are a family of bacterial superantigens produced by *Staphylococcus aureus*, and are associated with a number of serious diseases, including food poisoning, bacterial arthritis, and toxic shock syndrome (Fraser et al. 1976). Sugita-Konishi et al. (1996) found that specific IgY was capable of inhibiting the production of *S. aureus* enterotoxin A *in vitro*.

LeClaire et al. (2002) reported the production of IgY against *S. aureus* enterotoxin B (SEB), and found that systemically administered anti-SEB IgY provided both pre- and postexposure protection to SEB challenge, and protected monkeys against toxic shock syndrome in a rhesus monkey model of SEB-induced lethal shock. These results suggest that antienterotoxin IgY may provide protection as both a prophylactic and therapeutic agent against lethal doses of *S. aureus* enterotoxins, and could be used to reduce or eliminate enterotoxin-mediated disorders (LeClaire et al. 2002).

5.5.7.3. Prevention of Newcastle Disease Virus

Newcastle disease virus is one of the most virulent and lethal pathogenic viruses in chickens, resulting in instant death. Newcastle disease is officially designated as the most infectious disease of the chicken, and currently, it's obligatory to vaccinate (called *preventive vaccination*) all chickens against this disease. It takes usually 1–2 weeks to activate the immune system of a chicken by vaccination, by introducing a specific antibody as a curative dose into the chicken's body. Newcastle disease is a virus that often exists in a poultry farm, and the viral infectious disease sometimes spreads during the entire antibody induction period.

Stedman and coworkers (1969) prepared IgY as Newcastle disease virus–neutralizing antibody from egg yolk and performed a viral infection experiment by intramuscularly injecting a noninfected chicken with antibody obtained from a chicken previously infected by Newcastle disease. On injection of IgY antibody and circulation of the antibody, the titer of the chicken promptly rose, indicating the immediate effect of the antibody and confirming the prevention of Newcastle virus disease. Therefore, IgY antibody is helpful in preventing Newcastle virus disease epidemics.

5.5.7.4. Prevention of Rabies Virus

In an attempt to produce antirabies immunoglobulin affordable for people living in developing countries, we have immunized egg-laying hens with a part of the G protein of rabies virus expressed in *Escherichia coli*. Immunoglobulin (IgY) was purified from the yolks of eggs laid by immunized hens. It was revealed *in vitro* that the antibody specifically bound to *virions* as well as cells infected with rabies virus. Moreover, the antibody apparently neutralized rabies virus infectivity (Motoi et al. 2005a; 2005b). Inoculation of the antibody into mice infected with rabies virus reduced the mortality caused by the virus, suggesting that IgY directed to the part of the G protein expressed in *E. coli* could serve as a possible alternative to currently available antirabies human or equine immunoglobulins.

5.5.7.5. Use of IgY in Hyperacute Rejection in Xenotransplantation

Transplantation of pig organs into humans (xenotransplantation) is seriously considered due to the shortage of human donors for organ transplants. However, the problem with such xenografts is the risk of hyperacute rejection, mediated by natural antibodies in humans against pig antigens, complement fixation, and the rapid onset of intravascular coagulation (Sandrin and McKenzie 1994). The major target of these natural antibodies is the carbohydrate epitope Gala1–3Gal, which is expressed by all mammals except for humans, apes, and some monkeys. Besides humans and monkeys, birds, especially chickens, also lack Gala1–3Gal expression (Bouhours et al. 1998). And, since IgY does not bind human complement or Fc receptors, anti-aGal IgY is a good candidate for use in blocking antibodies to inhibit the interactions that may contribute to xenograft rejection. Fryer et al. (1999) demonstrated *in vitro* that anti-aGal IgY blocked human xenoreactive natural antibody binding to both porcine and rat tissues, as well as inhibiting cell lyses of porcine cells by human serum, suggesting that IgY could be of potential use in inhibiting pig-to-human xenograft rejection.

Anti-aGal IgY was also found to significantly reduce the infectivity of porcine endogenous retrovirus (PERV), an a-Gal-bearing virus, which has emerged as a potential zoonotic agent, with possible pig-to-human transmission (Leventhal et al. 2001).

5.5.8. Oral Administration of IgY

Although passive immunization is a relatively recent concept in human health, it is well established in animals, and the use of IgY for passive immunization was described as early as 1963, for the protection of chickens against Newcastle disease virus and Marek's disease virus (Wills and Luginbuhl 1963; Box et al. 1969; Kermani-Arab et al. 1975). The oral administration of IgY remains an attractive approach for the establishment of passive immunity (Larsson and Carlander 2003), and potential applications of IgY for the prevention and treatment of infections caused by pathogenic bacteria and viruses in both humans and animals have been studied at length; a nutshell summary is presented below.

Oral passive immunization by IgY is used to prevent rotaviral diarrhea, tooth decay, sterilization of *Helicobacter pylori*, cultured fish infection, and other disorders, discussed in the following sections.

5.5.8.1. Prevention of Rotaviral Diarrhea

In developing countries, millions of infant die every year from diarrhea due to human rotavirus (HRV) infection (Clark et al. 1999). In Japan, nearly 100,000 infants are also infected per year with HRV diarrhea. The infection adheres to the intestinal tract and particularly targets infants with immature immune systems. To prevent the HRV infection, it is necessary to administer an anti-HRV antibody orally to prevent or limit adherent infection of HRV in the intestinal tract (Bishop et al. 1973).

As an antibody specific to HRV infection, the thermal limitations and tolerance to pH in the digestive enzyme system are very important. The thermal denaturation temperature of IgY is around 73.9°C, and a structural change occurs at pH < 3.5 and IgY tends to be deactivated. However, IgY is relatively stable to trypsin and chymotrypsin. The perfect deactivation occurs at pH 2.0–pepsin (Hatta et al. 1993a, 1993b). It is clearly known that under the conditions assumed in the infant's stomach (pH 4.0, 4 h), antibody titer still remains at nearly 50%. In a typical experiment, HRV infection orally introduced to a young mouse caused loose bowels. An anti-HRV IgY was orally introduced to the infected mouse (22.5 μg/mouse) to neutralize the HRV infection. Thus, human rotaviral diarrhea could be prevented by administration of anti-HRV IgY.

The recombinant HRV coat protein VP8, a cleavage product of the rotavirus spike protein VP4, which is involved in viral infectivity and neutralization of the virus, has been reported for the induction of antibodies against HRV (Kovacs-Nolan et al. 2001). The resulting anti-VP8 IgY exhibited significant neutralizing activity *in vitro* against the Wa strain of HRV, suggesting its use for the prevention and treatment of HRV infection. Oral administration of anti-HRV egg yolk to children affected with HRV resulted in only a modest improvement in HRV-related symptoms; however further studies are required (Sarker et al. 2001).

Neonatal calf diarrhea, caused by bovine rotavirus (BRV), is a significant cause of mortality in cattle (Lee et al. 1995). Using a mouse model of BRV infection, Kuroki et al. (1993) observed protection against two strains of BRV using orally administered anti-BRV IgY. The passive protection of calves against BRV infection, using anti-BRV IgY, has also been demonstrated (Kuroki et al. 1994).

5.5.8.2. Prevention of Dental Caries

IgY antibodies are also found effective in prevention of dental caries. Dental caries is an infection caused by decayed tooth's carious bacteria, which are indigenous bacteria (*Streptpcoccus mutans*) in the mouth (Hamada and Slade 1980). Carious bacteria possess an enzyme called *glucosyltransferase*, which is responsible for forming adhesive polysaccharide at the surface of the bacterial cell. Adhesion, also termed *plaque formation*, enhances firmness of the cavity bacteria on the tooth surface by adhesive polysaccharide. Generally, in plaque formation, lactic acid forms via an action of a lactobacillus, which "melts" (erodes) a tooth and forms a cavity. Such adherent infection of carious bacteria on the tooth surface could be prevented using anticavity bacterial IgY. A preparation of an anti-carious bacterial IgY specific to the bacterial cell with an adhesive polysaccharide formation in human carious bacteria (formalin mortal bacteria) from egg yolks of immunized chickens has been reported. In an experiment rats previously infected with cavity bacteria in their teeth were separately doped with anticavity bacterial IgY antibodies as well as controlled IgY antibody. Rats were analyzed to estimate the degree of cavity formation in their teeth. With control IgY (derived from nonimmunized chickens) a severe cavity was observed while the degree of cavity formation could be restricted predominantly with anticarious bacterial IgY (Otake et al. 1991). The effectiveness of IgY antibody was also examined in humans using a mouth rinse test. A 10% sucrose solution was added with nearly 1% of anticarious bacterial IgY or controlled IgY, and a gargle test with a human was performed in another human subject. Saliva was extracted in 4 h later and the ratio of cavity bacteria occupied per the number of *Streptococcus* genus bacteria was estimated. The results revealed that in all volunteers the number of cavity bacteria increased by gargling with controlled IgY antibody solution but decreased by gargling with anticavity bacterial IgY antibody (Hatta et al. 1997).

IgY produced against the *S. mutans* glucan-binding protein B (GBP-B), which is believed to be involved in *S. mutans* biofilm development, has also been examined. Using a rat model of dental caries, Smith et al. (2001) observed a decrease in *S. mutans* accumulation in rats treated with anti-GBP-B IgY, as well as a decrease in the overall amount of dental caries, as compared to control rats. These studies suggest that IgY against *S. mutans*, or its components, may inhibit *S. mutans* accumulation and limit plaque and the subsequent oral health problems associated with plaque accumulation.

5.5.8.3. Sterilization of Helicobacter pylori

Helicobacter pylori bacteria is the main cause of a stomach ulcer and a stomach cancer. Approximately 95% of stomach cancer patients are reported positive with *H. pylori* bacteria (Dunn et al. 1997). Commonly, the *urease*, which is a bacterial surface enzyme protein, actively participates in causing *H. pylori* to adhere to the gastric mucosa. Therefore, daily sterilization by a particular food or antibiotic is necessary for sterilization of *H. pylori* to prevent stomach cancer. The effectiveness of anti–*H. pylori* urease IgY antibody has been reported and considered as a new sterilization method, which obstructs adhesion of *H. pylori* to the gastric mucosa simply when the subject ingests anti–*H. pylori* urease IgY (Fig. 5.13) (Horie et al. 2004).

In a planned study, the 16 *H. pylori*–positive subjects were fed yogurt supplemented with anti–*H. pylori urease* IgY antibody. An *H. pylori* antigen detection test in the feces as well as a urea expiration test (UBT) were performed after 4, 8, and 12 weeks. After 8 weeks, the amounts of *H. pylori* antigen in the feces and the UBT value were substantially lowered, while after 12 weeks of treatment the *H. pylori* bacteria were completely neutralized (Fig. 5.14) (Horie et al. 2004). Thus, a positive sterilization effect of the anti–*H. pylori* urease IgY administration specific to *H. pylori* disease was confirmed.

5.5.8.4. Prevention of Cystic Fibrosis

Respiratory infection is the major cause of morbidity and mortality in cystic fibrosis (CF) patients (Shale and Elborn 1996). Chronic *Pseudomonas aeruginosa* infections ultimately occur in virtually all patients. It is impossible to eradicate *P. aeruginosa* when a patient has been chronically colonized. Immunotherapy with specific IgY may be an alternative to administering antibiotics to prevent *P. aeruginosa* infections (Kollberg et al. 2003). Carlander et al. (1999b, 2000, 2002) found that a rise in IgY against *P. aeruginosa* was capable

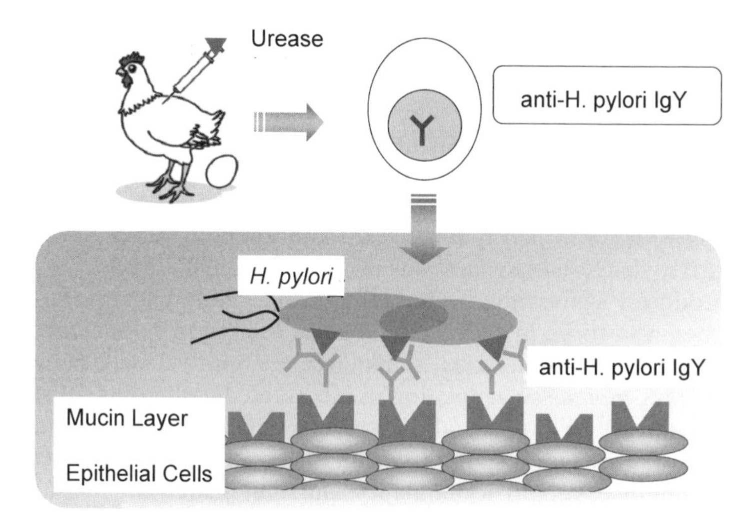

Figure 5.13. Competitive inhibition of *Helicobacter pylori* adherence by anti–*H. pylori* IgY. (See insert for color representation.)

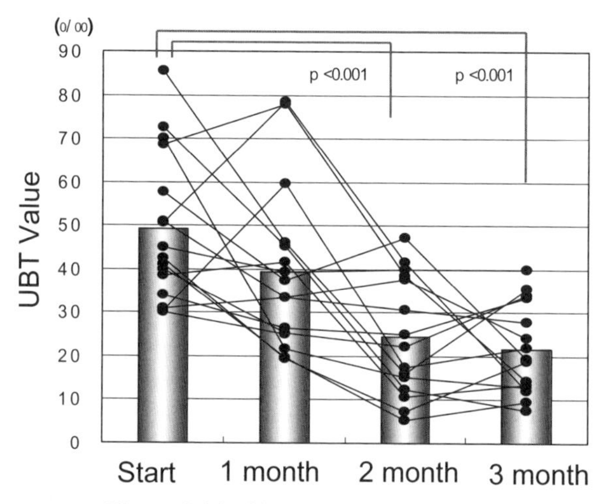

Figure 5.14. Changes of UBT values.

of inhibiting adhesion of *P. aeruginosa* to epithelial cells *in vitro*, but did not inhibit bacterial growth, suggesting that specific IgY might be capable of interfering with the bacterial infection process and preventing colonization in CF patients. CF patients gargled daily with an IgY antibody preparation, purified from eggs of hens immunized with *P. aeruginosa* bacteria. They were compared to a group of patients who did not gargle with this preparation. In both groups the first colonization with *P. aeruginosa* was eradicated by antibiotics. The basic treatment was essentially the same in both groups. In the initial study, the period between the first and second colonizations with *P. aeruginosa* was significantly prolonged for the treated versus the control groups.

This feasibility study shows that antipseudomonal IgY has the potential to effectively prevent *P. aeruginosa* colonization without any severe adverse effect (Kollberg et al. 2003). The stability of anti–*P. aeruginosa* IgY in saliva from healthy individuals, over time, following a mouth rinse with an aqueous IgY solution, was also examined (Carlander et al. 2002). Antibody activity in the saliva remained after 8 h. After 24 h, the antibody activity had decreased significantly, but was still detectable in some subjects, indicating that oral treatment with specific IgY for various local infections, such as the common cold and tonsillitis, might be possible.

5.5.8.5. Prevention of Crohn's Disease

Crohn's disease and ulcerative colitis are chronic inflammatory bowel diseases, and are an increasing burden to hospitals and society in terms of the cost of medication and treatment, and lost time from work due to illness (Hay and Hay 1992). Tumor necrosis factor (TNF) is implicated in the pathogenesis of inflammatory bowel disease (Worledge et al. 2000). Clinical trials indicate that intravenous infusion of anti-TNF antibody is an effective therapy for Crohn's disease. An oral anti-TNF therapy may be a preferred approach, reducing systemic side effects and eliminating the inconvenience and expense of intravenous infusions. Oral anti-TNF IgY antibody, in both the acute and chronic phases of the model, significantly decreased all inflammatory endpoints and proved to be more effective than standard anti-inflammatory drugs (sulfasalazine and dexamethasone). Oral delivery of anti-TNF IgY antibodies is an effective method for treating experimental colitis and may provide advantages over current parenteral administration of anti-TNF antibodies (Worledge et al. 2000).

5.5.8.6. Prevention of Cultured Fish Infection

Infectious paracola diseases of bacillus-infected eels caused by *Edwardjella tarda* invades the eel's intestinal tract and manifests symptoms. Presently, antibiotics are used for prevention and treatment of infections, but because of the resistance of some bacteria and the residual tendency of many antibiotics, large-volume application of antibiotics is limited. Therefore, a safer method for prevention of paracola must be developed (Hatta et al. 1994).

In recent years, a preparation of anti–*E. tarda* IgY antibody derived from whole-egg powder has been described and found effective in controlled prevention of the infectious paracola diseases of bacillus-infected eels. The eels were orally doped with anti–*E. tarda* IgY antibody, which was delivered into their intestinal tracts with hydrogen peroxide, and complete prevention of paracola disease was confirmed (Gutierrez et al. 1993). This antibody is now used commercially, after a field test on nearly 2,400,000 tails of cultured eels with confirmed paracolo disease prevention by anti–*E. tarda* IgY antibody.

Yersinia ruckeri is the causative agent of enteric redmouth disease, a systemic bacterial septicemia of salmonid fish (Stevenson et al. 1993), and the persistence of *Y. ruckeri* in carrier fish and shedding of bacteria in feces can present a continuing source of infection. The passive immunization of rainbow trout against infection with *Y. ruckeri* using IgY has been studied (Lee et al. 2000). The fish fed anti–*Y. ruckeri* IgY prior to challenge with *Y. ruckeri* showed a lower mortality rate, and, on the basis of organ and intestine culture, demonstrated a lower *Y. ruckeri* infection rate. This infection rate appeared lower regardless of whether IgY was administered prior to or after *Y. ruckeri* challenge.

The oral administration of specific IgY against fish pathogens would provide an alternative to antibiotic and chemotherapy treatment for the prevention of fish diseases in fishfarms, and would present a cost-effective alternative to slaughtering a stock of fish, which would also pose a health risk.

5.5.8.7. Prevention of Other Diseases

In addition, specific IgY has been shown to be effective in preventing and treating several other pathogens, including the passive protection of chicks against infectious bursal disease virus (IBDV) (Eterradossi et al. 1997), the protection of piglets against porcine epidemic virus (PEDV) (Kweon et al. 2000), the protection of calves against bovine coronavirus (BCV) (Ikemori et al. 1997), and the prevention of cryptosporidiosis due to *Cryptosporidium* infection (Cama and Sterling 1991). Finally, IgY has also been produced against *Bordetella bronchiseptica, Pasteurella multocida*, and *Actinobacillus pleuropneumoniae*, causative agents of swine respiratory diseases, and have been proposed as an alternative method to control infectious respiratory diseases in swine (Ling et al. 1998; Shin et al. 2002). Yang et al. (1997) described the production of IgY against P110, a protein purified from human stomach cancer cells, and suggested its use as a carrier for antitumorigenic drugs, to target gastrointestinal cancers.

5.5.9. Practical Use of IgY in Passive Immunizaton

In summary, large quantities of specific antibodies are required for practical use of adherent infection prevention by oral passive immunity. In comparison to specific antibody preparation using a conventional mammal (rabbit, goat, etc.), large-scale production of antibodies is possible using egg yolk, which provides the most suitable antibody via oral passive immunity in diverse applications. In addition, when considering oral administration, the specific antibody derived from food–grade egg yolks is highly recommended.

The observations and data discussed and summarized in this chapter indicate that the egg yolk-derived IgY antibody is effective in preventing dental cavities, in sterilization of *Helicobacter pylori*, and in other diverse applications.

Additionally, the use of IgY antibody for the prevention of human rotaviral (HRV) diarrhea, in serum treatments, and as detoxicants indicates potential for the safe and practical utility and effectiveness of IgY in advanced applications.

5.6. CONCLUDING REMARKS

From the beginning of ancient culture, to the present, the hen egg has been used as a food throughout the world regardless of religion, race, or culture. The nutritious and health-promoting functions of the egg are valuable for humankind to maintain and increase recovery of health. Raw egg is also useful for preventing fatigue and the common cold and in beauty health management and other applications. Overall, the functional ingredients such as egg yolk lipids and egg yolk–derived antibody help maintain body functions and health. Therefore, considering the physiological functions described in this chapter, inclusion of an egg in the daily diet is highly recommended for prolonged life expectancy and healthy living.

REFERENCES

Akita EM, Li-Chan EC (1998). Isolation of bovine immunoglobulin G subclasses from milk, colostrums, and whey using immobilized egg yolk antibodies. J Dairy Sci 81:54–63.

Akita EM, Nakai S (1993). Comparison of four purification methods for the production of immunoglobulins from eggs laid by hens immunized with an enterotoxigenic E. coli strain. J Immunol Meth 160:207–214.

Akita EM, Nakai S (1992). Immunoglobulins from egg yolk: Isolation and purification. J Food Sci 57:629–634.

Al-Haddad S, Zhang Z, Leygue E, Snell L, Huang A, Niu Y, Hiller-Hitchcock T, Hole K, Murphy LC, Psoriasin LC (1999). (S100A7) expression and invasive breast cancer. Am J Pathol 155:2057–2066.

Almeida CM, Kanashiro MM, Rangel Filho FB, Mata MF, Kipnis TL (1998). Development of snake antivenom antibodies in chickens and the purification from yolk. Vet Res 143:579–584.

Altschuh D, Hennache G, von Regenmortel MH (1984). Determination of IgG and IgM levels in serum by rocket immunoelectrophoresis using yolk antibodies from immunized chickens. J Immunol Meth 69:1–7.

Amoresanno A, Brancaccio A, Andolfo A, Perduca M, Monaco HL, Marino G (1999). The carbohydrates of the isoforms of three avian riboflavin-binding proteins. Eur J Biochem 263(3):849–858.

Baker C (1989). Lecithin in cosmetics. In: Szuhaj BF, ed. Lecithins. Champaign, IL: American Oil Chemists Society; pp 253–260.

Baker ME (1988). Invertebrate vitellogenin is homologous to human von Willebrand factor (letter). Biochem J 256(3):1059–1061.

Bauwens RM, Devos MP, Kint JA, De Leenheer AP (1988). Chicken egg yolk and rabbit serum compared as sources of antibody for radioimmunoassay of 1,25-dihydroxyvitamin D in serum or plasma. Clin Chem 34(10):2153–2154.

Beeley JG (1976). Location of the carbohydrate groups of ovomucid. Biochem J 159:335–345.

Best CH, Huntsman ME (1932). The effects of the components of lecithin upon deposition of fat in the liver. J Physiol 75:405–412.

Bishop RF, Davidson GP, Holmes IH, Ruck BJ (1973). Virus particles in epithelial cells of duodenal mucosa from children with acute nonbacterial gastroenteritis. Lancet 2:1281–1283.

Bouhours JF, Richard C, Ruvoen N, Barreau N, Naulet J, Bouhours D (1998). Characterization of a polyclonal anti-Gala1-3Gal antibody from chicken. Glycoconj J 15: 93–99.

Box PG, Stedman RA, Singleton L (1969). Newcastle disease. I. The use of egg yolk derived antibody for passive immunization of chickens. J Compar Pathol 79:495–506.

Brockbank RL, Vogel HJ (1990). Structure of oligosaccharide of hen phosvitin as determined by two-dimensional 1H-NMR of the intact glycoprotein. Biochemistry 29:5574–5583.

Burger D, Ramus MA, Schapira M (1990). Antibodies to human plasma kallikrein from egg yolks of an immunized hen: preparation and characterization. Thromb Res 40(2):283–288.

Burley RW, Vadehra DV (1989). The Avian Egg, Chemistry and Biology. New York: Wiley.

Cama VA, Sterling CR (1991). Hyperimmune hens as a novel source of anti-Cryptosporidium antibodies suitable for passive immune transfer. J Protozool 38:42S–43S.

Carlander D, Larsson A (2001). Avian antibodies can eliminate interference due to complement activation in EUSA. Upsala J Med Sci 106:189–195.

Carlander D, Kollberg H, Larsson A (2002). Retention of specific yolk IgY in the human oral cavity. BioDrugs 16:433–437.

Carlander D, Kollberg H, Wejaker PE, Larsson A (2000). Prevention of chronic *Pseudomonas aeruginosa* colonization by gargling with specific antibodies: A preliminary report. In: Sim JS, Nakai S, Guenter W, eds. Egg Nutrition and Biotechnology. New York: CABI Publishing; pp 371–374.

Carlander D, Stalberg J, Larsson A (1999a). Chicken antibodies: A clinical chemistry perspective. Ups J Med Sci 104:179–190.

Carlander D, Sundstrom J, Berglund A, Larsson A, Wretlind B, Kollberg HO (1999b). Immunoglobulin Y (IgY)—a new tool for the prophylaxis against *Pseudomonas aeruginosa* in cystic fibrosis patients. Pediatr Pulmonol (Suppl) 19:241 (abstract).

Carroll SB, Stollar BD (1983). Antibodies to calf thymus RNA polymerase II from egg yolks of immunized hens. J Biol Chem 258(1):24–26.

Chang CH, Jensen LS (1975). Inefficiency of carnitine as a substitute for choline for normal reproduction in Japanese quail. Poultry Sci 54(5):1718–1720.

Choi I, Jung C, Seog H, Choi H (2004). Purification of phosvitin from egg yolk and determination of its physiochemical properties. Food Sci Biotechnol 13:434–437.

Chong YS, Liang Y, Tan D, Gazzard G, Stone RA, Saw SM (2005). Association between breastfeeding and likelihood of Myopia in children. J Am Med Assoc 24:293.

Chou HH, Hayakawa T, Diaz S, Krings M, Indriati E, Leakey M, Paabo, SY, Takahata N, Varki A (2002). Inactivation of CMP-N-acetylneuraminic acid hydroxylase occurred prior to brain expansion during human evolution. Proc Natl Acad Sci USA 99(18):11736–11741.

Cipolla A, Cordeviola J, Terzolo H, Combessies G, Bardon J, Ramon N, Martinez A, Medina D, Morsella C, Malena R (2001). Campylobacter fetus diagnosis: Direct immunofluorescence comparing chicken IgY and rabbit IgG conjugates. ALTEXx 18:165–170.

Clark HF, Glass RI, Offit PA (1999). Rotavirus vaccines. In: Plotkin SA, Orentstein WA, eds. Vaccines, 3rd eds. Philadelphia: Saunders; pp 987–1005.

Cook WH, Martin WV (1969). Egg lipoproteins. In: Tria E, Scanu AM, eds. Structural and Functional Aspects of Lipoproteins in Living Systems. New York: Academic Press.

Davies P, Maloney AJ (1976). Selective loss of central cholinergic neurons in Alzheimer's diseases (letter). Lancet 2(8000):1403–1403.

De Ceuninck F, Pastoureau P, Agnellet S, Bonnet J, Vanhoutte PM (2001). Development of an enzyme-linked immunoassay for the quantification of YKL-40 (cartilage gp-39) in guinea pig serum using hen egg yolk antibodies. J Immunol Meth 252:153–161.

DeMol P, Zissis G, Butzler JP, Mutwewingabo A, André FE (1986). Failure of live, attenuated oral rotavirus vaccine. Lancet 2(8498):108–108.

Di Lonardo A, Luisa Marcante M, Poggiali F, Hamsøikovà E, Venuti A (2001). Egg yolk antibodies against the E7 oncogenic protein of human papillomavirus type 16. Arch Virol 146:117–125.

Dunn BE, Cohen H, Blaser MJ (1997). Helicobacter pylori. Clin Microbiol Rev 10:720–741.

Du Plessis DH, Van Wyngaardt W, Romito M, Du Plessis M, Maree S (1999). The use of chicken IgY in a double sandwich ELISA for detecting African horsesickness virus. Onderstepoort J Vet Res 66:25–28.

Dziedzic SZ, Hudson BJF (1984). Phosphatidyethanolamine as a synergist for primary antioxidants in edible oils. J Am Oil Chem Soc 61(6):1042–1045.

Dziedzic SZ (1986). Fate of propyl gallate and diphosphatidyethanolamine in lard during autoxidation at 120°C. J Agric Food Chem 34:1027–1029.

Eterradossi N, Toquin D, Abbassi H, Rivallan G, Cotte JP, Guittet M (1997). Passive protection of specific pathogen free chicks against infectious bursal disease by in-ovo injection of semi-purified egg-yolk antiviral immunoglobulins. J Vet Med B 44:371–383.

Evans AJ, Burley RW (1987). Proteolysis of apoprotein B during the transfer of very low density lipoprotein from hen's blood to egg yolk. J Biol Chem 262(2):501–504.

Fertel R, Yetiv JZ, Coleman MA, Schwarz RD, Greenwald JE, Bianchine JR (1981). Formation of antibodies to prostaglandins in the yolk of chicken eggs. Biochem Biophys Res Commun 102(3):1028–1033.

Fishman PH, Brady RO (1976). Biosynthesis and function of gangliosides. Science 194(4268):906–915.

Fraser J, Arcus V, Kong P, Baker E, Proft T (1976). Superantigens—powerful modifiers of the immune system. Mol Med Today 6:125–132.

Fryer JP, Firca J, Leventhal JR, Blondin B, Malcolm A, Ivancic D, Gandhi R, Shah A, Pao W, Abecassis M, Kaufman DB, Stuart F, Anderson B (1999). IgY antiporcine endothelial cell antibodies effectively block human antiporcine xenoantibody binding. Xenotransplantation 56:98–109.

Fukudome K, Yoshie O, Konno T (1999). Comparison of human, simian and bovine rotaviruses for requirement of sialic acid in hemagglutination and cell adsorption. Virology 172(1):196–205.

Gardner PS, Kaye S (1982). Egg globulins in rapid virus diagnosis. J Virol Meth 4(4):257–262.

Gottstein B, Hemmeler E (1985). Egg yolk immunoglobulin Y as an alternative antibody in the serology of echinococcosis. Z Parasitenkd 71(2):273–276.

Grebenschikov N, Geurts-Moespot A, De Witte H, Heuvel J, Leake R, Sweep F, Benraad T (1997). A sensitive and robust assay for urokinase and tissue-type plasminogen activators (uPA and tPA) and their inhibitor type I (PAI-1) in breast tumor cytosols. Int J Biol Markers 12:6–14.

Griffith WH, Wade NJ (1939). Choline metabolism: part I. The occurrence and prevention of hemorrhagic degeneration in young rats on low choline diet. J Biol Chem 131:567–569.

Groche D, Rashkovetsky LG, Falchuk KH, Auld DS (2000). Subunit composition of the zinc proteins alpha- and beta-lipovitellin from chicken. J Protein Chem 19(5): 379–387

Gutierrez Calzado E, Garcia Garrido RM, Schade R (2001). Human haemoclassification by use of specific yolk antibodies obtained after immunisation of chickens against human blood group antigens. Alter Lab Anim 29:717–726.

Gutierrez MA, Miyazaki T, Hatta H, Kim M (1993). Protective properties of egg yolk IgY containing anti-Edwardsiella tarda antibody against paracolo disease in the Japanese eel, Anguilla japanica Temminck & Schlegel. J Fish Dis 16:113–122.

Hamada S, Horikoshi T, Minami T, Kawabata S, Hiraoka J, Fujiwara T, Ooshima T (1991). Oral passive immunization against dental caries in rats by use of hen egg yolk antibodies specific for cell-associated glucosyltransferase of *Streptococcus mutans*. Infect Immun 59:4161–4167.

Hamada S, Slade HD (1980). Biology, immunology, and cariogenicity of *Streptococcus mutans*. Microbiol Rev 44:331–384.

Hassl A, Aspock H (1988). Purification of egg yolk immunoglobulins. A two-step procedure using hydrophobic interaction chromatography and gel filtration. J Immunol Meth 110:225–228.

Hatta H, Kim M, Yamamoto T (1990). A novel isolation method for hen egg yolk antibody "IgY." Agric Biol Chem 54:2531–2535.

Hatta H, Mabe K, Kim M, Yamamoto T, Gutierrez MA, Miyazaki T (1994). Prevention of fish disease using egg yolk antibody In: Sim JS, Nakai S, eds. Egg Uses and Processing Technologies, New Developments. Oxon, UK: CAB International; pp 241–249.

Hatta H, Ozeki M, Tsuda K (1997). Egg yolk antibody IgG and its application. In: Yamamoto T, Juneja LR, Hatta H, Kim M, eds. Hen Eggs, Their Basic and Applied Science. New York: CRC Press; pp 151–178.

Hatta H, Sim JS, Nakai S (1988). Separation of phospholipids from egg yolk and recovery of water-soluble proteins. J Food Sci 53:425–431.

Hatta H, Tsuda K, Akachi S, Kim M, Yamamoto T, Ebina T (1993a). Oral passive immunization effect of anti-human rotavirus IgY and its behavior against proteolytic enzymes. Biosci Biotechnol Biochem 57:1077–1081.

Hatta H, Tsuda K, Akachi S, Kim M, Yamamoto T (1993b). Productivity and some properties of egg yolk antibody (IgY) against human rotavirus compared with rabbit IgG. Biosci Biotechnol Biochem 57:450–454.

Hatta H, Tsuda K, Ozeki M, Kim M, Yamamoto T, Otake S, Hirosawa M, Katz J, Childers NK, Michalek SM (1997). Passive immunization against dental plaque formation in humans: Effect of a mouth rinse containing egg yolk antibodies (IgY) specific to Streptococcus mutans. Caries Res 31:268–274.

Hay JW, Hay AR (1992). Inflammatory bowel disease: costs-of-illness. J Clin Gastroenterol 14:309–317.

Hirabayashi Y, Suzuki T, Suzuki Y, Taki T, Matumoto M, Higashi H, Kato S (1983). A new method for purification of anti-glycosphingolipid antibody. Avian antihematoside (NeuGc) antibody. J Biochem 94(1):327–330.

Hirsch MJ, Wurtman RJ (2002). Lecithin consumption increases acetylcholine concentrations in rat brain and adrenal gland. Science 202(4364):223–225.

Holmgren J, Elwing H, Fredman P, Strannegård O, Svennerholm L (1980). Gangliosides as receptors for bacterial toxins and Sendai virus. Adv Exp Med Biol 125:453–467.

Horie K, Horie N, Abdou AM, Yang JO, Yun SS, Park CK, Kim M, Hatta H (2004). Suppressive effect of functional drinking yogurt containing specific egg yolk immunoglobulin on Helicobacter pylori in humans. J Dairy Sci 87(12):4073–4079.

Huttunen JK (1966). Neuraminic acid-containing oligosaccharides of human urine. Isolation and identification of di-N-acetylneuraminyl-3-galactosyl-N-acetyl galactosamine 62-N-acetylneuraminyl-lactose 62-N-acetylneuraminyl-N-acetyl lactosamine and 32-N-acetylneuraminyllactose. Ann Med Exp Biol Fenn 44(12):1–60.

Ikemori Y, Ohta M, Umed K, Icatlo Jr FC, Kuroki M, Yokoyama H, Kodama Y (1997). Passive protection of neonatal calves against bovine coronavirus-induced diarrhea by administration of egg yolk or colostrum antibody powder. Vet Microbiol 58:105–111.

Inglis AS, Strike PM, Burley RW (1982). Two low-molecular weight apoproteins (apovitellenins I and II) from a lipoprotein of goose's egg yolk: A comparison with related species. Austral J Biol Sci 35(3):263–269.

Ishikawa S, Yano Y, Arihara K, Itoh M (2004). Egg yolk phosvitin inhibits hydroxyl radical formation from the Fenton reaction. Biosci Biotechnol Biochem 68:1324–1331.

Ito Y, Fujii T (1962). Chemical compositions of the egg yolk lipoproteins. J Biochem (Tokyo) 52:221.

Jenike MA, Albert MS, Heller H, LoCastro S, Gunther J (1986). Combination therapy with lecithin and ergoloid mesylates for Alzheimer's disease. J Clin Psychiatr 47:249–251.

Jensenius JC, Anderson I, Hau J, Crone M, Koch C (1981). Eggs: Conveniently packaged antibodies. Method for purification of yolk IgG. J Immunol Meth 46:63–68.

Jiang B, Mine Y (2001). Phosphopeptides derived from hen egg yolk phosvitin: Effect of molecular size on the calcium-binding properties. Biosci Biotechnol Biochem 65:1187–1190.

Jiang B, Mine Y (2000). Preparation of novel functional oligophos-phopeptides from hen egg yolk phosvitin. J Agric Food Chem 48:990–994.

Jukes YH (1940). Prevention of perosis by choline. J Biol Chem 134:789–790.

Juneja LR (1999). Biological characteristics of egg components, specifically sialyloligosaccharides in egg yolk. In: Sim JS, Nakai S, Guenter W, eds, Egg Nutrition and Biotechnology. Wallingford, UK: CABI Publishing.

Juneja LR, Kazuoka T, Goto N, Yamane T, Shimizu, S (1989). Conversion of phosphatidycholine to phosphatidylserine by various phospholipases D in the presence of L- or D-serine. Biochem Biophys Acta 1003:277–283.

Juneja LR, Koketsu M, Nishimoto K, Kawanami H, Kim M, Yamamoto T, Itoh T (1991). Large-scale preparation of sialic acid from chalaza and egg yolk membrane. Carbohydr Res 214(1):179–186.

Juneja LR, Sugino H, Fujiki M, Kim M, Yamamoto T (1994). Preparation of pure phospholipids from egg yolk. In: Sim, JS, Nakai, S, eds. Egg Uses and Processing Technologies—New Developments. Oxon, UK: CAB International; pp 139–149.

Juneja LR (1997). Egg yolk lipids. In: Yamamoto T, Juneja LR, Hatta H, Kim M, eds. Hen Eggs, Their Basic and Applied Science. New York: CRC Press; pp 73–98.

Kannagi R, Kitahara A, Itai S, Zenita K, Shigeta K, Tachikawa T, Noda A, Hirano H, Abe M, Shin S (1988). Quantitative and qualitative characterization of human cancer-associated serum glycoprotein antigens expressing epitopes consisting of sialyl or sialyl-fucosyl type 1 chain. Cancer Res 48(13):3856–3861.

Kato I, Schrode J, Kohr WJ, Laskowski MJ (1987). Chicken ovomucoid: Determination of its amino acid sequence, determination of the trypsin reactive site, and preparation of all three of its domains. Biochemistry 26:193–201.

Kawano T, Koyama S, Takematsu H, Kozutsumi Y, Kawasaki H, Kawashima S, Kawasaki T, Suzuki A (1995). Molecular cloning of cytidine monophospho-N-acetylneuraminic acid hydroxylase. Regulation of species- and tissue-specific expression of N-glycolylneuraminic acid. J Biol Chem 270(27):16458–16463.

Kelm S, Schauer R (1997). Sialic acids in molecular and cellular interactions. Int Rev Cytol 175:137–240.

Kermani-Arab V, Moll T, Davis WC, Cho BR, Lu YS, Leslie GA (1975). Immunoglobulins and anti-Marek's disease virus antibody synthesis in chickens after passive immunization with immunoglobulin Y anti-Marek's disease virus antibody. Am J Vet Res 36:1655–1661.

Kim H, Durance TD, Li-Chan EC (1999). Reusability of avidin-biotinylated immunoglobulin Y columns in immunoaffinity chromatography. Anal Biochem 268:383–397.

Kim H, Nakai S (1998). Simple separation of immunoglobulin from egg yolk by ultrafiltration. J Food Sci 63:485–490.

Klemperer F (1893). Ueber naturliche Immunitat und ihre Verwertung fur die Immunisierungstherapie. Archiv Exp Pathol Pharmakol 31:356–382.

Koketsu M, Enoki Y, Juneja LR, Kim M, Yamamoto T (1996). Isolation of sialogosaccharides from egg yolk using enzymes and some biofunctional activities of the oligosaccharides isolated. Oyo Toshitsu Kagaku 43:283–287.

Koketsu M, Juneja LR, Kawanami H, Kim M, Yamamoto T (1992). Preparation of N-acetylneuraminic acid from delipidated egg yolk. Glycoconj J 9:70–74.

Koketsu M, Juneja LR, Kim M, Ohta M, Matsuura F, Yamamoto T (1993). Sailyloligosaccharides of egg yolk fraction. J Food Sci 58:743–747.

Koketsu M, Nakata K, Juneja LR, Kim M, Yamamoto T (1995a). Learning performance of egg yolk sialyloligosaccharides. Oyatoshitsu Kagaku 9:15–18.

Koketsu M, Nitoda T, Juneja LR, Kim M, Kashimura N, Yamamoto T (1995b). Sialylglycopeptides from egg yolk as an inhibitor of rotaviral infection. J Agric Food Chem 43:858–861.

Koketsu M, Seko A, Juneja LR, Kim M, Kashimure N, Yamamoto T (1995c). An efficient preparation and structural characterization of sialylglycopeptides from protease treated egg yolk. J Carbohydr Chem 14(6):833–841.

Koketsu M (1997). Glycochemistry of hen eggs. In: Yamamoto T, Juneja LR, Hatta H, Kim M, eds. Hen Eggs—Their Basics and Applied Science. New York: CRC Press; pp 99–115.

Kollberg H, Carlander D, Olesen H, Wejaker PE, Johannesson M, Larsson A (2003). Oral administration of specific yolk antibodies (IgY) may prevent *Pseudomonas aeruginosa* infections in patients with cystic fibrosis: A phase I feasibility study. Pediatr Pulmonol 35:433–440.

Kovacs-Nolan J, Sasaki E, Yoo D, Mine Y (2001). Cloning and expression of human rotavirus spike protein, VP8*, in *Escherichia coli*. Biochem Biophys Res Commun 282:1183–1188.

Kozutsumi Y, Kawano T, Yamakawa T, Suzuki, A (1990). Participation of cytochrome b5 in CMP-N- acetylneuraminic acid hydroxylation in mouse liver cytosol. J Biochem 108(5):704–706.

Kurisaki J, Yamauchi K, Isshiki H, Ogiwara S (1964). Difference between α- and β-lipovitellin from hen egg yolk. Agric Biol Chem 45:699.

Kuroki M, Ikemori Y, Yokoyama H, Peralta RC, Icatlo Jr FC, Kodama Y (1993). Passive protection against bovine rotavirus-induced diarrhea in murine model by specific immunoglobulins from chicken egg yolk. Vet Microbiol 37:135–146.

Kuroki M, Ohta Y, Ikemori Y, Peralta RC, Yokoyama H, Kodama Y (1994). Passive protection against bovine rotavirus in calves by specific immunoglobulins from chicken egg yolk. Arch Virol 138:143–148.

Kweon CH, Kwon BJ, Woo SR, Kim JM, Woo GH, Son DH, Hur W, Lee YS (2000). Immunoprophylactic effect of chicken egg yolk immunoglobulin (IgY) against porcine epidemic diarrhea virus (PEDV) in piglets. J Vet Med Sci 62:961–964.

Larsson A, Balow R, Lindahl TL, Forsberg P (1993). Chicken antibodies: taking advantage of evolution; a review. Poultry Sci 72:1807–1812.

Larsson A, Carlander D (2003). Oral immunotherapy with yolk antibodies to prevent infections in humans and animals. Ups J Med Sci 108:129–140.

Larsson A, Karlsson-Parra A, Sjöquist J (1991). Use of chicken antibodies in enzyme immunoassays to avoid interference by rheumatoid factors. Clin Chem 37:411–414.

Larsson A, Mellstedt H (1992). Chicken antibodies: A tool to avoid interference by human anti-mouse antibodies in ELISA after in vivo treatment with murine monoclonal antibodies. Hybridoma 11:33–39.

LeClaire RD, Hunt RE, Bavari S (2002). Protection against bacterial superantigen staphylococcal enterotoxin B by passive immunization. Infect Immun 70:2278–2281.

Lee J, Babiuk LA, Harland R, Gibbons E, Elazhary Y, Yoo D (1995). Immunological response to recombinant VP8* subunit protein of bovine rotavirus in pregnant cattle. J Genet Virol 76:2477–2483.

Lee K, Ametani A, Shimizu M, Hatta H, Yamamoto T, Kaminogawa S (1991). Production and characterization of anti-human insulin antibodies in the hen's egg. Agric Biol Chem 55:2141–2413.

Lee SB, Mine Y, Stevenson RMW (2000). Effects of hen egg yolk immunoglobulin in passive protection of rainbow trout against *Yersinia ruckeri*. J Agric Food Chem 48:110–115.

Lemamy GJ, Roger P, Mani JC, Robert M, Rochefort H, Brouillet JP (1999). High-affinity antibodies from hen's egg-yolks against human mannose-6-phosphate/insulin-like growth-factor-II receptor (M6P/IGFII-R): Characterization and potential use in clinical cancer studies. Int J Cancer 80:896–902.

Leslie GA, Clem LW (1969). Phylogen of immunoglobulin structure and function. 3. Immunoglobulins of the chicken. J Exp Med 130(6):1337–1352.

Leslie GA, Martin LN (1973). Studies on the secretory immunologic system of fowl. 3. Serum and secretory IgA of the chicken. J Immunol 110:1–9.

Leventhal JR, Su A, Kaufman DB, Abecassis MI, Stuart FP, Anderson B, Fryer JP (2001). Altered infectivity of porcine endogenous retrovirus by "protective" avian antibodies: implications for pig-to-human xenotransplantation. Transplant Proc 33:690.

Levine AS, Doscherholmen A (1983). Vitamin B12 bioavailability from egg yolk and egg white: Relationship to binding proteins. Am J Clin Nutr 38(3):436–439.

Li SC, Chien JL, Wan CC, Li YT (1978). Occurrence of glycosphingolipids in chicken egg yolk. Biochem J 173:697–699.

Li X, Nakano T, Sunwoo HH, Paek BH, Chae HS, Sim JS (1997). Effects of egg and yolk weights on yolk antibody (IgY) production in laying chickens. Poultry Sci 77:266–270.

Li-Chan EC, Ler SS, Kummer A, Akita EM (1998). Isolation of lactoferrin by immunoaffinity chromatography using yolk antibodies. J Food Biochem 22:179–195.

Li-Chan EC (2000). Applications of egg immunoglobulins in immunoaffinity chromatography. In: Sim JS, Nakai S, Guenter W, eds. Egg Nutrition and Biotechnology. New York: CAB International; pp 323–339.

Li-Chan ECY, Powrie WD, Nakai S (1995). The chemistry of eggs and egg products. In: Stadelman WJ, Cotterill OJ, eds. Egg Science and Technology, 4th ed. New York: Haworth Press; pp 105–175.

Lieber C, Robins SJ, Li J, Decarli LM, Mak M, Fasulo JM, Leo MA (1994). Phosphatidylcholine protects against fibrosis and cirrhosis in the baboon. Gastroenterology 106(1):152–159.

Lin Y, Feeney RE (1972). Glycoproteins. Amsterdam: Elsevier/North-Holland Biomedical Press.

Ling YS, Guo YJ, Li JD, Yang LK, Luo YX, Yu SX, Zhen LQ, Qiu SB, Zhu GF (1998). Serum and egg yolk IgG antibody titres from laying chickens vaccinated with *Pasteurella multocida*. Avian Dis 42:186–189.

Loeken MR, Roth TF (1983). Analysis of maternal IgG subpopulations which are transported into the chicken oocyte. Immunology 49:21–28.

Lu CL, Baker R (1986). Characteristics of egg yolk phosvitin as an antioxidant for inhibiting metal-catalyzed phospholipid oxidations. Poultry Sci 65:2065–2070.

Lungberg B (1973). Isolation and characterization of egg lecithin. Acta Chem Scand 27(9):3535–3549.

Maddison SE, Reimer CB (1976). Normative values of serum immunoglobulins by single radial immunodiffusion: A review. Clin Chem 22(5):594–601.

Makrides M, Hawkes JS, Neumann MA, Gibson RA (2002). Nutritional effect of including egg yolk in the weaning diet of breast-fed and formula-fed infants: A randomized controlled trial. Am J Clin Nutr 75:1084–1092.

Martin WG, Augstyniak J, Cook WH (1964). Fractionation and characterization of the low density lipoproteins of hen's egg yolk. Biochem Biophs Acta 84:714–720.

Matsuda T (1998). Biosynthesis of egg. In: Nakamura R, ed. Tamago no kagaku. Tokyo, Asakura Shoten; pp 30–41.

Matsurra F, Imaoka A (1988). Chromatographic separation of asparagines-linked oligosaccharides labeled with an ultraviolet absorbing compounds, p-aminobenzoic acid ethyl ester. Glycoconj J 5(1):13–26.

Matsuura F, Ohta M, Murakami K, Hirano K, Sweeley CC (1992). The combination of normal phase with reserved phase high performance liquid chromatography for the analysis of aparagine-linked neutral oligosaccharides labeled with p-aminobenzoic acid ethyl ester. Biomed Chromatogr 6(2):77–83.

McLaren RD, Prosser CG, Grieve RCJ, Borissenko M (1994). The use of caprylic acid for the extraction of the immunoglobulin fraction from egg yolk of chickens immunised with bovine α-lactalbumin. J Immunol Meth 177:175–184.

McReynolds L, O'Malley BW, Nisbet AD, Fothergill JE, Givol D, Fields S, Robertson M, Brownlee GC (1978). Sequence of chicken ovalbumin mRNA. Nature 273:723–728.

Michael UF, Cookson SL, Chvez R, Pardo V (1975). Renal function in the choline deficient rat. Proc Soc Exp Biol Med 150:672–676.

Mine Y, Kovacs-Nolan J (2002). Chicken egg yolk antibodies as therapeutics in enteric infectious disease: A review. J Med Food 5(3):159–169.

Morrison SL, Mohammed MS, Wims LA, Trinh R, Etches R (2002). Sequences in antibody molecules important for receptor-mediated transport into the chicken egg yolk. Mol Immunol 38:619–625.

Morschhäuser J, Hoschützky H, Jann K, Hacker J (1990). Functional analysis of the sialic acid binding adhesion SfaS of pathogenic *Escherichia coli* by site-specific mutagenesis. Infect Immun 58(7):2133–2138.

Motoi Y, Inoue S, Hatta H, Sato K, Morimoto K, Yamada A (2005a). Detection of rabies-specific antigens by egg yolk antibody (IgY) to the recombinant rabies virus proteins produced in Escherichia coli. Jpn J Infect Dis 58(2):115–118.

Motoi Y, Sato K, Hatta H, Morimoto K, Inoue S, Yamada A (2005b). Production of rabies neutralizing antibody in hen's eggs using a part of the G protein expressed in *Escherichia coli*. Vaccine 23(23):3026–3032.

Muchmore EA, Mileewski M, Varki A, Diaz S (1989). Biosynthesis of N-glycolyneuraminic acid. The primary site of hydroxylation of N-acetylneuraminic acid is the cytosolic sugar nucleotide pool. J Biol Chem 264(34):20216–20223.

Muniyappa K, Adiga PR (1981). Nature of the thiamin-binding protein from chicken egg yolk. Biochem J 193(3):679–685.

Nakane S, Tokumura A, Wahu K, Sugira T (2001). Hen egg yolk and white contain high amounts of lysophosphatidic acids, growth factor-like lipids: Distinct molecule species compositions. Lipid 36(4):413–419.

Nisbet AD, Saundry RH, Moir AJG, Fothergill LA, Fothergill JE (1981). The complete amino acid sequence of hen ovalbumin. Eur J Biochem 115(2):335–345.

Niskanen EA, Hytonen VP, Grapputo A, Nordlund HR, Kuloma MS, Laitinen OH (2005). Chicken genome analysis reveals novel genes encoding biotin-binding proteins related to avidin family. BMC Genomics 6(1):41.

Noack F, Helmecke D, Rosenberg R, Thorban S, Nekarda H, Fink U, Lewald J, Stich M, Schutze K, Harbeck N, Magdolen V, Graeff H, Schmitt M (1999). CD87-positive tumor cells in bone marrow aspirates identified by confocal laser scanning fluorescence microscopy. Int J Oncol 15:617–623.

Ntakarutimana V, Demedts P, Sande MV, Scharpe S (1992). A simple and economical strategy for downstream processing of specific antibodies to human transferrin from egg yolk. J Immunol Meth 153:133–140.

Ohta M, Kobatake M, Matsumura A, Matsuura F (1990). Separation of Asn-linked sialyloligosaccharides labeled with p-aminobenzoic acid ethyl ester by high performance liquid chromatography. Agric Biol Chem 54(4):1045–1047.

Ostro MJ (1987). Liposomes. Sci Am 256:102–111.

Otake S, Nishihara Y, Makimura M, Hatta H, Kim M, Yamamoto T, Hirasawa M (1991). Protection of rats against dental caries by passive immunization with hen egg yolk antibody (IgY). J Dent Res 70:162–166.

Pangborn MC (1951). A simplified purification of lecithin. J Biol Chem 188(2):471–476.

Parkinson TL (1966). The chemical composition of eggs. J Sci Food Agric 17: 101–111.

Patterson JM, McHenry EW (1944). Choline and prevention of hemorrhsgic kidneys in the rat. J Biol Chem 156:265–269.

Paulson JC, Rogers GN, Carroll SM, Higa HH, Pritchett T, Milks G, Sabesan S (1984). Selection of influenza virus variants based on sialyoligosaccharide receptor specificity. Pure Appl Chem 56:797–805.

Pawelczyk T, Lowenstein JM (1993). Inhibition of phospholipase delta by hexadecylphosphoryl choline and lysophospholipids with antitumor activity. Biochem Pharmacol 45:493–497.

Phillips ML, Nudelman E, Gaeta FCA, Perez M, Singhal AK, Hakomori S, Paulson JC (1990). ELMA-1 mediates cell adhesion by recognition of a carbohydrate ligand, siayl-lex. Science 250(4984):1130–1132.

Polson A, von Wechmar MB, van Regenmortel MH (1980). Isolation of viral IgY antibodies from yolks of immunized hens. Immunol Commun 9:475–493.

Prosise WE (1985). Commercial lecithin products: Food use of soyabean lecithin. In: Szuhaj BF, List GR et al, eds. Lecithins. Champaign, IL: American Oil Chemists Society; pp 163–182.

Rabinowich H, Lyte M, Steiner Z, Klajman A, Ahinitzky M (1987). Augmentation of mitogen responsiveness in the aged by a special lipid diet Al 721. Mech Age Devel 40:131–138.

Rose ME, Orlans E, Buttress N (1974). Immun globulin classes in the hen's eggs: Their segregation in yolk and white. Eur J Immunol 4:521–523.

Sagarin E (1957). Cosmetics Science and Technology. New York: Interscience.

Sandrin MS, McKenzie IF (1994). Gal alpha(1,3)Gal, the major xenoantigen(s) recognised in pigs by human natural antibodies. Immunol Rev 141:169–190.

Sarker SA, Casswall TH, Juneja LR, Hoq E, Hossain E, Fuchs GJ, Hammarstrom L (2001). Randomized, placebo-controlled, clinical trial of hyperimmunized chicken egg yolk immunoglobulin in children with rotavirus diarrhea. J Pediatr Gastroenterol Nutr 32:19–25.

Sattar Khan MA, Nakamura S, Ogawa M, Akita E, Azakami H, Kato A (2000). Bactericidal action of egg yolk phosvitin against *Escherichia coli* under thermal stress. J Agric Food Chem 48:1503–1506.

Schade R, Pfister C, Halatsch R, Henklein P (1991). Polyclonal IgY antibodies from chicken egg yolk-an alternative to the production of mammalian IgG type antibodies in rabbits. Altern Lab Anim 19:403–419.

Schauer R (1982a). Chemistry, metabolism, and biological function of sialic acids. Adv Carbohydr Chem Biochem 40:131–234.

Schauer R (1982b). Sialic Acids: Chemistry, Metabolism and Function. Cell Biology Monographs, vol 10. New York: Springer-Verlag.

Scholfield CR (1985). Occurrence, structure, composition and nomenclature. In: Szuhaj BF, List, GR, eds. Lecithins. Champaign IL: American Oil Chemists Society; pp 1–64.

Schultze B, Herrler G (1994). Recognition of cellular receptors by bovine corona virus. Arch Virol 9:451–459.

Seko A, Koketsu M, Nishizono M, Enoki Y, Ibrahim HR, Juneja LR, Kim M, Yamamoto T (1997). Occurrence of a sialylglycopeptide and free sialylglycans in hen's egg yolk. Biochim Biophys Acta 1335(1–2):23–32.

Shale DJ, Elborn JS (1996). Lung injury. In: Shale DJ, ed. Cystic Fibrosis. London: BMJ Publishing Group; pp 62–78.

Sharon N (1975). Complex Carbohydrates: Their Chemistry, Biosynthesis and Functions. New York: Addison-Wesley.

Shaw L, Schauer R (1988). The biosynthesis of N-glycoloylneuraminic acid occurs by hydroxylation of the CMP-glycoside of N-acetylneuraminic acid. Biol Chem Hoppe-Seyler 369(6):477–486.

Shin NR, Choi IS, Kim JM, Hur W, Yoo HS (2002). Effective methods for the production of immunoglobulin Y using immunogens of *Bordatella bronchiseptica, Pasteurella multocida* and *Actinobacillus pleuropneumoniae*. J Vet Sci 3:47–57.

Sillanaukee P, Pönniö M, Jääskeläinen IP (1999). Occurrence of sialic acids in healthy humans and different disorders. Eur J Clin Invest 29:413–425.

Smith DJ, King WF, Godiska R (2001). Passive transfer of immunoglobulin Y to Streptococcus mutans glucan binding protein B can confer protection against experimental dental caries. Infect Immun 69:3135–3142.

Smith H, Gaastra W, Kamerling JP, Vliegenthart JFG, de Graaf FK (1984). Isolation and structural characterization of the equine erthrocyte receptor for enterotoxigenic *Escherichia coli* K99 fimbrial adhesion. Infect Immun 46:578–581.

Stedman RA, Singlenton L, Box PG (1969). Purification of Newcastle disease virus antibody from the egg yolk of the hen. J Compar Pathol 79(4):507–516.

Stevenson RMW, Flett D, Raymond BT (1993). Enteric redmouth (ERM) and other enterobacterial infections of fish. In: Inglis V, Roberts RJ, Bromage NR, eds. Bacterial Diseases of Fish. Oxford: Blackwell Scientific Publications; pp 80–105.

Stryer L (1988). Biochemistry, 3rd ed. New York: Freeman.

Sugano M, Hirahara F (2000). Polyunsaturated fatty acds in the food chain in Japan. Am J Clin Nutr 71(Suppl):189S–196S.

Sugino H, Nitoda T, Juneja LR (1997). General chemical composition of hen eggs. In: Yamamoto T, Juneja LR, Hatta H, Kim M, eds. Hen Eggs, Their Basic and Applied Science. New York: CRC Press; pp 13–24.

Sugita-Konishi Y, Shibata K, Yun SS, Yukiko HK, Yamaguchi K, Kumagai S (1996). Immune functions of immunoglobulin Y isolated from egg yolk of hens immunized with various infectious bacteria. Biosci Biotechnal Biochem 60:886–888.

Svendsen L, Crowley A, Ostergaard LH, Stodulski G, Haul J (1995). Development and comparison of purification strategies for chicken antibodies from egg yolk. Lab Anim Sci 45:89–93.

Tabassum S, Shears P, Hart CA (1994). Genomic characterization of rotavirus strains obtained from hospitalized children with diarrhea in Bangladesh. J Med Virol 43(1):50–56.

Tesedo J, Barrado E, Sanz MA, Tesedo A, de la Rosa F (2006). Fatty acid profiles of processed chicken egg yolks. J Agric Food Chem 54(17):6255–6260.

Thalley BS, Carroll SB (1990). Rattlesnakes and scorpion antivenoms from the egg yolk of immunized hens. Biotechnology 8:934–937.

Tyrrell DA, Heath TD, Colley CM, Ryman BE (1976). New aspects of liposomes. Biochim Biophys Acta 457(3–4):259–302.

Van Nieuwenhuyzen W (1976). Lecithin production and properties. J Am Oil Chem Soc 53:425–427.

Varki A, Cumming R, Esko D, Freeze H, Hart G, Marth J, eds (1999). Essentials of Glycobiology. Cold Spring Harbor, NY: Cold Spring Harbor Laboratory.

Varki A (1997). Sialic acids as ligands in recognition phenomena. FASEB J 11: 248–255.

Vasudevan N, Bahadur U, Kondaiah P (2001). Characterization of chicken riboflavin carrier protein gene structure and promoter regulation by estrogen. J Biosci 26(1):39–46.

Vieira AV, Sanders EJ, Schneider WJ (1995). Transport of serum transthyretin into chicken oocytes. A receptor-mediated mechanism. J Biol Chem 270(7):2952–2956.

von Itzstein M, Wu WY, Kok GB, Pegg MS, Dyason JC, Jin B, Van Phan T, Smythe ML, White HF et al (1993). Rational design of potent sialidase-based inhibitors of influenza virus replication. Nature 363:418–423.

Wallance RA, Morgan JP (1986). Chromatographic resolution of chicken phosvitin. Multiple macromolecular species in a classic vitellogenin-derived phosphoprotein. Biochem J 240(3):871–878.

Warren L (1959). Sialic acid in human semen and in the male genital tract. J Clin Invest 38(5):755–761.

Weiner N, Martin F, Riaz M (1989). Liposomes as a drug delivery system. Drug Devel Indust Pharm 15:1523.

Williams J (1962). Serum proteins and for livetins of hen's egg yolk. Biochem J 82:346–355.

Wills FK, Luginbuhl RE (1963). The use of egg yolk for passive immunization of chickens against Newcastle disease. Avian Dis 7:5–12.

Worledge KL, Godiska R, Barrett TA, Kink JA (2000). Oral administration of avian tumor necrosis factor antibodies effectively treats experimental colitis in rats. Digest Dis Sci 45:2298–2305.

Wu C, Yang R, Zhou J, Bao S, Zou L, Zhang P, Mao Y, Wu J, He Q (2003). Production and characterisation of a novel chicken IgY antibody raised against C-terminal peptide from human thymidine kinase 1. J Immunol Meth 277:157–169.

Yamamura J, Adachi T, Aoki N, Nakajima H, Nakamura R, Matsuda T (1995). Precursor-product relationship between chicken vitellogenin and the yolk proteins: The 40 kDa yolk plasma glycoprotein is derived from the C-terminal cysteine-rich domain of vitellogenin II. Biochim Biophys Acta 1244(2–3):384–394.

Yang J, Jin Z, Yu Q, Yang T, Wang H, Liu L (1997). The selective recognition of antibody IgY for digestive system cancers. Chin J Biotechnol 13:85–90.

Yano N, Fukinbara I, Takano M (1979). A Process for Obtaining Yolk Lecithin from Raw Egg Yolk. US Patent 4,157,404.

Yoshimura S, Watanabe K, Suemizu H, Onozawa T, Mizoguchi J, Tsuda K, Hatta H, Moriuchi T (1991). Tissue specific expression of the plasma glutathione peroxidase gene in rat kidney. J Biochem (Tokyo) 109:918–923.

Zeisel SH, Da Costa KA, Franklin PD, Alexander EA, Lamont JT, Sheard NF, Beiser A (1991). Choline, an essential nutrient for human. FASEB J 5(7):2093–2098.

Zeisel SH (1981). Dietary choline: Biochemistry, physiology and pharmacology. Ann Rev Nutr 1:95–121.

6

EGG ALLERGENS

MARIE YANG AND YOSHINORI MINE
Department of Food Science, University of Guelph, Guelph, Ontario, Canada

6.1. INTRODUCTION

Hen egg allergy is one of the most important causes of childhood food allergy. There is currently no efficient approach for the prevention and/or treatment of egg-induced allergy. The only causative treatment is complete avoidance of the offending food. However, the omnipresence of minute amounts of egg-derived components in processed and prepared food put egg-allergic individuals at high risk. There is therefore an urgent need for the development of novel approaches for the treatment and prevention of egg allergy.

In order to develop accurate diagnostic tools, efficient therapeutic approaches, and preventive interventions, molecular characterization of the allergenic components present in eggs, combined with understanding of the immunologic mechanisms underlying egg hypersensitivity, is an absolute prerequisite. After a brief definition and discussion of food allergy, this chapter presents an overview on the epidemiologic and clinical aspects of egg allergy. It then reviews the molecular properties of egg proteins with regard to their antigenicity and allergenicity, and discusses how this information can be utilized for the investigation of safe and efficient forms of specific immunotherapy. The concept of hypoallergenicity will be discussed in the context of clinical applications (e.g., hypoallergenic vaccines) as well as in industrial applications (e.g., potential preparation of hypoallergenic egg products).

Egg Bioscience and Biotechnology Edited by Yoshinori Mine

6.2. FOOD ALLERGY OVERVIEW

6.2.1. Definition and Prevalence of Food Allergy

An allergic reaction to food is believed to result from an overt response of the mucosal immune system to innocuous dietary antigens, mainly food proteins (Bischoff and Crowe 2005). According to a recently revised nomenclature, food allergies should be distinguished from food intolerances, which are non-immunological in nature (Fig. 6.1). The best understood and most common form of food allergy, also known as a type I hypersensitivity reaction (Ebo and Stevens 2001), is mediated by a class of antibodies secreted by B lymphocytes, called *immunoglobulin E* (IgE).

Food-induced allergies are now recognized as a worldwide medical problem, especially in industrialized countries. They are estimated to occur in 6% of young children and often persist beyond childhood to reach 3–4% of adults according to an epidemiologic study completed in the United States (Sampson

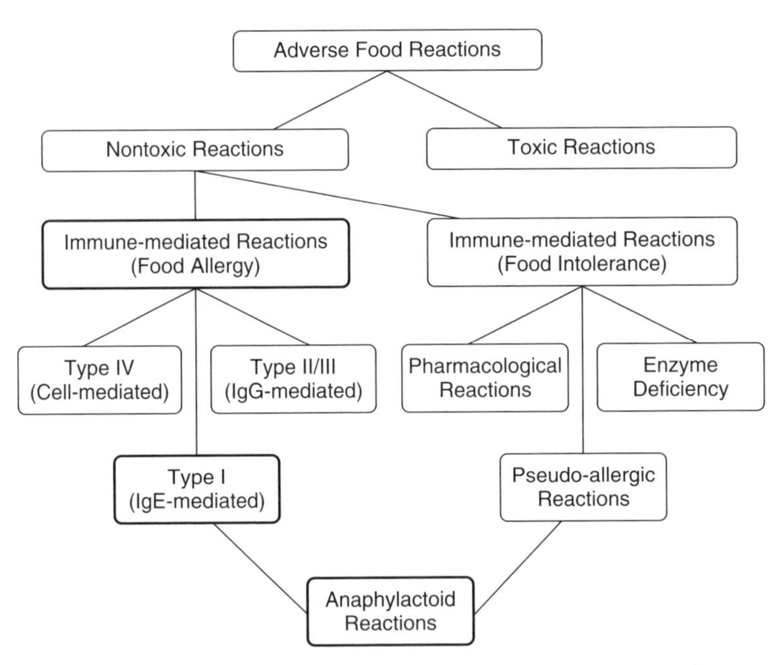

Figure 6.1. Classification of adverse food reactions. (Adapted from Vieleuf et al. 2002). Adverse reactions to food can be classified as either toxic or nontoxic. Nontoxic reactions are further divided into immune-mediated and non-immune-mediated. For immune-mediated reactions, the term *food allergy* is recommended, while non-immune-mediated reactions should be referred to as *food intolerance*. Food allergies are either IgE-mediated or non-IgE-mediated. The term *pseudoallergic reactions* refers to cases in which clinical symptoms mimic those seen in allergic reactions, with no apparent immunologic sensitization.

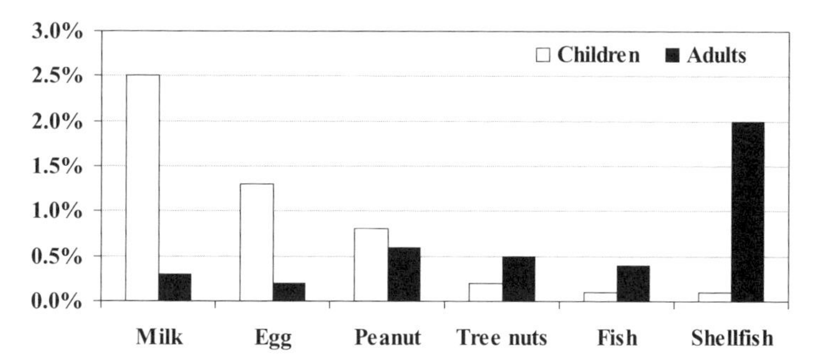

Figure 6.2. Prevalence of food allergies in children and adults in the United States [adapted from Zuercher et al. (2006)].

2004) (Fig. 6.2). Approximately 80% of allergic reactions in children are due to milk, egg, and peanut alone (Burks et al. 2004), while adults are more likely to develop sensitivity to shellfish products (Sampson 2004).

The development of an allergic disease seems to result from complex interactions between multiple influences, including genetic (e.g., atopic heredity), environmental (e.g., early exposure to indoor and outdoor allergens), and cultural factors (e.g., dietary habits) (Host 1995; Halken 2004; Bischoff and Crowe 2005).

6.2.2. Clinical Symptoms of Food Allergy

Food-induced allergic symptoms usually occur within minutes to hours after ingestion of the offending food (Williams and Bock 1999). They range in severity from mild to life-threatening reactions in some cases, and may affect multiple organs, including the skin (urticaria, angioedema, eczema), the gastrointestinal tract (nausea, vomiting, abdominal pain, diarrhea), the respiratory tract (cough, asthma, and rhinitis), and the cardiovascular system (tachycardia, hypotension, anaphylactic shock) (Sicherer 2002) (Table 6.1). Systemic anaphylaxis is a potentially life-threatening manifestation of food allergy. It is characterized by a sudden onset of dyspnea, cyanosis, angina, hypotension, and shock. Cutaneous manifestations, such as urticaria and angiodema, may also occur simultaneously. If untreated, systemic anaphylaxis can be fatal.

6.2.3. Major Food Allergens: The "Big Eight"

The term "Big Eight" was coined to designate peanuts, tree nuts (e.g., walnuts, cashews, Brazil nuts), cow's millk, soy, wheat, hen's egg, fish, and crustaceans with respect to IgE-mediated food allergy, both internationally and in the United States (Teuber et al. 2006). These eight foods account for more than 90% of food allergies of group studies completed in the United States (Sampson and McCaskill 1985; Burks et al. 1998; Ellman et al. 2002). Among these foods,

TABLE 6.1. Clinical Symptoms of Food Allergy

Skin reactions	Reactions of gastrointestinal tract
Contact urticaria	Abdominal cramps and distension
Urticaria/angioedema	Nausea, vomiting
Flush, pruritus	Gastritis, gastroenteritis with diarrhea
Oral allergy syndrome[a]	Anorexia
Protein contact dermatitis	Flatulence
Atopic eczema	Colitis
Reactions of the respiratory system	Reactions of cardiovascular system
Rhinoconjonctivitis	Cardiac dysrhythmia
Bronchial Asthma	Hypotension
Laryngeal and/or pharyngeal edema	Vascular collapse
Hoarseness	Anaphylactic shock

[a]Local itching and tingling and/or edema of lips, tongue, palate, and pharynx.

Source: Adapted from Vieleuf et al. (2002).

peanut and tree nuts have been accounted for the most severe anaphylactic responses, leading to fatal or near-fatal reactions in Europe and in the United States (Yunginger et al. 1988; Sampson et al. 1992; Bock et al. 2001). The Codex Alimentarius Commission (1999) recommended that member countries adopt this list of eight common foods and ensure that food manufacturers list these eight foods or ingredients derived from them on their labels. Around the world, subsequent legislation has flourished to help and protect allergic consumers against these "Big Eight" (Gowland 2001).

6.2.4. Molecular and Cellular Mechanisms of Food Allergy

The etiology of type I food allergies classically involves two phases: (1) the sensitization phase, without symptoms, during which the immune system is primed, resulting in the production of antibodies belonging to the immunoglobulin E class; and (2) the elicitation phase, with specific clinical responses (i.e., allergic symptoms), which occurs after subsequent exposure to the offending compound. During the onset of an allergic reaction, it has been shown that antigen could cross the intestinal barrier, be captured by the underlying immune cells, processed, and presented by the underlying antigen-presenting cells (Fig. 6.3).

It is generally accepted that an immune response results from a balance between two subsets of T lymphocytes: CD4+ T helper (Th) 1 and Th2 lymphocytes (Mosmann et al. 1986; Mosmann and Coffman 1989). In general, Th1 cells play a prominent role in cell-mediated immunity with the secretion of cytokines such as interferon-(IFN)-γ, interleukin (IL)-2, and tumor necrosis factor (TNF)-β, which promotes the development of cytotoxic T lymphocytes (CTL) and macrophages (Untersmayr and Jensen-Jarolim 2006). On the other hand, Th2 cells promote primarily humoral immune responses, with the expression of cytokines such as IL-4 and IL-13 (promoting IgE synthesis), IL-5 (eosinophil proliferation), and IL-9 (mast cell activation). A dominant Th2

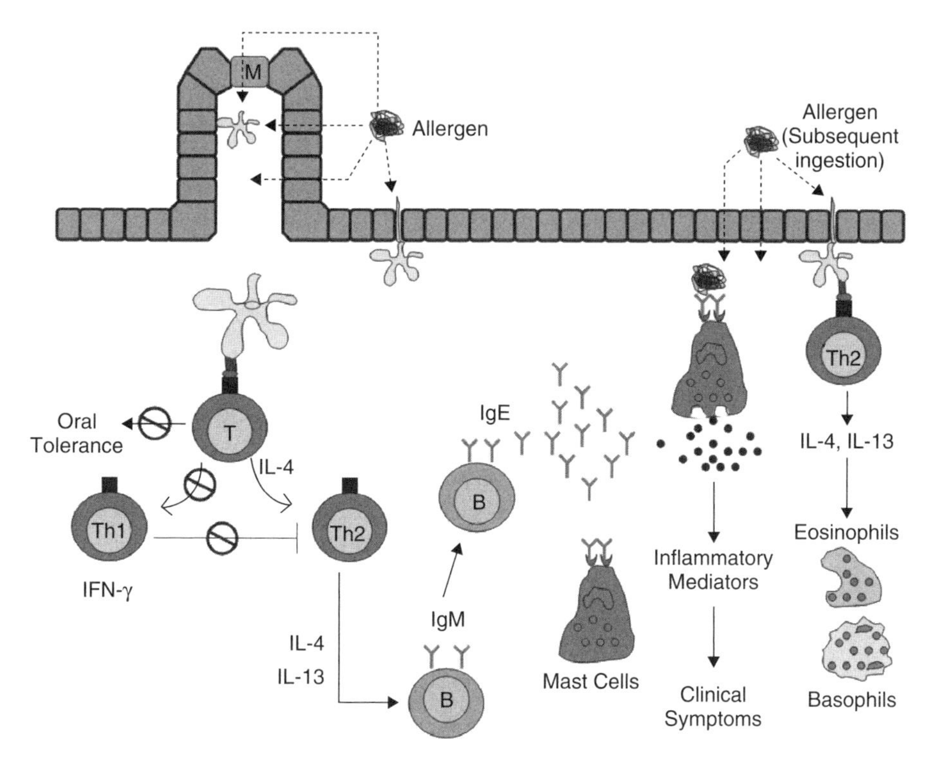

Figure 6.3. Cellular and molecular mechanisms of food allergy [adapted from Prioult and Nagler-Anderson (2005)]. (See insert for color representation.)

pattern of cytokine expression has been associated with allergic immune responses. A significant feature of the Th1/Th2 paradigm is that Th1 and Th2 are capable of cross-regulating one another. The cytokine environment encountered by a naive T helper cell plays a prominent role in determining whether the naive T cell will develop into a Th1 or Th2 cell. Cytokines such as IL-4 will deviate naive T cells toward a Th2 phenotype, while IL-12 and IFN-γ are important in the development of Th1 cells.

In healthy individuals, ingestion of innocuous antigens usually leads to a status of oral tolerance, whereas allergic individuals will develop an IL-4 dominant cytokine microenvironment in the gastrointestinal mucosa. When food antigens are presented to naive T cells by antigen-presenting cells (e.g., dendritic cells), antigen-specific Th2 cells will be generated. The activation of Th2 cells will lead to the production of IL-4 and IL-13, which both promote immunoglobulin E (IgE) production by B cells. The allergen-specific IgE will then bind to their high-affinity receptor (FcεRI) present at the surface of mast cells (Fig. 6.3). On subsequent ingestion of the food allergen, antigen presentation will lead to a rapid activation of Th2 cells, followed by the activation of eosinophils and basophils. In the meantime, allergenic fragments may also bind to receptor-bound IgE on mast cells, triggering the aggregation of

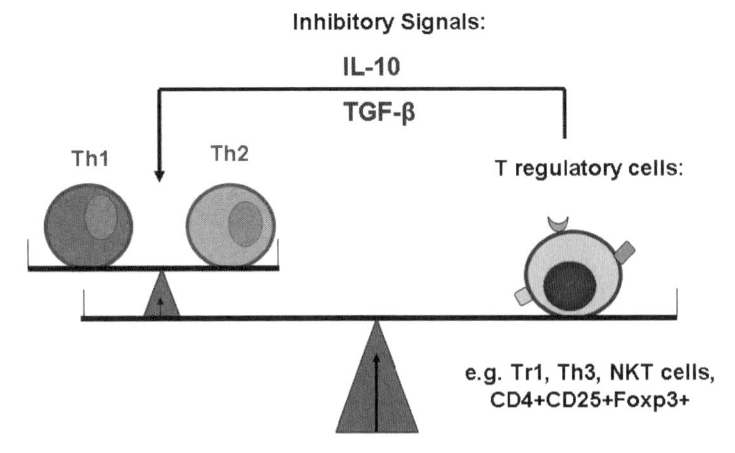

Figure 6.4. Currently accepted view on T-cell regulation. Regulatory T-cell populations exert inhibitory effects on the immune balance of Th1 and Th2, thereby preventing overt immune response by either type of cell. Several populations of Treg have been reported, including IL-10-producing inducible Treg cells, TGF-β-producing Th3 cells, naturally occurring CD4+CD25+Foxp3 Tr1 cells, and natural-killer T cells (NKT cells). Either TGF-β or IL-10 would exert direct and/or indirect inhibitory effects on neighboring cells, and both are believed to play an important role in the induction of immune tolerance [adapted from Jeurinks et al. (2006)]. (See insert for color representation.)

the receptors and the subsequent release of inflammatory and vasoactive mediators such as histamine, which are directly responsible for the clinical symptoms of food allergy (Table 6.1).

Previously, the immunologic scene was dominated by the concept that the Th1/Th2 balance was a central player in the regulation of immune responses (Romagnani 1991). This led to a very fruitful period during which it was shown that an efficient therapy would be one that would modulate this balance toward a Th1-leading response and therefore a reduction in allergic disease. Currently, the Th1/Th2 model has evolved to introduce a third subset of T lymphocytes, the regulatory T cells (Tregs). Tregs produce suppressive cytokines, such as IL-10 and transforming-growth factor (TGF)-β, to keep inflammatory T cells (both Th1 and Th2) and their downstream effectors under control (Stock et al. 2006) (Fig. 6.4). It is suggested that the lack of Treg cells activity would result in a Th2-skewed immune response, with elevated levels of cytokines such as IL-4, IL-5, and IL-13.

6.3. EGG ALLERGY

6.3.1. Epidemiologic and Clinical Aspects of Egg Allergy

The hen's egg is an important cause of childhood allergy. In fact, egg hypersensitivity has been reported in two-thirds of children and adolescents with atopic dermatitis (Sampson 1997). Egg allergy usually develops within the first 2 years of life, and usually resolves by school age (Heine et al. 2006; Teuber et al. 2006). Sensitization to egg white, as with that to milk, might develop very early in exclusively breastfed children. Sensitization may result from the transfer of small doses of the antigen into breast milk. Infants sensitized to egg through this route can develop allergic symptoms to their first ingestion of egg (Caffarelli et al. 1995).

In the United States, the prevalence of egg allergy was estimated at about 1.3% (Fig. 6.2) and represents the second most common form of food allergy in young children, while in adults the estimate is 0.2% (Sampson 2004). In other countries, such as Japan and Spain, hen egg represents the major cause of food allergy in children less than 2 years old, exceeding the prevalence of cow's milk allergy (Crespo et al. 1995; Imai and Iikura 2003; Teuber et al. 2006). In Korea, for instance, a study in 2004 showed that hen's egg hypersensitivity significantly surpassed other food allergen sources in children with atopic dermatitis. In this latter study, CAP-FEIA tests showed that 87 out of 266 children had specific IgE levels highly suggestive of clinical allergy to egg, against only 12 out of 266 for cow's milk, 8 for peanut, and 3 for soy (Han et al. 2004). Similarly, a prospective study was recently conducted in Sweden where 100 children with atopic dermatitis were followed up to the age of 7 (Gustafsson et al. 2003). The report showed that hen egg was the most common food sensitizer, with 46 out of 58 food allergic children showing detectable levels of hen egg–specific IgE. By the age of 7, 74% of these children had resolved their IgE sensitization to hen egg, as measured by radioallergosorbent (RAST) assays.

Other studies suggested that early childhood sensitization to hen egg was associated with the subsequent development of respiratory allergies (Nickel et al. 1997; Tariq et al. 2000). In adults, occupational asthma has been associated with the inhalation of aerosolized dried egg powder, leading to the development of IgE-mediated response on egg ingestion (Leser et al. 2001; Escudero et al. 2003).

In egg-allergic patients, the clinical symptoms involve various organs such as the skin (e.g., urticaria, angioedema, and atopic dermatitis), the respiratory system (e.g., asthma, rhinoconjonctivitis), and/or the gastrointestinal system (e.g., vomiting, diarrhea) (Martorell Aragones et al. 2001). Severe anaphylactic reactions can also develop in patients with acute IgE-mediated sensitivity to hen egg (Novembre et al. 1998). However, in infancy, atopic dermatitis represents the main clinical manifestation (Heine et al. 2006). It has been reported that egg allergy tends to persist in children with more severe reactions (e.g., respiratory symptoms, angioedema, multisystemic reactions) or with positive skin tests (Ford and Taylor 1982). In some patients, although ingestion is tolerated, contact with eggs can cause urticaria. It has been reported that these patients have IgE antibodies that recognize egg white epitopes unstable to the action of digestive enzymes (Yamada et al. 2000).

Diagnosis of egg allergy is usually determined by skin prick tests or RAST assays, but the gold standard remains the double-blind placebo controlled food challenge, which will confirm the clinical diagnosis. Diagnostic cutoff values have been suggested for food-specific serum IgE antibody levels and skin prick test wheal diameters, allowing for the prediction of adverse oral challenge outcomes (Heine et al. 2006). Such knowledge would indeed decrease the need for formal food challenges. Recent studies have provided supportive data to the reliability of skin prick tests results in the diagnosis of IgE-mediated egg allergy (Knight et al. 2006).

Egg hypersensitivity has raised other issues related to the potential development of anaphylaxis subsequent to viral vaccine administration. Indeed, the triple MMR (measles, mumps, and rubella) vaccines, as well as influenza and yellow fever vaccines, classically use egg in their manufacturing processes. The MMR vaccines are classically prepared from viral strains multiplied in chick embryo fibroblast and may contain only minute amounts of hen egg allergenic components. On the other hand, yellow fever and influenza vaccines are prepared directly from egg embryos and may therefore have an immunologically more significant content in egg allergens (Teuber et al. 2006). More recent studies have provided evidence that MMR vaccine is safe in children with anaphylactic sensitivity to egg (James et al. 1995; Goodyear-Smith et al. 2005). The risks of developing anaphylaxis to yellow fever and influenza vaccines have also been the object of several studies (Chino et al. 1999; Zeiger 2002; Kursteiner et al. 2006).

6.3.2. Thresholds for Clinical Reactions to Egg Allergens

In a trial involving 124 patients, 16% of the subjects were reactive to <6.5 mg of hen egg proteins (Morisset et al. 2003). The same study reported that the lowest threshold dose in food challenges was <0.2 mg of hen egg proteins. A different study also involving oral food challenges, conducted in France, reported an even lower threshold, equal to only 0.13 mg of hen egg proteins (Taylor et al. 2004).

6.4. ANTIGENIC AND ALLERGENIC PROPERTIES OF EGG ALLERGENS

6.4.1. Major Egg Allergens

As early as 1912, egg proteins were implicated in food allergy (Schloss 1912). Reports have documented that the major egg allergens were contained mainly in the egg white (Anet et al. 1985). Early studies involving a cohort of 342 patients reported that the major egg allergens were, in increasing order

lysozyme > ovomucin > ovalbumin > ovomucoid,

based on skin tests (Miller and Campbell 1950). Using 13 egg-allergic patients, further studies documented ovomucoid (OVM) as the dominant egg allergen

(Bleumink and Young 1971). Radioallergosorbent test (RAST) and crossed radioimmunoelectrophoresis (CRIE) studies completed in the 1980s established that ovalbumin (OVA), ovomucoid (OVM), and ovotransferrin (OVT) were the major egg allergens (Langeland 1982; Hoffman 1983); later, lysozyme (LYS) was also demonstrated to be a significant egg allergen (Holen and Elsayed 1990). In an effort to establish which egg proteins were major allergens, another study investigated the binding of specific IgE to eight purified egg white and yolk proteins by RAST, using sera from 40 egg-sensitive children (Walsh et al. 2005). The study confirmed that the major egg allergens originated primarily from egg white, and include ovomucoid (Gal d1), ovalbumin (Gal d2), ovotransferrin (Gal d3), and lysozyme (Gal d4). The same reports confirmed previous results that egg yolk also contains allergenic proteins, which were identified as apovitellenins I and VI and phosvitin (Anet et al. 1985; Walsh et al. 1987, 1988). In previous studies, other proteins such as ovoflavoprotein and ovoinhibitor were identified as antigenic, but they lacked allergenic activity (Langeland 1982; Hoffman 1983; Anet et al. 1985).

The two major allergens ovomucoid and ovalbumin constitute about 11% and 54% of egg white proteins, respectively (Kovacs-Nolan et al. 2005). Both proteins are glycosylated, with as high as 25% of the mass of ovomucoid containing carbohydrates. Debate flourished over the immunodominance of ovalbumin as the major egg allergen; however, studies showed that use of contaminated commercial ovalbumin led to an overestimation of its dominance as a major egg allergen in egg-sensitive patients (Bernhisel-Broadbent et al. 1994). Using egg-allergic patients' sera, further studies confirmed that OVM was the dominant allergenic component in egg white (Cooke and Sampson 1997; Urisu et al. 1997; Zhang and Mine 1998, 1999).

A large number of studies have been conducted to characterize both the antigenic (IgG-binding) and allergenic (IgE-binding) properties of the major egg allergens. Variability in results have been attributed to (1) the varying degree of purity of the different egg proteins, (2) the use of egg allergic patients' sera with different clinical profiles, and (3) the use of animal antisera raised by parenteral administration of egg proteins (Matsuda et al. 1986; Mine and Zhang 2002). One study in particular compared the binding activity of IgG and IgE antibodies from eight egg-allergic patients sera versus rabbit anti–egg white IgG, to physically and chemically modified egg allergens (Mine and Zhang 2002). Native OVM and OVT were the dominant antigens in rabbit, while the human patients' IgG preferentially bound to native forms of OVM and LYS. The four major egg allergens (OVM, OVA, LYS, OVT) were thus thermally treated for 15 min at 95°C, or chemically modified with either urea 6 *M* or by carboxymethylation. Urea-treated OVA, LYS, and OVT led to an increase in human IgG-binding activity, while carboxymethylation and thermal treatment of OVM and OVT led to a significant drop of their antigenicity. Treatment with urea also increased the binding activity of rabbit IgG antibodies to OVT. On the other hand, the carboxymethylation of OVM, OVT, and LYS decreased their allergenicity, but did not affect IgE-binding activity to OVA. Altogether, these results suggest that OVA contains sequential IgE epitopes, while OVM and LYS may contain both sequential and conformational epitopes.

The same study further reported that the sensitization profile may have a significant impact on the nature and severity of the clinical symptoms. Indeed, the authors found that OVT and OVM were the dominant allergens recognized by IgE antibodies from patients with egg-induced anaphylaxis history, while atopic dermatitis patients' sera showed dominant binding activities to OVA and OVM.

6.4.2. Ovomucoid (Gal d1)

Ovomucoid (OVM) has a molecular weight of 28,000 Da, representing 11% of egg white proteins and is noncoagulable by heat. Furthermore, it has been shown that the protein is not denatured by trichloric acid–acetase procedures or 8 *M* urea (Bernhisel-Broadbent et al. 1994). Its molecular properties are described in detail in Chapter 1 (this book). Essentially, the molecule consists of three structurally independent tandem homologous domains (domains I, II, and III), nine intramolecular disulfide bridges, and 20–25% of carbohydrate moieties (Kato et al. 1987). Its main function is based on its trypsin inhibitory activity.

As mentioned earlier, a large number of studies have reported OVM as the major allergenic component in the hen egg. Previous studies using inhibition ELISA, T-cell proliferation assay, and epitope mapping tools (SPOTs membranes, sigma-Genosys) have determined that the profile of OVM sensitization could be quite variable among individuals with respect to the sites of IgE and IgG binding (Cooke and Sampson 1997). This latter study also suggested for the first time that recognition of linear versus conformational epitopes may be associated with the persistence of egg allergy. Two independent studies also found that ovomucoid was the major allergenic component in egg white and also proposed that ovomucoid-dominant sensitization played a determinant role in the prognosis of egg allergy. Thus, it has been suggested that the outgrowth of egg allergy could be predicted by the absence or the decrease of OVM-specific IgE titers (Bernhisel-Broadbent et al. 1994).

In its native form, OVM has been shown to be resistant to extreme denaturing temperature. Thus, early studies revealed that the heating of OVM at 100°C only had little effect on its antibody-binding properties (Deutsch and Morton 1956). The midpoint thermal transition of OVM was determined at 74°C, and the reversibility of its conformational changes has been studied in association with its high heat stability (Matsuda et al. 1981; Swint-Kruse and Robertson 1995). Early studies reported that OVM was more resistant to heat denaturation in acidic rather than alkaline environments (Lineweaver and Murray 1947). Further reports demonstrated that its trypsin inhibitory activity was indeed pH-dependent (Konishi et al. 1985). The allergenic and antigenic properties of ovomucoid were shown to be maintained after peptic digestion (Matsuda et al. 1985; Kovacs-Nolan et al. 2000; Takagi et al. 2005). This phenomenon was later partly attributed to the presence of its nine disulfide bonds. Indeed, the chemical reduction of OVM enhanced its digestibility and reduced its allergenicity (Kovacs-Nolan et al. 2000). Similarly, the chemical oxidation of OVM did not alter egg-allergic patient's specific IgE-binding activity, while its reduced form displayed a significant decrease (Cooke and Sampson 1997). In contrast, other studies reported that reduced forms of OVM did not affect its allergenicity (Djurtoft et al. 1991) or lead to an increase in IgE-binding activity in some patients' sera (Zhang and Mine 1998).

In a human study involving 72 subjects it was reported that OVM was present in an immunologically active form in cooked eggs, as evidenced by double-blind placebo controlled food challenges (Urisu et al. 1997). Evidence was provided that subjects with high IgE-binding activity to pepsin-treated ovomucoid were unlikely to outgrow their egg white allergies (Urisu et al. 1999). This suggests that linear epitopes play a significant role in the persistence of food allergy, since they represents the fragments that have survived the degradation process in the gastrointestinal environment.

6.4.2.1. Allergenicity of Ovomucoid Carbohydrate Moieties

Many food proteins undergo post-translational modifications such as glycosylation during their passage through the endoplasmic reticulum. Since the existence of *N*-glycan-specific IgE was initially reported (van Ree and Aalberse 1995; van Ree 2002), the allergenicity of carbohydrate moieties has been a matter of active debate. There are two isoforms of ovomucoid: nonglycosylated and glysosylated (Kato et al. 1987). The main saccharides present on OVM are *N*-acetylglucosamine (NAcGlc), mannose (Man), galactose (Gal), and *N*-acetylneuraminic acid (NAcNeu) (Beeley 1971), which are usually present as multiantennary structures. The antigenic properties of ovomucoid carbohydrate chains have been investigated (Gu et al. 1986, 1989). *In vitro* studies showed that the removal of its carbohydrate moiety led to a decreased stability against tryptic hydrolysis and heat denaturation (Gu et al. 1989). It is known that *N*-glycosylation can have a significant stabilizing or rigidifying effect on protein structure, conferring an enhanced resistance to denaturation (Pedrosa et al. 2000). In the particular case of highly glycosylated ovomucoid, it was initially reported that the carbohydrate moiety contributed to its IgE-binding activity in human sera (Matsuda et al. 1985), but subsequent studies reported that it did not participate to the protein allergenicity (Besler 1999a). In fact, the carbohydrate chains may have an inhibitory effect on its IgG- and IgE-binding activities (Zhang and Mine 1998). Cooke and Sampson (1997) reported that carbohydrate moieties present on the ovomucoid molecules had no contribution to its IgE- and IgG-binding activity, but the carbohydrates may, in counterpart, play a role in its resistance to gastrointestinal digestion. The potential role and effects of carbohydrates on egg protein allergenicity therefore remained a controversial issue.

In order to shed some light in this area, recent studies were conducted. In these experiments, recombinant isoforms of OVM domain III (P-Gly) have been cloned and expressed in the eukaryotes yeast system *Pichia pastoris*, which has been documented as an ideal host for the large-scale expression of proteins from various sources (Cereghino and Cregg 2000; Schuster et al. 2000). This latter study compared the allergenicity of native OVM domain III versus P-Gly using specific IgE mouse antibodies (Rupa et al. 2007). Results showed a significantly lower IgE-binding activity to P-Gly than to native domain III, suggesting a phenomenon of steric hindrance due to the presence of the newly synthesized carbohydrate moieties.

These results were supported by immunization studies where two groups of BALB/c mice were respectively sensitized to P-Gly and native OVM domain III. In the mouse group sensitized to P-Gly, results showed significantly lower specific IgE levels accompanied by a Th1-skewed cytokine profile (IFN-γ vs. IL-4). This suggests that P-Gly may have the potential to modulate the Th1/Th2 balance toward a nonallergic immune response (Rupa et al. 2007).

Structural analysis determined that the *N*-glycan carbohydrates specifically located at residue 28 on OVM domain III were adjacent to a dominant IgE-binding epitope (Rupa and Mine 2006b). It was suggested that the partial masking of the IgE epitope by these *N*-glycans may have accounted for the hypoallergenicity inherent to P-Gly. Further investigations are being conducted in our laboratory to assess the potential use of recombinant glycosylated forms of major egg allergens, such as OVM and OVA, in immunotherapeutic approaches.

6.4.2.2. Differential Allergenicity of Ovomucoid Three Domains

Early studies reported that OVM domain II contained most of the allergenic determinants (Kurisaki et al. 1981). The same authors reported later that domains I and/or III were however essential for OVM to fully exhibit its allergenicity, otherwise resulting in a loss of 60% (Konishi et al. 1982). Subsequent studies also identified IgE- and IgG-binding regions in all three domains (Cooke and Sampson 1997). More recent reports, affirmed that OVM domain III was more allergenic than domain I or II (Zhang and Mine 1998; Rupa and Mine 2006c) in both humans and mice.

6.4.2.3. Ovomucoid Epitope Mapping

Ovomucoid B- and T-cell epitopes have been reported by a number of studies in both egg-allergic patients and murine models. Using enzymatic fragments and synthetic peptides, early studies identified human IgG- and IgE-binding epitopes present on the OVM molecule (Cooke and Sampson 1997; Besler 1999b). Five IgE-binding epitopes and seven IgG-binding regions were identified. Of particular importance, it has been shown that the major OVM epitopes were mainly linear epitopes and that OVM domain III displayed the highest IgG- and IgE-binding activities, as evidenced by *in vitro* tests using the sera of 18 egg-allergic children (Zhang and Mine 1998). Chemical modifications of the OVM domain III suggested that hydrophilic residues were more important for the binding activity of IgG, while hydrophobic residues were more critical for IgE binding (Zhang and Mine 1999). Also of interest was the finding that the genetic attachment of an undecane peptide to the *N*-terminal end of OVM domain III suppressed the production and binding activity of specific IgE and IgG in a murine model (Mine and Rupa 2003b). The fine IgG and IgE epitope mapping of the entire OVM molecule has subsequently been reported in humans (Mine and Zhang 2002). In this latter study, three IgE epitopes were identified in domain I, four IgE epitopes in domain II, and two IgE epitopes in domain III. Holen and coworkers also reported the identification of novel B and T cell epitopes (Holen et al. 2001). More recently, the B-cell epitope mapping of ovomucoid was reported, and residues 81–95 (domains I and II) and residues 161–175 (domain III) were identified as B-cell epitopes in BALB/c mouse (Rupa and Mine 2006b, 2006a). In accordance with the epitope results generated with egg-allergic patients' sera (Mine and Zhang 2002), it was shown that OVM domain III exhibited the highest IgE-binding signal intensity in BALB/c mouse.

Amino acid residues critical to the binding of egg-allergic patients IgE to OVM domain III have been identified. A mutant isoform (GMFA) containing a substitution of glycine at position 162 (from G to M) and phenylalanine at position 167 (from F to A) led to a complete abrogation of human IgE-binding activity as evidenced by ELISA and Western blot analysis (Mine et al. 2003). The potential use of GMFA as a hypoallergenic vaccine in the inhibition of egg-induced allergic responses has been explored recently (Rupa and Mine 2006b, 2006a).

A separate study reported the identification of OVM T-cell epitopes recognized by three different strains of mice (Mizumachi and Kurisaki 2003). T-cell clones specific to ovomucoid were also isolated from human egg-allergic patients (Eigenmann et al. 1996; Suzuki et al. 2002). However, further studies on these clones revealed that the T-cell clone phenotype did not reflect the patients' clinical manifestations (Kondo et al. 2005).

A summary of B- and T-cell epitope mapping results in OVM, as reported by different groups, is schematically represented in Figure 6.5.

6.4.3. Ovalbumin (Gal d2)

Ovalbumin (OVA) has a molecular weight of 45,000 Da, constitutes 54% of the protein in egg white, and is easily denatured by urea and guanidinium salts (Bernhisel-Broadbent et al. 1994). The molecule is classified as a member of the serpin family of protease inhibitors, but it is not known to inhibit any proteases (Huntington and Stein 2001). Although OVM has now been accepted as the major egg allergen, ovalbumin has been used most extensively in basic research as a model allergen. OVA is the most abundant protein in egg but is relatively heat-labile; therefore, hypersensitivity to cooked eggs is believed to be due mainly to heat-stable OVM (representing only 11% of egg white

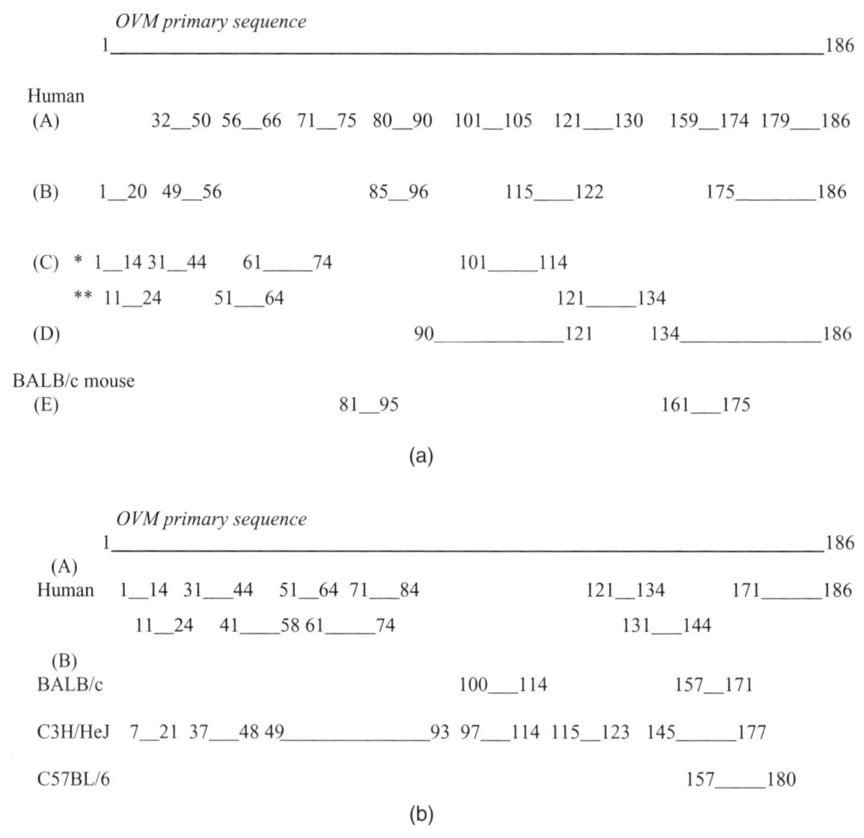

Figure 6.5. Schematic representation of IgE/T-cell epitope regions on ovomucoid, as reported by different groups. The numbers represent amino acid positions along the primary sequence of the protein. (a) Ovomucoid IgE/B-cell epitopes (human and mouse) [(A) Mine and Zhang 2002; (B) Cooke and Sampson 1997; (C) Holen et al. 2001; (D) Besler et al. 1999; (E) Rupa and Mine 2006a] (*IgE isotype; **IgG isotype); (b) ovomucoid allergenic T-cell epitope (human and mouse) [(A) Holen et al. 2001; (B) Mizumachi and Kurisaki 2003].

proteins). In its native form, ovalbumin is resistant to trypsin, while its heat-denatured form has shown an increased sensitivity to proteolysis (Kato et al. 1986). One study showed that the antigenicity of ovalbumin was resistant to heat treatment (Elsayed et al. 1986), while another report indicated a 90% decrease in its allergenicity after exposure to a temperature of 80°C for a duration of 3 min (Honma et al. 1994). To explain these variations, early studies suggested that sensitization to egg allergens was dependent on variation inherent in the group of patients selected, and not solely on the nature of the allergen (Miller and Campbell 1950). This was confirmed by subsequent reports (Walsh et al. 2005).

Several reports have documented OVA as a major allergen. In one study, 68 out of 68 sera from egg-sensitive individuals were determined positive to OVA in CRIE (Langeland 1982) The same authors determined in a subsequent study that 34 out of 34 sera from egg-allergic patients were positive to OVA in RAST and CRIE (Langeland 1983b).

6.4.3.1. Ovalbumin Epitope Mapping

A number of studies describe the allergenic and antigenic determinants of ovalbumin recognized by sera obtained from egg-allergic patients and murine models. A study in 2002 showed that the treatment of OVA by carboxymethylation, with 6 *M* urea or heat at 95°C had no significant effect on the IgE-binding capacity of human egg-allergic patients' sera (Mine and Zhang 2002). This suggests that patients' IgE antibodies mainly recognize linear epitopes on OVA.

The fine mapping of ovalbumin IgE epitope was initiated with identification of the N-terminal fragment, residues 1–10 by RAST using sera obtained from egg-allergic patients (Elsayed et al. 1988). Utilizing synthetic or cyanogen bromide (CNBr)-cleaved OVA peptides, subsequent studies demonstrated that the peptide regions 11–19, 34–70 (Elsayed and Stavseng 1994) and also fragments 41–172 and 301–385 (Kahlert et al. 1992) were reactive with patients' IgE antibodies. A different study also provided evidence that three overlapping haptenic peptides covering the sequence 347–385 were capable of specifically inhibiting histamine release from human basophil *in vitro* cultures, opening the door to innovative antiallergic strategies (Honma et al. 1996).

The OVA fragment 323–339 was initially identified as an immunodominant T-cell epitope (Sette et al. 1987). This fragment was later shown to present binding activities to human IgE antibodies (Buus et al. 1987), and to be responsible for up to 50% of OVA-specific IgE response in BALB/c mice sensitized intraperitoneally (Renz et al. 1993). The presence of both B- and T-cell epitopes (Janssen et al. 1999; Sun et al. 1999) in the fragment 323–339 has led to its widespread use as a hapten model for the induction of oral tolerance and the regulation of T-cell subsets in the context of peptide-based immunotherapy (Shimojo et al. 1994; Ohta et al. 1997). This same fragment also resulted in the creation of a specific BALB/c transgenic strain of mice coding for TCR-α and -β transgenes, which together specifically recognize the OVA peptide 323–339 (Sato et al. 1994).

Additional studies also identified the regions 41–172 and 301–385 as antigenic determinants of OVA, using specific anti-OVA monoclonal antibodies, containing IgG1 and IgM isotypes, produced from BALB/c mice (Kahlert et al. 1992). A further study showed that irradiation of OVA could induce the formation of neoepitopes by using IgG1 monoclonal antibodies from BALB/c. The fragment 173–196 was specifically identified as an IgG1-antigenic determinant found only on the irradiated form of OVA (Matsuda et al. 2000).

The mapping of IgE epitopes of the entire OVA molecule has been completed using egg-allergic patients' sera (Mine and Rupa 2003a) and in a mouse model (Mine and Yang 2007). The study involving egg-allergic patients' sera determined a total number of five IgE epitopes. The mouse study not only completed the entire map of IgE-binding epitopes present on the ovalbumin protein but also demonstrated the existence of a substantial heterogeneity in IgE-binding epitopes, between groups of BALB/c mice sensitized to ovalbumin via three different routes of exposure: oral, subcutaneous, and intraperitoneal routes. Thus, a total of eight prominent epitope regions were identified from orally sensitized mouse sera. In contrast, only two major IgE-binding epitopes were elicited by intraperitoneal immunization. Finally, results obtained with subcutaneously sensitized mice sera indicated five distinct binding regions.

Of particular interest were the results obtained from the orally challenged set of mice, as an oral challenge represents the natural route of human sensitization to food allergens. The elicitation of an allergic response in mouse by oral administration provides clear ties to the human condition; for this reason, the OVA IgE epitope map obtained in this study was compared to the one obtained in a previous study using human patients' sera (Mine and Rupa 2003a). Table 6.2 presents the similarities and differences between the two species.

TABLE 6.2. Comparison of Human versus BALB/c Mouse OVA IgE Sequential Epitopes

Variable Factor	Human Egg-Allergic Patients	BALB/c Mouse Oral Sensitization
Number of epitope regions	5	8
Epitope regions on OVA primary sequence	L38T49	I53D60
	D95A102	V77R84
	E191V200	S103E108
	V243E248	G127T136
	G251N260	E275V280
	G301F306	
	I323A332	
	A375S384	
Number of residues per epitope	6–12	6–10
Epitope physicochemical composition[a]	54.3% hydrophobic 23.9% polar 21.7% charged	53.1% hydrophobic 26.6% polar 20.3% charged
Main residues critical for IgE-binding activity	Hydrophobic and charged	Hydrophobic and charged
Epitope main secondary structures	β-Sheets and β-turns	β-Sheets and β-helices

[a]Percentage of total residues.

Source: Data in column 2 (Human Egg. Allergic Patients) from Mine and Rupa (2003a); data in column 3 (BALB/c Mouse Oral Sensitization) from Mine and Yang (2007).

Fragments resulting from tryptic digestion were tested for their ability to stimulate mouse hybridomas. The peptide 323–339 was identified for the first time as an immunodominant T-cell epitope in mouse (Shimonkevitz et al. 1984). Similarly, OVA-specific T-cell lines (TCL) have also been investigated. Thus studies have shown that the same peptide 323–339 was recognized by a T-cell line derived from a human egg-allergic patient (Shimonkevitz et al. 1984). The characterization of OVA-specific TCL was reported using PBMC from atopic dermatitis patients with egg allergy (Shimojo et al. 1994; Katsuki et al. 1996). It was shown that at least three peptide sequences were specifically recognized by T cells from these atopic dermatitis egg-allergic patients: residues 1–33, 198–213, and 261–277. Another study culturing PBMC from egg-allergic patients also determined that the peptide 323–339 and the novel region 105–122 both contained dominant T-cell epitopes of OVA, with no affinity to the patients' specific IgE, thereby suggesting their potential use in a context of peptide-based immunotherapy (Holen and Elsayed 1996).

Several studies have used ovalbumin as a model antigen for understanding T-cell immunity (Vidard et al. 1992b, 1992a, 1992c). One of these studies identified a total of nine antigenic regions recognized by mouse hybridomas. The peptide regions 225–240 and 233–248 were shown as the most dominant T-cell epitopes. From a structural point of view, it has been reported that these dominant T-cell epitopes occurred mainly between disordered loops, which may represent preferred sites of proteolytic cleavage (Landry 2006).

Ovalbumin B- and T-cell epitope mapping results, as reported by different groups, is schematically summarized in Figure 6.6.

6.4.4. Ovotransferrin (Gal d3)

Ovotransferrin (OVT), also known as *conalbumin*, has a molecular weight of 77,000 Da. It contains 12 disulfide bonds and 2.6% carbohydrate content (Mine and Rupa 2004). It has metal-binding activity and is capable of exerting antimicrobial effects (Kovacs-Nolan et al. 2005).

Along with OVM and OVA, ovotransferrin has long been accepted as a major allergen in egg white (Bleumink and Young 1969; Langeland 1982; Hoffman 1983; Langeland and Harbitz 1983). In one study, 35 out of 68 sera from egg-sensitive indivuduals were tested positive for OVT by CRIE (Langeland 1982) and 20 out of 34 sera had positive RAST and CRIE to OVT (Langeland 1983a). Another study using 34 egg-allergic adult patients sera indicated that OVT and OVM were immunodominant allergens compared to OVA and LYS (Aabin et al. 1996). However, contrary to ovomucoid and ovalbumin, the identification of ovotransferrin epitopes relevant to allergy has not yet been reported in humans or in animal models. Earlier reports documented however that there was allergenic cross-reactivity between ovotransferrin and whole egg yolk (Langeland 1983a).

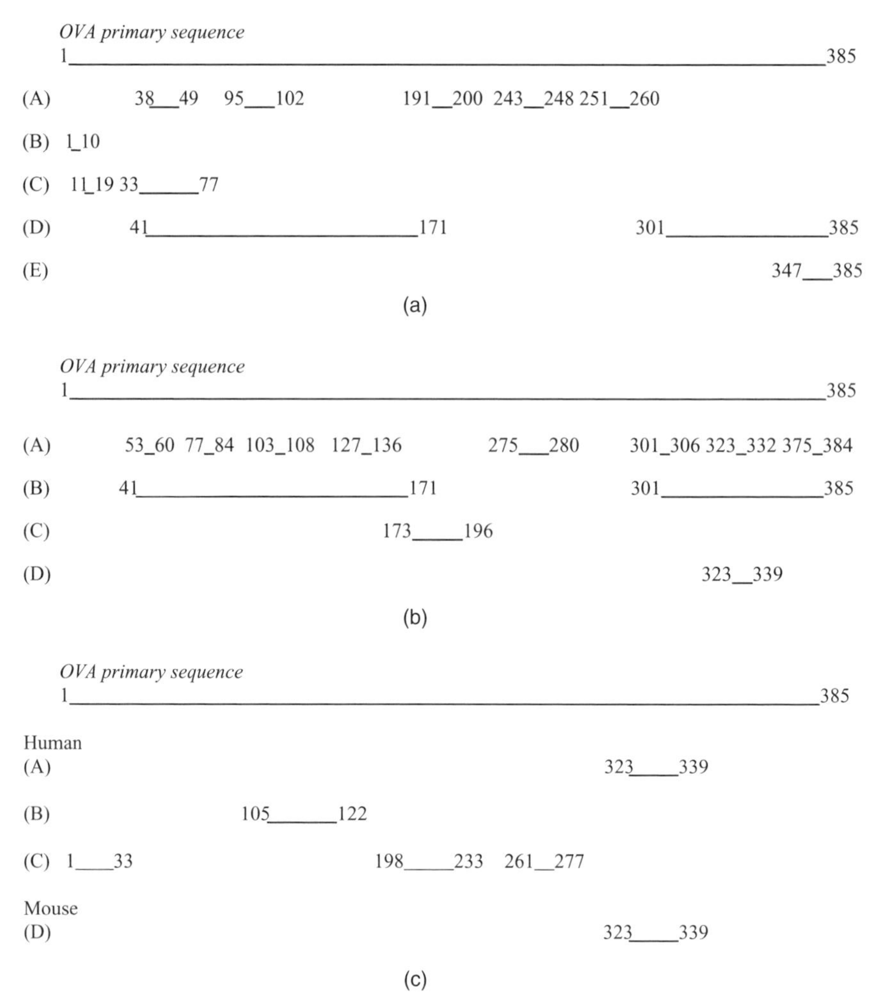

Figure 6.6. Schematic representation of IgE/T-cell epitope regions on ovalbumin, as reported by different groups. The numbers represent amino acid position along the primary sequence of the protein. (a) Ovalbumin allergenic (IgE) epitopes reported in human egg-allergic patients [(A) Mine and Rupa (2003a); (B) Elsayed et al. (1998); (C) Elsayed and Stavseng (1994); (D) Kahlert et al. (1992); (E) Honma et al. (1996); (b) ovalbumin allergenic (IgE) and antigenic (IgG1/IgM) epitopes reported in BALB/c mouse [(A) Mine and Yang (2007)—IgE isotype; (B) Kahlert et al. (1992)—IgG1/IgM isotypes (monoclonal); (C) Matsuda et al. (2000)—IgG1 isotype (monoclonal); (D) Renz et al. (1993)—IgE isotype]; (c) ovalbumin T-cell epitopes reported in human and BALB/c mouse [(A) Shimojo et al. (1994); (B) Holen and Elsayed (1996); (C) Katsuki et al. (1996); (D) Shimonkevitz et al. (1984)].

6.4.5. Lysozyme (Gal d4)

Hen egg white lysozyme (LYS) has a molecular weight of 14,300 Da and a single polypeptide chain crosslinked by four disulfide bridges (Mine and Rupa 2004). Early studies using high-purity egg proteins indicated that ovalbumin, ovomucoid, and ovotransferrin were important allergens, whereas lysozyme was only a weak allergen (Hoffman 1983), which was supported by other studies (Langeland 1982, Langeland and Harbitz 1983). Using egg-allergic patients' sera, a subsequent report also identified lysozyme as a minor egg allergen (Walsh et al. 1987). However, studies completed by different authors disputed these findings, and found that lysozyme could bind strongly to IgE antibodies in all egg-allergic sera tested. The authors concluded that lysozyme was also one of the major allergens of egg white (Holen and Elsayed 1990). Subsequent studies conducted in egg-allergic children brought support to the allergenic dominance of lysozyme (Yamada et al. 1993). To determine the prevalence of lysozyme sensitization in an egg-allergic population, specific IgE levels were determined by CAP-RAST (CAP, Pharmacia) using sera from 52 patients clinically allergic to eggs. Among these patients, 35% had anti-lysozyme-specific IgE antibodies (Fremont et al. 1997). The importance of lysozyme was further investigated, and a more recent study suggests that sensitization to lysozyme is associated predominantly with the development of occupational asthma (Leser et al. 2001; Escudero et al. 2003). In 2005, lysozyme was again documented as a significant egg allergen in human egg-sensitive individuals (Walsh et al. 2005). In light of this information, special attention should be drawn to the presence of lysozyme as a food additive, which may provoke allergic symptoms when unnoticed (Martorell Aragones et al. 2001). Indeed, egg lysozyme is used in drugs and in numerous foods as a bactericide to prevent the development of anaerobic bacteria (Fremont et al. 1997).

Hen egg lysozyme has been extensively used as an allergen model in the study of exercise-induced anaphylactic reactions to food in a B10. A mouse strain (Yano et al. 2002). Intestinal morphology, specific immunoglobulin levels, and intestinal permeability to lysozyme were examined. It has also served as a food allergen model for the investigation of airway allergic sensitization (Yoshino and Sagai 1999).

Among the four major allergens contained in egg white, lysozyme is found in the lowest concentration (3.4% of total egg white proteins). For this reason, the development of high-sensitivity detection assays for lysozyme in food products has been the objective of fairly recent investigations (Sato et al. 2001).

Hen egg lysozyme (HEL) has been a very popular model for the fine characterization of epitope–paratope interactions. A number of crystallization and X-ray diffraction studies have revealed the binding regions of Fab and monoclonal antibodies (Padlan et al. 1989; Kondo et al. 1999; Li et al. 2000). Similarly, HEL has been involved in the identification of peptide–MHC interactions (Moudgil et al. 1998; Weber et al. 1998). However, only a few investigations have been conducted toward the identification of lysozyme IgE epitopes relevant to egg allergy (Holen and Elsayed 1990, 1996; Batori et al. 2006).

A number of studies have investigated HEL-specific T-cell epitopes in BALB/c, B10.A, and CH3 and other inbred strains of mice (Adorini et al. 1988; Gammon et al. 1991; Moudgil et al. 1997). One report identified hen egg lysozyme-specific T-cell responses using PBMC from human egg-allergic patients (Holen and Elsayed 1996), but no human T-cell epitope has yet been characterized (Landry 2006).

6.4.6. Minor Egg Allergens in Egg White

Phosvitin represents 13.4% of total egg yolk protein content (Kovacs-Nolan et al. 2005). It was initially documented as a minor allergen in RAST from egg-allergic patients sera (Walsh et al. 1988). However, a more recent study by the same authors documented phosvitin as nonnegligible egg allergen (Walsh et al. 2005).

Ovomucin constitutes approximately 3.5% of total egg white proteins, while lysozyme also represents about 3.4%. Earlier studies have documented ovomucin as a significant allergen (Miller and Campbell 1950). However, skin test results were highly variable in the 342 patients studied. In patients who reacted to a single egg antigen, the frequency of allergen response was lysozyme > ovomucin > ovalbumin > ovomucoid. Subsequent reports suggested that the reagents used in this early study were undoubtedly contaminated with other protein fractions, and questioned whether the patients were truly allergic to eggs (Bernhisel-Broadbent et al. 1994). Although subsequent studies reported ovomucin as a minor allergen only (Walsh et al. 1988), a recent investigation suggests that ovomucin is a nonnegligible egg allergen (Walsh et al. 2005).

6.4.7. Minor Egg Allergens in Egg Yolk

Earlier reports indicated that egg yolk did not contain any allergenic components (Ankier 1969); therefore, egg yolk has long been ignored. However, subsequent studies have reported the existence of IgE binding activity against egg yolk components (Anet et al. 1985; Walsh et al. 1987, 1988, 2005).

In bird egg syndrome, found in groups of patients sensitized to egg through bird proteins (feathers, excrement, and bird serum), studies determined that sensitization was associated mainly with egg yolk components (Martorell Aragones et al. 2001). The main component responsible for this cross-sensitization is believed to be α-livetin (Szepfalusi et al. 1994). α-Livetin has been documented as the most relevant allergen in the sensitization to egg proteins via inhalation routes (van Toorenenbergen et al. 1994; Quirce et al. 1998).

Apovitellenins are the apoproteins present in the low-density lipoprotein fraction of the egg yolk (see Chapter 1, this book). Altogether, they represent up to 37% of total egg yolk proteins (Kovacs-Nolan et al. 2005). An earlier study determined that apovitellenins III, V, and VI were minor allergens in their human group of study (Anet et al. 1985), but a subsequent report documented apovitellenins I and VI as significant egg allergens by RAST analysis (Walsh et al. 1988). A more recent investigation conducted by the same authors confirmed that apovitellenins I and VI were nonnegligible egg allergens (Walsh et al. 2005).

6.4.8. Hen Egg Proteins Cross-Reactivity

Earlier reports determined that cross-reaction could take place between egg white proteins and egg yolk proteins, as well as between different bird eggs (hen, turkey, duck and seagull) (Langeland 1983a). In Western countries, eggs other than hen eggs are rarely consumed. However, in other countries, duck, quail, and goose eggs are sometimes consumed, but no report of cross-reactivity and severe anaphylactic response has been reported to these varieties of eggs (Landry 2006). A single case of duck and goose egg allergy has been reported, but the patient was tolerant to hen egg (Anibarro et al. 2000).

6.5. PROSPECTS FOR SPECIFIC IMMUNOTHERAPY FOR EGG ALLERGY

6.5.1. Current Treatment Options for Egg Allergy

6.5.1.1. Exclusion Diet

As mentioned in the introduction, there is currently no cure for egg allergy. Strict avoidance of the offending food is the only efficient approach. Once a diagnosis of hypersensitivity to egg proteins has been established, a strict exclusion diet should be initiated (Lever et al. 1998). However, the presence of "hidden allergens" represents a high risk for egg-sensitive individuals. Exclusion diets should be adequately supervised to eliminate egg derivatives and possible contamination of food products with these egg proteins. In cases of anaphylactic reactions, the administration of epinephrine remains the drug of choice.

6.5.1.2. Allergen-Specific Immunotherapy in Egg Allergy

Allergen-specific immunotherapy (SIT), commonly known as "allergy shot," exists for respiratory allergens such as grass pollen and house dust mite. However, SIT does not represent a viable alternative for egg and other food allergies, because of the potential severe-to-fatal side effects associated with such treatment. A successful SIT is commonly referred as *hyposensitization*, *desensitization*, or *specific oral tolerance induction* (SOTI). The terminology remains controversial, and a clear consensus has not been reached yet (Niggemann et al. 2006). Conventional SIT usually involves the administration of gradually increasing doses of allergen extracts, followed by a maintenance phase with doses given at fixed interval for several years, in an attempt to achieve clinical tolerance. The mechanisms underlying SIT are currently the focus of intensive investigations. A successful SIT is believed to involve diverse mechanism including (1) immune deviation from a Th2- to a Th1-skewed immune response, (2) activation of T regulatory cells, (3) pro-inflammatory T cell anergy, and (4) T cell apoptosis.

6.5.1.3. Induction of Oral Tolerance in Egg Allergy

On a daily basis, the gastrointestinal tract is exposed to a variety of dietary proteins, which seldom trigger an immune response. This phenomenon is commonly known as *oral tolerance* (Chehade and Mayer 2005).

During the onset of food allergy, it is believed that an abrogation of oral tolerance occurs. A study in 2005 involving two egg-allergic individuals determined that clinical tolerance was achievable with oral administration of dried egg proteins, but the beneficial effects were only transient if the ingestion of egg white proteins was discontinued (Rolinck-Werninghaus et al. 2005). On a larger scale, a similar study provided proof of concept that clinical tolerance by oral ingestion could be safely completed in egg-sensitive children without a history of anaphylaxis to egg (Buchanan et al. 2007). The study concluded that oral immunotherapy did not heighten sensitivity to egg and might in fact protect against reaction on accidental ingestion. However, the authors pointed out that a major limitation to the interpretation of the results was the absence of control groups.

Given the severity of egg hypersensitivity, the possibility of human testing is limited and potentially dangerous, which explains the need for use of animal models to study responses to egg allergens. A number of investigations were conducted to mimic allergic sensitization processes and create murine model of egg allergy (van Halteren et al. 1997; Rask et al. 2000; Akiyama et al. 2001). In addition, many attempts to characterize oral tolerance induction to ovalbumin have been made using mouse models. Transgenic mice carrying the T-cell receptor specific for OVA residues 323–339 were submitted to continuous oral microfeeding of OVA (100 μg/day for 14 days). The study showed that OVA-specific Th2 responses, as well as specific IgE and IgG1 production, were significantly decreased in the groups of OVA-microfed mice. Tolerance induction to OVA is now well established in murine models and has been used as a reference against other major food allergens such as peanut proteins (Strid et al. 2004). Similarly, continuous feeding of nanograms and microgram doses of OVM was shown to induce oral tolerance to OVM in a mouse model (Kjaer and Frokiaer 2002). Oral tolerance to ovalbumin has also been investigated in the atopic dog (Zemann et al. 2003), which represents a closer animal model to the human immune system.

The oral administration of soluble proteins seems to be an effective way of inducing hyporesponsiveness in animal models. However, most animal studies investigating oral tolerance used single, highly purified proteins. In allergic patients, tolerance induction may therefore differ since antigenic mixtures are often utilized. It is also important to emphasize that different workers used different protocols leading to the same effects operationally, but the mechanisms underlying oral tolerance, as well as its safety in human patients, is still a matter of debate. The induction of clinical oral tolerance to egg allergens thus remains questionable, especially for highly sensitive individuals. There is therefore a need for the development of novel and safe immunotherapeutic approaches. This matter will be discussed in the following paragraphs.

6.5.2. Novel Immunotherapeutic Approaches for Egg Allergy

Knowledge gained from studies of allergic mechanisms, such as the importance of Th1 and Th2 cells, the immune regulation by cytokine networks, and the potential inhibition of allergic mechanisms by induction of T regulatory cells may be applicable to a variety of allergic disorders, including egg allergy.

Since conventional SIT does not represent a viable solution for the treatment of egg allergy, new immunomodulatory methods have been investigated. Modern DNA technology and protein chemistry have provided innovative ways to produce hypoallergenic forms of allergens. Engineered recombinant proteins, peptide-based immunotherapy, DNA-based vaccines, and chemically modified allergens have all been reported as promising strategies to tackle egg-induced allergic reactions.

6.5.2.1. Development of Recombinant "Vaccine" via Genetic Engineering

Since the mid-1980s, the introduction of molecular biology has allowed the cloning of major food allergens. Recombinant DNA technology permits the large-scale production of highly purified allergens for both diagnostic and therapeutic purposes. Clones of cDNA have been isolated for an increasing number of food allergens and expressed in prokaryotic and eukaryotic systems as recombinant allergens. Recombinant allergens are most commonly expressed in the bacteria *Escherichia coli* and in the yeasts *Saccharomyces* and *Pichia* (Chapman et al. 2000, 2002). A special issue of the journal Methods (Valenta and Kraft 2004) contains thorough review articles dedicated to the potential use of recombinant allergens for clinical applications. The purpose of using recombinant DNA technology is to produce recombinant molecules that mimic the wild-type food protein but differ in their immunologic properties (Chapman et al. 2002). For recombinant allergens to provide therapeutic value in a clinical setting, they need to display a reduced capacity to bind IgE bound to mast cells or basophils, thus ensuring a lower risk of IgE-mediated side effects, while retaining their T-cell epitopes (Ferreira et al. 1998) and their capacity to induce the production of so-called "blocking IgG" (Valenta and Kraft 2002). The manipulation of nucleotide bases allows for the introduction of substitutions or deletions altering the coding sequences of allergens, resulting in a mutated sequence on production of a recombinant protein

(Niederberger and Valenta 2006). Reduction of allergenicity by point mutation of the allergen IgE-binding site could in theory be applied to various allergens (Ferreira et al. 1998, Valenta 2002). Recombinant allergens have therefore been investigated in the context of specific immunotherapy for egg allergy.

OVM domain III was successfully cloned and expressed in *E. coli*, and its immunologic properties were shown to be very similar to those of its native counterpart, as evidenced by ELISA and Western blot analyses (Sasaki and Mine 2001). In study of five recombinant isoforms of OVM domain III (Mine et al. 2003), each mutant contained single or double amino acid substitutions in their IgE and IgG epitopes, and effects on their allergenicity and antigenicity were compared against native OVM domain III. Substitution of glycine by methionine at position 32 and substitution of phenylalanine by alanine at position 37 (or GMFA) resulted in a complete loss of IgE- and IgG-binding activity by human egg-allergic patients' sera. This suggests the importance of structural integrity in the antigenicity and allergenicity of egg proteins. Full ovomucoid and ovalbumin have also been successfully expressed as recombinant isoforms in *E. coli* (Rupa and Mine 2003b, 2003a). Immunoblot and ELISA assays using human egg-allergic patients' sera revealed the immunologic properties of the recombinant molecules to be similar to those of their native counterparts. Their potential use in the development of immunotherapeutic strategies has been confirmed by recent studies in a mouse model of egg allergy.

A first report has investigated the *in vivo* allergenicity of GMFA recombinant domain III, compared to its native counterpart (Rupa and Mine 2006c). Intraperitoneal injections of either GMFA recombinant or native OVM domain III were administered to groups of BALB/c mice. The titers of OVM-specific IgE levels were significantly lower in the mice groups administered the GMFA mutant isoform. The higher titers of OVM-specific IgG2a titers along with the significantly lower titers of IgE suggest that GMFA isoform may represent a hypoallergenic form of the antigen, and therefore led to the development of a Th1-skewed immune response. Increased levels of IFN-γ and low levels of IL-4 in splenocyte culture supernatants brought support to this hypothesis. These data were further investigated with the completion of two desensitization studies in mice. The OVM domain III GMFA isoform was administered to BALB/c mice previously sensitized to native OVM domain III (Rupa and Mine 2006b). Detailed analysis of the mice immune response showed that the GMFA mutant isoform could efficiently abolish OVM domain III–induced hypersensitivity in this mouse model. Indeed, the administration of GMFA mutant was accompanied by reduced plasma histamine levels, significantly decreased OVM-specific IgE and high titers of Th1-type cytokines (IFN-γ). Detection of increased concentrations of regulatory cytokine IL-10 in splenocyte culture supernatants suggested the ability of GMFA to deviate the mouse immune response from a Th2- to a Th1-dominant phenotype.

These encouraging results naturally led to the assessment of GMFA as a hypoallergenic desensitization agent in full OVM-sensitized BALB/c mouse (Rupa and Mine 2006a). In this study, BALB/c mice were first sensitized with native full ovomucoid and subsequently administered GMFA, the mutant isoform of the OVM domain III. Anaphylactic reactions, histamine levels, OVM-specific antibody levels, and cytokine secretion profiles were examined in detail. While the positive control group (OVM-sensitized) showed severe anaphylactic symptoms and high concentrations of serum histamine and OVM-specific IgE and IgG1, the mouse group exposed to GMFA showed significant improvement in their overall allergic response. The authors concluded that the engineered mutations of ovomucoid led to a structural disruption of the molecule, which was probably responsible for the desensitizing effects of GMFA in the OVM-sensitized mice. It is reasonable to envisage that the same concept could be applied to all food allergens. Such an approach has already been investigated for other major food allergens such as peanut proteins. Collectively, these data provide strong support to the potential use of hypoallergenic forms of recombinant egg proteins, in desensitization trials involving human egg-allergic patients (Fig. 6.7). There are currently no recombinant egg allergen-based vaccines commercially available, but the results presented above warrant thorough investigation.

1. Characterization, cloning and expression of egg allergens

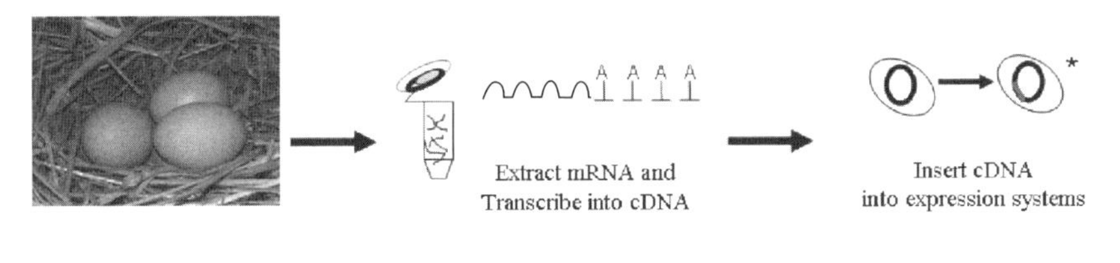

2. Identification of B-cell and T-cell epitopes

3. Production of recombinant allergens

IgE
IgE
Wild-type
recombinant

T-cell response
Overlapping egg peptides

IgE
IgE
Modify allergens
(e.g. alter IgE binding sites)

Use for egg allergens
detection and diagnostic
purposes

Use for peptide-based
immunotherapy in egg
allergic patients

Evaluate recombinant egg
allergens and derivatives in
human clinical studies

Figure 6.7. Potential immunotherapeutic strategies for the treatment of egg allergy [adapted from Niederberger and Valenta (2006) and de Leon et al. (2007)].

6.5.2.2. *Peptide-Based Immunotherapy*

The T-cell compartment and its effects in allergic responses represent a prime target for the intervention strategies in food allergy. Progress in synthetic peptide chemistry has allowed the identification of T-cell epitope of any allergen with a known primary sequence. The concept of downregulating allergic responses by using peptides containing T-cell epitopes originated from studies in murine models reporting that T cell could be rendered anergic (or nonresponsive) by administration of high doses of peptides representing immunodominant T-cell epitopes (Lamb et al. 1983; Hoyne et al. 1993). These initial findings were further supported by recent trials in cat-allergic patients. Patients receiving subcutaneous administration of T-cell-derived peptides showed significantly fewer symptoms than did the placebo control groups after exposure to cats (Norman et al. 1996; Larche 2005).

Peptides have been proposed as a therapeutic alternative to conventional SIT as they have the potential to inhibit T-cell function but do not induce the adverse reactions often associated with SIT because of the loss of their IgE binding activity (Francis and Larche 2005). A peptide-based approach for egg allergy requires the detailed characterization of the dominant T-cell epitopes present on the major egg allergens (Fig. 6.7). The mapping results of OVM and OVA T-cell epitopes in allergic patients and in mouse models were described above (Figs. 6.5 and 6.6). Earlier reports have shown that the orogastric administration of the OVA-peptidic fragment 173–196 could induce hyporesponsiveness to OVA in neonatal rats (Brown et al. 1994). The prophylactic effects of OVA peptides 323–339 were also investigated in a mouse model via intranasal treatment on the development of an allergic response (Janssen et al. 2000; Barbey et al. 2004).

A major limitation to peptide-based immunotherapy is that the variations brought by HLA genotype among individuals needs to be closely considered. The pool of tolerance-inducing peptides may need to be tailored to each individual (Prescott and Jones 2002). For this reason, the use of longer and in some cases overlapping peptides has been suggested as a solution (Fellrath et al. 2003).

With regard to the underlying mechanisms of peptide-based immunotherapy, it has been suggested that the induction of IL-10 and TGF-β-producing T cells (i.e., regulatory T cells) suppresses the activation of Th2 cells and the release of Th2 cell-derived cytokines. Interleukin-10 would also promote IgG4 isotype switching, leading to competition with IgE for allergen-binding sites.

The efficient dose, route of administration, frequency, and amount of short or long peptides to be administered have yet to be optimized, and the mechanisms underlying tolerance induction by peptides still remain obscure. To the best of our knowledge, no report on the use of peptide-based immunotherapy has been assayed in humans for egg or any other food allergy.

6.5.2.3. Plasmid DNA Vaccination

Among the innovative treatment options suggested for egg allergy immunotherapy, the use of gene or DNA vaccination has been investigated as well. Gene vaccination involves the intramuscular or intradermal administration of plasmid DNA encoding an allergen. How genes present in a DNA vaccine can be transcribed and translated, and how the protein product functions as a specific immunogen remains unclear (Satkauskas et al. 2001; Herweijer and Wolff 2003; Piccirillo and Prud'homme 2003). An excellent review discusses the potential of DNA vaccines for the treatment of allergic diseases (Weiss et al. 2005). A major advantage of using gene vaccination is that only a very small amount of allergen will be secreted and is therefore less likely to induce the anaphylactic reactions that can occur during a conventional SIT (Spiegelberg et al. 2002b). Chitosan nanoparticles carrying the gene for major allergens have been used to immunize murine models against subsequent sensitization with peanut allergens (Roy et al. 1999). Similar animal studies have focused on other clinically relevant allergens such as birch pollen allergen (Hartl et al. 1999), house dust mite allergen (Chew et al. 2003), and β-lactoglobulin (Adel-Patient et al. 2001). Promising results were obtained in all these studies. However, as yet, no human trials with allergen gene vaccination have been performed.

In a late-phase allergic response mouse model, a study reported that plasmid DNA immunization inhibited eosinophilic lung infiltration in OVA-sensitized mice (Raz and Spiegelberg 1999). The study discussed that gene vaccination had the ability to induce a Th1 immune response capable of down-regulating a preexisting Th2 response as well as inhibiting the formation of specific IgE antibody. In order to potentially enhance the efficacy of DNA vaccination, other studies have constructed fusion DNA plasmid containing the cDNA for a prototype allergen, such as OVA, combined to the cDNA of a potent immunomodulatory cytokine, such as IL-18 (Kim et al. 2001; Maecker et al. 2001).

In the context of egg allergy and other major food allergies, the development of DNA vaccination in humans has not been explored as yet. However, DNA-based vaccination is expected to offer a promising new approach to the desensitization of individuals with allergic hypersensitivities to foods, in both the prevention and reversal of the food allergic phenotype (Nguyen et al. 2001).

6.5.2.4. DNA Immunostimulatory Sequences or CpG Motifs

Recent discoveries have revealed the mechanism through which bacterial extracts can stimulate or enhance an immune response. Indeed, in order to detect the presence of infectious agents, the immune system has evolved to be equipped with recognition molecules commonly known as *pattern recognition receptors* (PRRs). The most widely characterized PRRs are the toll-like receptors (TLRs), more than 10 members of which have been identified in humans (Janeway and Medzhitov 2002; Takeda and Akira 2005). TLRs are expressed primarily on innate immune cells such as dendritic cells, macrophages, monocytes, and neutrophils and bind to conserved microbial motifs such as zymosan, lipopolysaccharides (LPS), and hypomethylated CpG dinucleotides (Janeway and Medzhitov 2002). It was observed that plasmid DNA containing immunostimulatory sequences (ISS), also known as *CpG motifs*, were capable of inducing a strong Th1 anti-allergic response (Spiegelberg et al. 2002a; Tsalik 2005). *CpG* refers to a cytosine followed by guanine and linked by a phosphodiester bond that forms the backbone of DNA (Krieg 2002). TLR-9 is the receptor used by the innate immune system for detecting CpG motifs, as shown by TLR-9 knockout mice (Hemmi et al. 2000). The activation of TLR-9 results in a response biased toward the expression and the production of IFN-γ, and therefore the possibility of triggering a strong Th1 adaptive response. The prospect of deliberately activating a TLR-9 pathway is of considerable interest for the treatment of immune disorders, such as food allergy.

A number of studies have used ovalbumin as a model allergen for the therapeutic use of immunostimulatory sequences in asthma mouse models (Ikeda et al. 2003; Youn et al. 2004). Ovalbumin–CpG conjugates were compared to ovalbumin mixed with CpG motifs and ovalbumin alone (Horner et al. 2002). The induction of mast cell degranulation was significantly lower with the conjugates than with ovalbumin, suggesting that the conjugate is less allergenic. Subsequent studies have also demonstrated that CpG sequences administered before or concurrently with allergen sensitization can attenuate the allergic response in mice (Kline et al. 2002; Santeliz et al. 2002).

Human studies with ragweed-sensitive patients have been performed to assess safety, tolerability, immunogenicity, and efficacy of CpG motifs coupled to allergen. Results suggest a potential for DNA-based immunotherapy to become an alternative to current immunotherapy protocols (Tulic et al. 2004). Other reports provided evidence that CpG motifs may also be useful in promoting immune tolerance in established atopic disorders (Jain et al. 2003). The use of CpG sequences or CpG motifs conjugated to a protein allergen entered phase I and II clinical trials in 2005 (Broide 2005).

6.5.2.5. Chemically Modified Allergens or Allergoids

Much effort has been made to chemically modify allergens. The resulting compounds were termed *allergoids*. Their most critical features are (1) their strongly reduced IgE-binding activity and (2) their retained immunogenicity, or T-cell reactivity, compared to their native counterpart. Formaldehyde and glutaraldehyde were among the first reagents to be used (HayGlass and Stefura 1990). The resulting compounds had lower allergenicity, but their molecular masses tended to be very high because of extensive crosslinking reactions. Carbamylation was then proposed as an alternative to modify allergen in a monomeric fashion (Mistrello et al. 1996; Bagnasco et al. 2001). Hypoallergenicity of ovalbumin protein conjugated with the copolymer *N*-vinylpyrrolidone and maleic anhydride were also investigated in a murine model (Babakhin et al. 1995). Allergoids have been established for allergen-specific immunotherapy against inhalant allergens for many years (Norman et al. 1982; Bousquet et al. 1987; Ria et al. 2004; von Baehr et al. 2005), and chemically modified allergens have now been licensed by the FDA for use in immunotherapy (perennial and seasonal inhalant allergies).

An interesting study using BALB/c mouse has shown that a chemically modified form of hen egg lysozyme (RCM-HEL) generated a skewed Th1 response compared to its native counterpart (Ria et al. 2004). These data suggest that the structure of a food allergen can have a significant impact on the immunologic outcome. Appropriate structural modifications of protein allergens can have a predefined impact on the efficacy of immunotherapeutic interventions in egg and other food allergies.

6.6. APPROACHES FOR PREPARATION OF HYPOALLERGENIC EGG PRODUCTS

The integrity of food allergen epitopes is determined mainly by their primary (B-cell sequential epitope) structure, and sometimes tertiary (B-cell conformational epitope) structure. The presence of B-cell linear epitopes explains why most food protein allergenicity can be quite resistant to the effects of food processing or gastrointestinal digestion. The concept of hypoallergenicity is based on the knowledge of these epitope structures and the factors that determine their chemical and physical stability. The aim of preparing hypoallergenic food is to provide allergic individuals with a source of food that is safe. Possibilities of developing hypoallergenic formulations via food processing have been investigated.

Egg proteins constitute a unique model system for studying the effects of physical and chemical processing, as the egg is used as an ingredient in a large variety of processed foods (dry, heated, or pasteurized) with varying physicochemical states (from intact egg to an ingredient incorporated into a food product). The main challenge in the development of hypoallergenic food through elimination or denaturating of the food allergen is to maintain the nutritional and functional properties identical to "normal" untreated food.

There are only a few examples of hypoallergenic foods that were introduced in the market, and proteolytically hydrolyzed milk proteins for infant formulations is probably the best known example (Hays and Wood 2005). One review described various processing technologies focusing on reducing allergenicity in food products (Wichers et al. 2003). These technologies include chemical, biochemical (using proteases or oxidases), and physical (such as heat or extraction) methods. Some of these are discussed below.

6.6.1. Thermal Processing

Thermal processing is usually carried out to enhance texture and flavor, or to ensure microbiological safety, but not to reduce allergenicity. Previous reports have described studies of the effect of thermal processing on the nutritional properties of egg white proteins (Kato et al. 1986; Mine et al. 1990), but only a few studies have considered heat application as a fast and efficient procedure for the production of hypoallergenic foods. During heat denaturation, many of the sites recognized by IgE antibodies present on the native molecule can be destroyed; cooking can eliminate the allergenicity of some food proteins. The basis for this approach was demonstrated in a clinical study in which patients allergic to freeze-dried egg white did not react to cooked egg white (Urisu et al. 1997). However, hypoallergenic products obtained by thermal processing may lead to clinical tolerance, only in individuals reacting to heat-labile allergens (Fiocchi et al. 2003). *In vitro* studies have shown that the combination of gamma radiations and heat treatment could be an efficient method in reducing egg hypersensitivity due to OVM allergenicity (Lee et al. 2002). The allergenicity of the four major egg allergens, native versus heat-treated forms, were examined. Results suggested that human OVA-specific IgE recognized more sequential epitopes, and that OVM- and LYS-specific IgE antibodies bound to both conformational and sequential epitopes (Mine and Zhang 2002). Nevertheless, thermal processing may not represent a viable alternative for egg proteins, since egg proteins are used as an ingredient in food products for their unique functional properties, such as foaming and gelling.

6.6.2. Enzymatic Fragmentation

Enzymatic processing of food materials represents another potential alternative for the reduction of their allergenicity. It offers a more specific approach than thermal processing. However, epitopes of most food allergens are known to be sequential, so that reduction in allergenicity would be expected only when allergenic epitopes are eliminated. Cleaving allergenic proteins by enzymatic digestion may result in a marked reduction of allergenic activity due to either (1) the generation of small fragments that will be unable to crosslink receptor-bound IgE present at the surface of mast cells or (2) their potential haptenic activity, which will inhibit IgE-mediated response (Tanabe et al. 1996).

One of the best known examples is derived from extensively hydrolyzed cow milk–based formula (Sampson et al. 1991; Hill et al. 1995; Hoffman and Sampson 1997). This approach was shown to be efficient in the prevention of sensitization to cow milk proteins (Marini et al. 1996). In Japan, the enzymatic treatment of rice, which leads to the decomposition of globulin, a major rice allergen, gave birth to one of the first FOSHU ("food for specialized health uses") applications in 1993 (Shimada et al. 2005). Further information on rice allergens and investigations on hypoallergenic rice products by enzymatic treatment are available from the Internet Symposium on Food Allergens (Besler et al. 1999). For a staple food such as wheat-containing food products, preparation of hypoallergenic wheat flour has also been investigated in a murine model using enzymatic fragmentation with cellulose and actinase preparations (Watanabe et al. 2000, 2004). Again enzymatic hydrolysis may not represent a viable alternative for egg proteins, as egg proteins are used as an ingredient in food products for their unique functional properties, for example, foaming and gelling.

6.6.3. Other Food Processing Methods

One review suggests that novel food-processing methods, such as high-pressure processing, pulse electric field processing, combined with physical and biochemical processing, also hold promises for the development of hypoallergenic foods (Wichers et al. 2003).

The production of egg food products with low ovomucoid content by solvent extraction has been investigated. Ethanol (20%) precipitation successfully removed 70% of the OVM content from egg white powder, without affecting the whipping ability and foam stability of the remaining egg white proteins (Tanabe et al. 2000).

Radiation technology has also been explored in a number of studies for the modification of food allergen, such as major shrimp and milk allergens, in order to reduce the IgE-binding activity of these antigens (Byun et al. 2000; Lee et al. 2001). One study in particular has investigated the use of γ-irradiation in the confection of hypoallergenic egg-based cakes (Seo et al. 2004). ELISA assays using egg-allergic patients' sera and OVA-specific rabbit IgG revealed a significant decrease in OVA content after γ-irradiation treatment. The same authors investigated the potential use of irradiation technology in the prevention or treatment of food allergy (Seo et al. 2007). In this latter study, the immunologic properties of γ-irradiated OVA versus native OVA were compared using ELISA and immunoblot analyses with OVA-specific monoclonal and polyclonal antibodies. Results showed that γ-irradiation exceeding 10 kGy altered the structure of OVA and led to a decreased antigenicity. Immunization trials in BALB/c mouse revealed that the titers of OVA-specific immunoglobulins and OVA-specific T-cell-mediated response were significantly lower in groups of mice sensitized to irradiated OVA. Again, these results suggest that appropriate changes in food allergen structures may offer great opportunity for the development of desensitization approaches in the context of egg and other food allergies.

6.6.4. Development of Hypoallergenic Products via Genetic Engineering

Since the mid-1990s, molecular biology tools have been widely used to enhance beneficial effects or eliminate negative effects carried out by food (Lehrer 2004). In this context, genetic engineering has also been investigated in the making of hypoallergenic foods. Techniques such as antisense RNA or RNA interference, have allowed the production of hypoallergenic soybean (Ogawa et al. 2000; Herman et al. 2003) or rice (Nakamura and Matsuda 1996). The use of RNA antisense technology led to reduction in content of the major rice allergen from 300 μg/seed to 60–70 μg/seed in control nontransgenic plants.

More recently, production of hypoallergenic apple using RNA interference technology has also successfully reduced the expression of Mal d1, a major allergen (Gilissen et al. 2005). Finally, antisense strategy was shown to be efficient for the silencing of Ara h2, a major peanut allergen (Dodo et al. 2005). Such approach has thus become popular and may be accepted for the making of hypoallergenic plant-derived allergens, but it would doubtlessly represent a challenge for animal-derived allergens such as egg proteins.

6.7. IMMUNOMODULATORY COMPOUNDS IN MANAGEMENT OF EGG ALLERGY

6.7.1. Use of Probiotics for Inhibition of Egg Hypersensitivity

The term *probiotic* was introduced about forty years ago to describe "any organism or substance contributing to the intestinal microbial balance in the host animal" (Lilly and Stillwell 1965; Prescott 2003). The definition has evolved since and probiotics are currently defined as "nonpathogenic microorganisms that exert a positive influence on the health and physiology of the host" (Isolauri et al. 2004). They most often belong to the genera *Bifidobacterium* or *Lactobacillus*, both members of the commensal microflora of a healthy human intestine. The intestinal microbiota constitutes a vital component of the mucosal barrier through their metabolic and trophic effects, their resistance to colonization by microbial invaders, and their proven immunomodulatory properties (Rautava et al. 2005).

The use of probiotics in the treatment or prevention of food-allergic disorders is a relatively new concept. A correlation between the increase of allergic disease incidence and highly sanitized areas of the world suggested the possible benefit of stimulating the gut immune system with probiotics (Renz et al. 1993). This correlation was termed the "hygiene hypothesis" (Strachan 1989). According to this hypothesis, prolonged exposure to pathogens during early childhood has a crucial role in the education and maturation of the immune system (Yazdanbakhsh et al. 2002).

Egg hypersensitivity was documented in two-thirds of children with severe atopic dermatitis (Sampson 1997). The use of probiotics in the prevention or the treatment of atopic dermatitis has been investigated. For example, one study involved 56 children with moderate to severe atopic dermatitis. Among them, 71% tested positive for specific IgE RAST to food mixture containing egg white, milk, cod, wheat, peanut, and soybean (Prescott et al. 2005). Participants received *Lactobacillus* fermentum VRI003 PCCTM (Protract Probiomics, Eveleigh, NSW, Australia) freeze-dried powder twice daily for 8 weeks. The ingestion of the probiotics strain led to a significant improvement in atopic dermatitis severity but no changes were observed on the food allergen-specific responses. The study concluded that the effects of probiotics were mediated by mechanisms that still needed to be explored.

More recently, a study has investigated the potential role of the natural intestinal microflora of allergic eczema/dermatitis children, but failed to demonstrate a quantitative relationship between IgE sensitization against important food allergens, such as egg and cow's milk, and the composition of fecal microflora (Kendler et al. 2006).

Strains of *Lactococcus* microorganisms are commonly used as starter bacteria in the manufacture of many types of fermented dairy products. One mouse study investigated the potential immunomodulatory properties of 15 strains of the genus *Lactococcus* (Kimoto et al. 2004). Among these strains, the *Lactococcus lactis* subspecies *lactis G50* (Napiergrass) was selected for further *in vivo* investigations. Ovomucoid-sensitized BALB/c mice were orally challenged with the bacteria G50 strain in a daily fashion. The total IgE antibody level in the G50-treated group was significantly lower than that of the control group. The same observations were found for the titers of OVM-specific IgG1 and IgE antibodies, suggesting that *L. lactis* subspecies *lactis G50* is a potential probiotic strain for the suppression of egg hypersensitivity mediated by a Th2-skewed response. Similar conclusions were reached in studies where ovalbumin-sensitized BALB/c mice were administered orally with heat-killed *Lactobacillus* casei strain Shirota (Matsuzaki et al. 1998; Shida et al. 2002).

A prospective intervention study, modifying the gut flora starting at birth, already has yielded encouraging results, in terms of primary prevention of allergy (Kalliomaki et al. 2003). However, further prospective studies are probably needed to examine in detail the therapeutic and preventive potential of probiotics in egg and other food allergies.

6.7.2. Other Immunomodulatory Food Components

Besides probiotics, a large number of food components have been assessed for their efficacy in allergy prevention or therapy. For instance, encouraging results were obtained from animal studies using oral administration of seaweed polysaccharide carrageenan to β-lactoglobulin-sensitized mice (Frossard et al. 2001) or the administration of Chinese herbal medicine to peanut-sensitized mice (Srivastava et al. 2005). Similarly, it has been shown that dietary lipids such as *n*-3 and *n*-6 PUFA could result in immune deviation with a reduction of specific IgE and a downregulation of IL-4,5,13 cytokines (Albers et al. 2002; Dunstan et al. 2003).

It was also demonstrated that *Quillaja* saponin could suppress IgE-mediated allergic response in a mouse model presensitized to ovalbumin (Katayama and Mine 2006). Oral administration of *Quillaja* saponin to OVA-sensitized BALB/c mice suppressed the anaphylactic reactions observed in control groups. Titers of IgE and IgG1 were markedly inhibited by oral administration of *Quillaja* saponin, while serum IgG2a levels were increased. These results suggest the occurrence of an immune deviation from a Th2- to a Th1-skewed response.

6.8. CONCLUSIONS

With the rising prevalence of atopic diseases, food allergy is today perceived as an important public health issue and is associated with great economic and social impacts (Primeau et al. 2000; Cohen et al. 2004).

Hen egg allergy represents the second most common food allergy in the pediatric population. The major egg allergens are contained in the egg white and are ovomucoid, ovalbumin, ovotransferrin, and lysozyme. Minor allergens are also found in egg yolk. A large number of studies sought to characterize the antigenic and allergenic properties of egg allergens. Results were variable and have been attributed to factors such as the varying degree of purity of the different egg proteins, the use of egg-allergic patients' sera with different clinical profiles, or the use of animal antisera raised after parenteral administration of egg proteins, which do not reflect the natural route of oral sensitization in humans.

Avoiding the offending compound is a logical way to treat or prevent egg-induced allergy, but the existence of hidden allergens makes it a difficult, not to say impossible, approach. Currently, conventional specific immunotherapy only can affect the course of allergy development, while pharmacological therapy only alleviates its symptoms. However, conventional SIT is not applicable to egg allergy because of its potential side effects. Therefore, beyond egg allergen avoidance and conventional SIT, novel approaches are being investigated.

Egg allergens have been cloned, immunodominant epitopes have been characterized, and modified forms of egg allergens are being assessed for the development of novel forms of immunotherapy. For example, promising results have been generated with use of recombinant OVM in desensitization trials conducted in animal models. While immunotherapeutic strategies for egg allergy are being explored, there is a growing awareness that the implementation of preventive methods is also required. For example, the production of hypoallergenic egg products—a food source with reduced or no egg allergens that can be safely ingested by egg allergic individuals—represents an alternative. However, the modification of egg proteins using conventional processing methods, such as heat and enzymatic processing, is not always feasible since it may affect their nutritional and functional values. On the other hand, the more recently discovered immunomodulatory properties of probiotics and other food components hold great promises for the management of egg allergy.

REFERENCES

Aabin B, Poulsen LK, Ebbehoj K, Norgaard A, Frokiaer H, Bindslev-Jensen C, Barkholt V (1996). Identification of IgE-binding egg white proteins: Comparison of results obtained by different methods. Int Arch Allergy Immunol 109(1):50–57.

Adel-Patient K, Boquet D, Creminon C, Wal JM, Chatel JM (2001). Genetic immunisation with bovine beta-lactoglobulin cDNA induces a preventive and persistent inhibition of specific anti-BLG IgE response in mice. Int Arch Allergy Immunol 126(1):59–67.

Adorini L, Appella E, Doria G, Nagy ZA (1988). Mechanisms influencing the immunodominance of T cell determinants. J Exp Med 168(6):2091–2104.

Akiyama H, Teshima R, Sakushima JI, Okunuki H, Goda Y, Sawada JI, Toyoda M (2001). Examination of oral sensitization with ovalbumin in brown Norway rats and three strains of mice. Immunol Lett 78(1):1–5.

Albers R, Bol M, Bleumink R, Willems A, Blonk C, Pieters R (2002). Effects of dietary lipids on immune function in a murine sensitisation model. Br J Nutr 88(3):291–299.

Anet J, Back JF, Baker RS, Barnett D, Burley RW, Howden ME (1985). Allergens in the white and yolk of hen's egg. A study of IgE binding by egg proteins. Int Arch Allergy Appl Immunol 77(3):364–371.

Anibarro B, Seoane FJ, Vila C, Lombardero M (2000). Allergy to eggs from duck and goose without sensitization to hen egg proteins. J Allergy Clin Immunol 105(4):834–836.

Ankier SI (1969). Reactivity of rat and man to egg-white. Progress Drug Res 13:340–394.

Babakhin AA, DuBuske LM, Wheeler AW, Stockinger B, Nolte H, Andreev S, Gushchin IS, Khaitov RM, Petrov RV (1995). Immunological properties of allergen chemically modified with synthetic copolymer of N-vinylpyrrolidone and maleic anhydride. Allergy Proc 16(5):261–268.

Bagnasco M, Passalacqua G, Villa G, Augeri C, Flamigni G, Borini E, Falagiani P, Mistrello G, Mariani G (2001). Pharmacokinetics of an allergen and a monomeric allergoid for oromucosal immunotherapy in allergic volunteers. Clin Exp Allergy 31(1):54–60.

Barbey C, Donatelli-Dufour N, Batard P, Corradin G, Spertini F (2004). Intranasal treatment with ovalbumin but not the major T cell epitope ovalbumin 323–339 generates interleukin-10 secreting T cells and results in the induction of allergen systemic tolerance. Clin Exp Allergy 34(4):654–662.

Batori V, Friis EP, Nielsen H, Roggen EL (2006). An in silico method using an epitope motif database for predicting the location of antigenic determinants on proteins in a structural context. J Mol Recognit 19(1):21–29.

Beeley JG (1971). The isolation of ovomucoid variants differing in carbohydrate composition. Biochem J 123(3):399–405.

Bernhisel-Broadbent J, Dintzis HM, Dintzis RZ, Sampson HA (1994). Allergenicity and antigenicity of chicken egg ovomucoid (Gal d III) compared with ovalbumin (Gal d I) in children with egg allergy and in mice. J Allergy Clin Immunol 93(6):1047–1059.

Besler M (1999a). Hen's egg white. Internet Symp Food Allergens 1(1):13–33.

Besler M (1999b). Identification of IgE-Binding Peptides from OM. Internet Symp Food Allergens 1(1):1–12.

Besler M, Helm RM, Ogawa T (1999). Allergen data base collection—rice. Internet Symp Food Allergens 1(4):147–160.

Bischoff S, Crowe SE (2005). Gastrointestinal food allergy: New insights into pathophysiology and clinical perspectives. Gastroenterology 128(4):1089–1113.

Bleumink E, Young E (1969). Studies on the atopic allergen in hen's egg. I. Identification of the skin reactive fraction in egg-white. Int Arch Allergy Appl Immunol 35(1):1–19.

Bleumink E, Young E (1971). Studies on the atopic allergen in hen's egg. II. Further characterization of the skin-reactive fraction in egg-white; immuno-electrophoretic studies. Int Arch Allergy Appl Immunol 40(1):72–88.

Bock SA, Munoz-Furlong A, Sampson HA (2001). Fatalities due to anaphylactic reactions to foods. J Allergy Clin Immunol 107(1):191–193.

Bousquet J, Hejjaoui A, Skassa-Brociek W, Guerin B, Maasch HJ, Dhivert H, Michel FB (1987). Double-blind, placebo-controlled immunotherapy with mixed grass-pollen allergoids. I. Rush immunotherapy with allergoids and standardized orchard grass-pollen extract. J Allergy Clin Immunol 80(4):591–598.

Broide DH (2005). Immunostimulatory sequences of DNA and conjugates in the treatment of allergic rhinitis. Curr Allergy Asthma Rep 5(3):182–185.

Brown W, Le M, Lee E (1994). Induction of systemic immunologic hyporesponsiveness to ovalbumin in neonatal rats by the enteric administration of peptic fragments of ovalbumin. Immunol Invest 23(2):73–83.

Buchanan AD, Green TD, Jones SM, Scurlock AM, Christie L, Althage KA, Steele PH, Pons L, Helm RM, Lee LA, Burks AW (2007). Egg oral immunotherapy in nonanaphylactic children with egg allergy. J Allergy Clin Immunol 119(1):199–205.

Burks AW, James JM, Hiegel A, Wilson G, Wheeler JG, Jones S, Zuerlein N (1998). Atopic dermatitis and food hypersensitivity reactions. J Pediatr 132(1):132–136.

Burks W, Lehrer SB, Bannon GA (2004). New approaches for treatment of peanut allergy: Chances for a cure. Clin Rev Allergy Immunol 27(3):191–196.

Buus S, Sette A, Grey HM (1987). The interaction between protein-derived immunogenic peptides and Ia. Immunol Rev 98:115–141.

Byun MW, Kim JH, Lee JW, Park JW, Hong CS, Kang IJ (2000). Effects of gamma radiation on the conformational and antigenic properties of a heat-stable major allergen in brown shrimp. J Food Protect 63(7):940–944.

Caffarelli C, Cavagni G, Giordano S, Stapane I, Rossi C (1995). Relationship between oral challenges with previously uningested egg and egg-specific IgE antibodies and skin prick tests in infants with food allergy. J Allergy Clin Immunol 95(6):1215–1220.

Cereghino JL, Cregg JM (2000). Heterologous protein expression in the methylotrophic yeast Pichia pastoris. FEMS Microbiol Rev 24(1):45–66.

Chapman MD, Smith AM, Vailes LD, Arruda LK, Dhanaraj V, Pomes A (2000). Recombinant allergens for diagnosis and therapy of allergic disease. J Allergy Clin Immunol 106(3):409–418.

Chapman MD, Smith AM, Vailes LD, Pomes A (2002). Recombinant allergens for immunotherapy. Allergy Asthma Proc 23(1):5–8.

Chehade M, Mayer L (2005). Oral tolerance and its relation to food hypersensitivities. J Allergy Clin Immunol 115(1):3–12.

Chew JL, Wolfowicz CB, Mao HQ, Leong KW, Chua KY (2003). Chitosan nanoparticles containing plasmid DNA encoding house dust mite allergen, Der p 1 for oral vaccination in mice. Vaccine 21(21–22):2720–2729.

Chino F, Oshibuchi S, Ariga H, Okuno Y (1999). Skin reaction to yellow fever vaccine after immunization with rabies vaccine of chick embryo cell culture origin. Jpn J Infect Dis 52(2):42–44.

Cohen BL, Noone S, Munoz-Furlong A, Sicherer SH (2004). Development of a questionnaire to measure quality of life in families with a child with food allergy. J Allergy Clin Immunol 114(5):1159–1163.

Cooke SK, Sampson HA (1997). Allergenic properties of ovomucoid in man. J Immunol 159(4):2026–2032.

Crespo JF, Pascual C, Burks AW, Helm RM, Esteban MM (1995). Frequency of food allergy in a pediatric population from Spain. Pediatr Allergy Immunol 6(1):39–43.

de Leon MP, Rolland JM, O'hehir RE (2007). The peanut allergy epidemic: Allergen molecular characterization and prospects for specific therapy. Expert Rev Mol Med 9(1):1–18.

Deutsch HF, Morton JI (1956). Immunochemical properties of heated ovomucoid. Arch Biochem Biophys 64(1):19–25.

Djurtoft R, Pedersen HS, Aabin B, Barkholt V (1991). Studies of food allergens: Soybean and egg proteins. Adv Exp Med Biol 289:281–293.

Dodo H, Konan K, Viquez O (2005). A genetic engineering strategy to eliminate peanut allergy. Curr Allergy Asthma Rep 5(1):67–73.

Dunstan JA, Mori TA, Barden A, Beilin LJ, Taylor AL, Holt PG, Prescott SL (2003). Fish oil supplementation in pregnancy modifies neonatal allergen-specific immune responses and clinical outcomes in infants at high risk of atopy: A randomized, controlled trial. J Allergy Clin Immunol 112(6):1178–1184.

Ebo DG, Stevens WJ (2001). IgE-mediated food allergy—extensive review of the literature. Acta Clin Belg 56(4):234–247.

Eigenmann PA, Huang SK, Sampson HA (1996). Characterization of ovomucoid-specific T-cell lines and clones from egg-allergic subjects. Pediatr Allergy Immunol 7(1):12–21.

Ellman LK, Chatchatee P, Sicherer SH, Sampson HA (2002). Food hypersensitivity in two groups of children and young adults with atopic dermatitis evaluated a decade apart. Pediatr Allergy Immunol 13(4):295–298.

Elsayed S, Hammer AS, Kalvenes MB, Florvaag E, Apold J, Vik H (1986). Antigenic and allergenic determinants of ovalbumin. I. Peptide mapping, cleavage at the methionyl peptide bonds and enzymic hydrolysis of native and carboxymethyl OA. Int Arch Allergy Appl Immunol 79(1):101–107.

Elsayed S, Holen E, Haugstad MB (1988). Antigenic and allergenic determinants of ovalbumin. II. The reactivity of the NH2 terminal decapeptide. Scand J Immunol 27(5):587–591.

Elsayed S, Stavseng L (1994). Epitope mapping of region 11–70 of ovalbumin (Gal d I) using five synthetic peptides. Int Arch Allergy Immunol 104(1):65–71.

Escudero C, Quirce S, Fernandez-Nieto M, Miguel J, Cuesta J, Sastre J (2003). Egg white proteins as inhalant allergens associated with baker's asthma. Allergy 58(7):616–620.

Fellrath JM, Kettner A, Dufour N, Frigerio C, Schneeberger D, Leimgruber A, Corradin G, Spertini F (2003). Allergen-specific T-cell tolerance induction with allergen-derived long synthetic peptides: Results of a phase I trial. J Allergy Clin Immunol 111(4):854–861.

Ferreira F, Ebner C, Kramer B, Casari G, Briza P, Kungl AJ, Grimm R, Jahn-Schmid B, Breiteneder H, Kraft D, Breitenbach M, Rheinberger HJ, Scheiner O (1998). Modulation of IgE reactivity of allergens by site-directed mutagenesis: Potential use of hypoallergenic variants for immunotherapy. FASEB J 12(2):231–242.

Fiocchi A, Bouygue GR, Sarratud T, Terracciano L, Martelli A, Restani P (2003). Clinical tolerance of processed foods. Ann Allergy Asthma Immunol 93(5 Suppl 3):S38–46.

Ford RP, Taylor B (1982). Natural history of egg hypersensitivity. Arch Dis Child 57(9):649–652.

Francis JN, Larché M (2005). Peptide-based vaccination: Where do we stand? Curr Opin Allergy Clin Immunol 5(6):537–543.

Fremont S, Kanny G, Nicolas JP, Moneret-Vautrin DA (1997). Prevalence of lysozyme sensitization in an egg-allergic population. Allergy 52(2):224–228.

Frossard CP, Hauser C, Eigenmann PA (2001). Oral carrageenan induces antigen-dependent oral tolerance: Prevention of anaphylaxis and induction of lymphocyte anergy in a murine model of food allergy. Pediatr Res 49(3):417–422.

Gammon G, Geysen HM, Apple RJ, Pickett E, Palmer M, Ametani A, Sercarz EE (1991). T cell determinant structure: Cores and determinant envelopes in three mouse major histocompatibility complex haplotypes. J Exp Med 173(3):609–617.

Gilissen LJ, Bolhaar ST, Matos CI, Rouwendal GJ, Boone MJ, Krens FA, Zuidmeer L, Van Leeuwen A, Akkerdaas J, Hoffmann-Sommergruber K, Knulst A, Bosch D, Van de Weg WE, Van Ree R (2005). Silencing the major apple allergen Mal d 1 by using the RNA interference approach. J Allergy Clin Immunol 115(2):364–369.

Goodyear-Smith F, Wong F, Petousis-Harris H, Wilson E, Turner N (2005). Follow-up of MMR vaccination status in children referred to a pediatric immunization clinic on account of egg allergy. Hum Vaccin 1(3):118–122.

Gu J, Matsuda T, Nakamura R (1986). Antigenicity of ovomucoid remaining in boiled shelled eggs. J Food Sci 51:1448–1450.

Gu JX, Matsuda T, Nakamura R, Ishiguro H, Ohkubo I, Sasaki M, Takahashi N (1989). Chemical deglycosylation of hen ovomucoid: Protective effect of carbohydrate moiety on tryptic hydrolysis and heat denaturation. J Biochem (Tokyo) 106(1): 66–70.

Gustafsson D, Sjoberg O, Foucard T (2003). Sensitization to food and airborne allergens in children with atopic dermatitis followed up to 7 years of age. Pediatr Allergy Immunol 14(6):448–452.

Halken S (2004). What causes allergy and asthma? The role of dietary factors. Pediatr Pulmonol Suppl 26:223–224.

Han DK, Kim MK, Yoo JE, Choi SY, Kwon BC, Sohn MH, Kim KE, Lee SY (2004). Food sensitization in infants and young children with atopic dermatitis. Yonsei Med J 45(5):803–809.

Hartl A, Kiesslich J, Weiss R, Bernhaupt A, Mostbock S, Scheiblhofer S, Ebner C, Ferreira F, Thalhamer J (1999). Immune responses after immunization with plasmid

DNA encoding Bet v 1, the major allergen of birch pollen. J Allergy Clin Immunol 103(1 Pt 1):107–113.

HayGlass KT, Stefura W (1990). Isotype-selective abrogation of established IgE responses. Clin Exp Immunol 82(3):429–434.

Hays T, Wood RA (2005). A systematic review of the role of hydrolyzed infant formulas in allergy prevention. Arch Pediatr Adolesc Med 159(9):810–816.

Heine RG, Laske N, Hill DJ (2006). The diagnosis and management of egg allergy. Curr Allergy Asthma Rep 6(2):145–152.

Hemmi H, Takeuchi O, Kawai T, Kaisho T, Sato S, Sanjo H, Matsumoto M, Hoshino K, Wagner H, Takeda K, Akira S (2000). A toll-like receptor recognizes bacterial DNA. Nature 408(6813):740–745.

Herman EM, Helm RM, Jung R, Kinney AJ (2003). Genetic modification removes an immunodominant allergen from soybean. Plant Physiol 132(1):36–43.

Herweijer H, Wolff JA (2003). Progress and prospects: Naked DNA gene transfer and therapy. Gene Ther 10(6):453–458.

Hill DJ, Cameron DJ, Francis DE, Gonzalez-Andaya AM, Hosking CS (1995). Challenge confirmation of late-onset reactions to extensively hydrolyzed formulas in infants with multiple food protein intolerance. J Allergy Clin Immunol 96(3):386–394.

Hoffman DR (1983). Immunochemical identification of the allergens in egg white. J Allergy Clin Immunol 71(5):481–486.

Hoffman KM, Sampson HA (1997). Serum specific-IgE antibodies to peptides detected in a casein hydrolysate formula. Pediatr Allergy Immunol 8(4):185–189.

Holen E, Bolann B, Elsayed S (2001). Novel B and T cell epitopes of chicken ovomucoid (Gal d 1) induce T cell secretion of IL-6, IL-13, and IFN-gamma. Clin Exp Allergy 31(6):952–964.

Holen E, Elsayed S (1990). Characterization of four major allergens of hen egg-white by IEF/SDS-PAGE combined with electrophoretic transfer and IgE-immunoautoradiography. Int Arch Allergy Appl Immunol 91(2):136–141.

Holen E, Elsayed S (1996). Specific T cell lines for ovalbumin, ovomucoid, lysozyme and two OA synthetic epitopes, generated from egg allergic patients' PBMC. Clin Exp Allergy 26(9):1080–1088.

Honma K, Kohno Y, Saito K, Shimojo N, Horiuchi T, Hayashi H, Suzuki N, Hosoya T, Tsunoo H, Niimi H (1996). Allergenic epitopes of ovalbumin (OVA) in patients with hen's egg allergy: Inhibition of basophil histamine release by haptenic ovalbumin peptide. Clin Exp Immunol 103(3):446–453.

Honma K, Kohno Y, Saito K, Shimojo N, Tsunoo H (1994). Specificities of IgE, IgG and IgA antibodies to ovalbumin. Comparison of binding activities to denatured ovalbumin or ovalbumin fragments of IgE antibodies with those of IgG or IgA antibodies. Int Arch Allergy Immunol 103(1):28–35.

Horner AA, Takabayashi K, Beck L, Sharma B, Zubeldia J, Baird S, Tuck S, Libet L, Spiegelberg HL, Liu F, Raz E (2002). Optimized conjugation ratios lead to allergen immunostimulatory oligodeoxynucleotide conjugates with retained immunogenicity and minimal anaphylactogenicity. J Allergy Clin Immunol 110(3):413–420.

Host A (1995). Adverse reactions to foods: Epidemiology and risk factors. Pediatr Allergy Immunol 6(Suppl 8):20–28.

Hoyne GF, O'Hehir RE, Wraith DC, Thomas WR, Lamb J (1993). Inhibition of T cell and antibody responses to house dust mite allergen by inhalation of the dominant T cell epitope in naive and sensitized mice. J Exp Med 178(5):1783–1788.

Huntington JA, Stein PE (2001). Structure and properties of ovalbumin. J Chromatogr B Biomed Sci Appl 756(1–2):189–198.

Ikeda RK, Nayar J, Cho JY, Miller M, Rodriguez M, Raz E, Broide DH (2003). Resolution of airway inflammation following ovalbumin inhalation: Comparison of ISS DNA and corticosteroids. Am J Respir Cell Mol Biol 28(6):655–663.

Imai T, Iikura Y (2003). The national survey of immediate type of food allergy. Arerugi 52(10):1006–1013.

Isolauri E, Salminen S, Ouwehand AC (2004). Probiotics. Microbial-gut interactions in health and disease. Best Pract Res Clin Gastroenterol 18(2):299–313.

Jain VV, Kitagaki K, Kline JN (2003). CpG DNA and immunotherapy of allergic airway diseases. Clin Exp Allergy 33(10):1330–1335.

James JM, Burks AW, Roberson PK, Sampson HA (1995). Safe administration of the measles vaccine to children allergic to eggs. New Engl J Med 332(19):1262–1266.

Janeway CA, Medzhitov R (2002). Innate immune recognition. Annu Rev Immunol 20:197–216.

Janssen EM, van Oosterhout AJ, Nijkamp FP, van Eden W, Wauben MH (2000). The efficacy of immunotherapy in an experimental murine model of allergic asthma is related to the strength and site of T cell activation during immunotherapy. J Immunol 165(12):7207–7214.

Janssen EM, Wauben MH, Jonker EH, Hofman G, Van Eden W, Nijkamp FP, Van Oosterhout AJ (1999). Opposite effects of immunotherapy with ovalbumin and the immunodominant T-cell epitope on airway eosinophilia and hyperresponsiveness in a murine model of allergic asthma. Am J Respir Cell Mol Biol 21(1): 21–29.

Jeurink PV, Savelkoul HFJ (2006). Induction and regulation of allergen-specific IgE. In: Gilissen LJWJ, Wichers HJ, Savelkoul HFJ, Bogers RJ, eds. Allergy Matters: New Approaches to Allergy Prevention and Management. Dordrecht, The Netherlands: Springer; pp 13–27.

Kahlert H, Petersen A, Becker WM, Schlaak M (1992). Epitope analysis of the allergen ovalbumin (Gal d II) with monoclonal antibodies and patients' IgE. Mol Immunol 29(10):1191–1201.

Kalliomaki M, Salminen S, Poussa T, Arvilommi H, Isolauri E (2003). Probiotics and prevention of atopic disease: 4-year follow-up of a randomised placebo-controlled trial. Lancet 361(9372):1869–1871.

Katayama S, Mine Y (2006). Quillaja saponin can modulate ovalbumin-induced IgE allergic responses through regulation of Th1/Th2 balance in a murine model. J Agric Food Chem 54(9):3271–3276.

Kato A, Fujimoto K, Matsudomi N, Kobayashi K (1986). Protein flexibility and functional properties of heat-denatured ovalbumin and lysozyme. Agric Biol Chem 50:417–420.

Kato I, Schrode J, Kohr WJ, Laskowski M (1987). Chicken ovomucoid: determination of its amino acid sequence, determination of the trypsin reactive site, and preparation of all three of its domains. Biochemistry 26(1):193–201.

Katsuki T, Shimojo N, Honma K, Tsunoo H, Kohno Y, Niimi H (1996). Establishment and characterization of ovalbumin-specific T cell lines from patients with egg allergy. Int Arch Allergy Immunol 109(4):344–351.

Kendler M, Uter W, Rueffer A, Shimshoni R, Jecht E (2006). Comparison of fecal microflora in children with atopic eczema/dermatitis syndrome according to IgE sensitization to food. Pediatr Allergy Immunol 17(2):141–147.

Kim SH, Cho D, Hwang SY, Kim TS (2001). Efficient induction of antigen-specific, T helper type 1-mediated immune responses by intramuscular injection with ovalbumin/interleukin-18 fusion DNA. Vaccine 19(30):4107–4114.

Kimoto H, Mizumachi K, Okamoto T, Kurisaki J (2004). New Lactococcus strain with immunomodulatory activity: Enhancement of Th1-type immune response. Microbiol Immunol 48(48):75–82.

Kjaer TM, Frokiaer H (2002). Induction of oral tolerance with micro-doses of ovomucoid depends on the length of the feeding period. Scand J Immunol 55(4): 359–365.

Kline JN, Kitagaki K, Businga TR, Jain VV (2002). Treatment of established asthma in a murine model using CpG oligodeoxynucleotides. Am J Physiol Lung Cell Mol Physiol 283(1):L170–L179.

Knight AK, Shreffler WG, Sampson HA, Sicherer SH, Noone S, Mofidi S, Nowak-Wegrzyn A (2006). Skin prick test to egg white provides additional diagnostic utility to serum egg white-specific IgE antibody concentration in children. J Allergy Clin Immunol 117(4):842–847.

Kondo H, Shiroishi M, Matsushima M, Tsumoto K, Kumagai I (1999). Crystal structure of anti-hen egg white lysozyme antibody (HyHEL-10) Fv-antigen complex. Local structural changes in the protein antigen and water-mediated interactions of Fv-antigen and light chain-heavy chain interfaces. J Biol Chem 274(39):27623–27631.

Kondo M, Suzuki K, Inoue R, Sakaguchi H, Matsukuma E, Kato Z, Kaneko H, Fukao T, Kondo N (2005). Characterization of T-cell clones specific to ovomucoid from patients with egg-white allergy. J Invest Allergol Clin Immunol 15(2):107–111.

Konishi Y, Kurisaki J, Kaminogawa S, Yamauchi K (1982). Localization of allergenic reactive sites on hen ovomucoid. Agric Biol Chem 46:305–307.

Konishi Y, Kurisaki J, Kaminogawa S, Yamauchi K (1985). Determination of antigenicity by radioimmunoassay and of trypsin inhibitory activities in heat or enzyme denatured ovomucoid. J Food Sci 50:1422–1426.

Kovacs-Nolan J, Phillips M, Mine Y (2005). Advances in the value of eggs and egg components for human health. J Agric Food Chem 53(22):8421–8431.

Kovacs-Nolan J, Zhang JW, Hayakawa S, Mine Y (2000). Immunochemical and structural analysis of pepsin-digested egg white ovomucoid. J Agric Food Chem 48(12): 6261–6266.

Krieg AM (2002). CpG motifs in bacterial DNA and their immune effects. Annu Rev Immunol 20:709–760.

Kurisaki J, Konishi Y, Kaminogawa S, Yamauchi K (1981). Studies on the allergenic structure of hen ovomucoid by chemical and enzymatic fragmentation. Agric Biol Chem 45:879–886.

Kursteiner O, Moser C, Lazar H, Durrer P (2006). Inflexal V—the influenza vaccine with the lowest ovalbumin content. Vaccine 24(44–46):6632–6635.

Lamb JR, Skidmore BJ, Green N, Chiller JM, Feldmann M (1983). Induction of tolerance in influenza virus-immune T lymphocyte clones with synthetic peptides of influenza hemagglutinin. J Exp Med 157(5):1434–1447.

Landry SJ (2006). The relatioship of T-cell epitopes and allergen structure. In: Maleki SJ, Burks W, Helm RM, eds. Food Allergy. Washington, DC: ASM Press; pp 123–259.

Langeland T (1982). A clinical and immunological study of allergy to hen's egg white. III. Allergens in hen's egg white studied by crossed radio-immunoelectrophoresis (CRIE). Allergy 37(7):521–530.

Langeland T (1983a). A clinical and immunological study of allergy to hen's egg white. VI. Occurrence of proteins cross-reacting with allergens in hen's egg white as studied in egg white from turkey, duck, goose, seagull, and in hen egg yolk, and hen and chicken sera and flesh. Allergy 38(6):399–412.

Langeland T (1983b). A clinical and immunological study of allergy to hen's egg white. IV. Specific IGE-antibodies to individual allergens in hen's egg white related to clinical and immunological parameters in egg-allergic patients. Allergy 38(7):493–500.

Langeland T, Harbitz O (1983). A clinical and immunological study of allergy to hen's egg white. V. Purification and identification of a major allergen (antigen 22) in hen's egg white. Allergy 38(2):131–139.

Larche M (2005). Peptide therapy for allergic diseases: Basic mechanisms and new clinical approaches. Pharmacol Ther 108(3):353–361.

Lee JW, Kim JH, Yook HS, Kang KO, Lee SY, Hwang HJ, Byun MW (2001). Effects of gamma radiation on the allergenic and antigenic properties of milk proteins. J Food Protect 64(2):272–276.

Lee JW, Lee KY, Yook HS, Lee SY, Kim HY, Jo C, Byun MW (2002). Allergenicity of hen's egg ovomucoid gamma irradiated and heated under different pH conditions. J Food Protect 65(7):1196–1199.

Lehrer SB (2004). Genetic modification of food allergens. Ann Allergy Asthma Immunol 93(5 Suppl 3):S19–S25.

Leser C, Hartmann AL, Praml G, Wuthrich B (2001). The "egg-egg" syndrome: Occupational respiratory allergy to airborne egg proteins with consecutive ingestive egg allergy in the bakery and confectionery industry. J Invest Allergol Clin Immunol 11(2):89–93.

Lever R, MacDonald C, Waugh P, Aitchison T (1998). Randomised controlled trial of advice on an egg exclusion diet in young children with atopic eczema and sensitivity to eggs. Pediatr Allergy Immunol 9(1):13–19.

Li Y, Li H, Smith-Gill SJ, Mariuzza RA (2000). Three-dimensional structures of the free and antigen-bound Fab from monoclonal antilysozyme antibody HyHEL-63. Biochemistry 39(21):6296–6309.

Lilly DM, Stillwell RH (1965). Probiotics: Growth-promoting factors produced by microorganisms. Science 147:747–748.

Lineweaver H, Murray CW (1947). Identification of the trypsin inhibitor of egg white with ovomucoid. J Biol Chem 171:565–581.

Maecker HT, Hansen G, Walter DM, DeKruyff RH, Levy S, Umetsu DT (2001). Vaccination with allergen-IL-18 fusion DNA protects against, and reverses established, airway hyperreactivity in a murine asthma model. J Immunol 166(2): 959–965.

Marini A, Agosti M, Motta G, Mosca F (1996). Effects of a dietary and environmental prevention programme on the incidence of allergic symptoms in high atopic risk infants: Three years' follow-up. Acta Paediatr Suppl 114:1–21.

Martorell Aragones A, Bone Calvo J, Garcia Ara MC, Nevot Falco S, Plaza Martin AM (2001). Allergy to egg proteins. Food Allergy Committee of the Spanish Society of Pediatric Clinical Immunology and Allergy. Allergol Immunopathol (Madrid) 29(2):72–95.

Matsuda T, Koseki SY, Yasumoto K, Kitabatake N (2000). Characterization of anti-irradiation-denatured ovalbumin monoclonal antibodies. Immunochemical and structural analysis of irradiation-denatured ovalbumin. J Agric Food Chem 48(7): 2670–2674.

Matsuda T, Nakamura R, Nakashima I, Hasegawa Y, Shimokata K (1985). Human IgE antibody to the carbohydrate-containing third domain of chicken ovomucoid. Biochem Biophys Res Commun 129(2):505–510.

Matsuda T, Nakashima I, Nakamura R, Shimokata K (1986). Specificity to ovomucoid domains of human serum antibody from allergic patients: Comparison with anti-ovomucoid antibody from laboratory animals. J Biochem (Tokyo) 100(4):985–988.

Matsuda T, Wanatabe K, Sato Y (1981). Temperature-induced structural changes in chicken egg white ovomucoid. Agric Biol Chem 45:1609.

Matsuzaki T, Yamazaki R, Hashimoto S, Yokokura T (1998). The effect of oral feeding of *Lactobacillus casei* strain Shirota on immunoglobulin E production in mice. J Dairy Sci 81(1):48–53.

Miller H, Campbell DH (1950). Skin test reactions to various chemical fractions of egg white and their possible clinical significance. J Allergy Clin Immunol 21(6): 522–524.

Mine Y, Noutomi T, Haga N (1990). Thermally induced changes in egg white proteins. J Agric Food Chem 38:2122–2125.

Mine Y, Rupa P (2003a). Fine mapping and structural analysis of immunodominant IgE allergenic epitopes in chicken egg ovalbumin. Protein Eng 16(10):747–752.

Mine Y, Rupa P (2003b). Genetic attachment of undecane peptides to ovomucoid third domain can suppress the production of specific IgG and IgE antibodies. Biochem Biophys Res Commun 311(1):223–228.

Mine Y, Rupa P (2004). Immunological and biochemical properties of egg allergens. World Poultry Sci J 60:321–330.

Mine Y, Sasaki E, Zhang JW (2003). Reduction of antigenicity and allergenicity of genetically modified egg white allergen, ovomucoid third domain. Biochem Biophys Res Commun 302(1):133–137.

Mine Y, Zhang JW (2002). Identification and fine mapping of IgG and IgE epitopes in ovomucoid. Biochem Biophys Res Commun 292(4):1070–1074.

Mine Y, Yang M (2007). Epitope characterization of ovalbumin in BALB/c mice using different entry routes. Biochim Biophys Acta 1774(2):200–212.

Mine Y, Zhang JW (2002). Comparative studies on antigenicity and allergenicity of native and denatured egg white proteins. J Agric Food Chem 50(9):2679–2683.

Mistrello G, Brenna O, Roncarolo D, Zanoni D, Gentili M, Falagiani P (1996). Monomeric chemically modified allergens: Immunologic and physicochemical characterization. Allergy 51(1):8–15.

Mizumachi K, Kurisaki J (2003). Localization of T cell epitope regions of chicken ovomucoid recognized by mice. Biosci Biotechnol Biochem 67(4):712–719.

Morisset M, Moneret-Vautrin DA, Kanny G, Guenard L, Beaudouin E, Flabbee J, Hatahet R (2003). Thresholds of clinical reactivity to milk, egg, peanut and sesame in immunoglobulin E-dependent allergies: Evaluation by double-blind or single-blind placebo-controlled oral challenges. Clin Exp Allergy 33(8):1046–1051.

Mosmann TR, Cherwinski H, Bond MW, Giedlin MA, Coffman RL (1986). Two types of murine helper T cell clone. I. Definition according to profiles of lymphokine activities and secreted proteins. J Immunol 136(7):2348–2357.

Mosmann TR, Coffman RL (1989). TH1 and TH2 cells: Different patterns of lymphokine secretion lead to different functional properties. Annu Rev Immunol 7:145–173.

Moudgil KD, Sekiguchi D, Kim SY, Sercarz EE (1997). Immunodominance is independent of structural constraints: Each region within hen eggwhite lysozyme is potentially available upon processing of native antigen. J Immunol 159(6):2574–2579.

Moudgil KD, Wang J, Yeung VP, Sercarz EE (1998). Heterogeneity of the T cell response to immunodominant determinants within hen eggwhite lysozyme of individual syngeneic hybrid F1 mice: Implications for autoimmunity and infection. J Immunol 161(11):6046–6053.

Nakamura R, Matsuda T (1996). Rice allergenic protein and molecular-genetic approach for hypoallergenic rice. Biosci Biotechnol Biochem 60(8):1215–1221.

Nguyen MD, Cinman N, Yen J, Horner AA (2001). DNA-based vaccination for the treatment of food allergy. Allergy 56(Suppl 67):127–130.

Nickel R, Kulig M, Forster J, Bergmann R, Bauer CP, Lau S, Guggenmoos-Holzmann I, Wahn U (1997). Sensitization to hen's egg at the age of twelve months is predictive for allergic sensitization to common indoor and outdoor allergens at the age of three years. J Allergy Clin Immunol 99(5):613–617.

Niederberger V, Valenta R (2006). Molecular approaches for new vaccines against allergy. Exp Rev Vaccines 5(1):103–110.

Niggemann B, Staden U, Rolinck-Werninghaus C, Beyer K (2006). Specific oral tolerance induction in food allergy. Allergy 61(7):808–811.

Norman PS, Lichtenstein LM, Kagey-Sobotka A, Marsh DG (1982). Controlled evaluation of allergoid in the immunotherapy of ragweed hay fever. J Allergy Clin Immunol 70(4):248–260.

Norman PS, Ohman JL, Long AA, Creticos PS, Gefter MA, Shaked Z, Wood RA, Eggleston PA, Hafner KB, Rao P, Lichtenstein LM, Jones NH, Nicodemus CF (1996). Treatment of cat allergy with T-cell reactive peptides. Am J Respir Crit Care Med 154(6 Pt 1):1623–1628.

Novembre E, Cianferoni A, Bernardini R, Mugnaini L, Caffarelli C, Cavagni G, Giovane A, Vierucci A (1998). Anaphylaxis in children: Clinical and allergologic features. Pediatrics 101(4):E8.

Ogawa A, Samoto M, Takahashi K (2000). Soybean allergens and hypoallergenic soybean products. J Nutr Sci Vitaminol (Tokyo) 46(6):271–279.

Ohta A, Sato N, Yahata T, Ohmi Y, Santa K, Sato T, Tashiro H, Habu S, Nishimura T (1997). Manipulation of Th1/Th2 balance in vivo by adoptive transfer of antigen-specific Th1 or Th2 cells. J Immunol Meth 209(1):85–92.

Padlan EA, Silverton EW, Sheriff S, Cohen GH, Smith-Gill SJ, Davies DR (1989). Structure of an antibody-antigen complex: Crystal structure of the HyHEL-10 Fab-lysozyme complex. Proc Natl Acad Sci USA 86(15):5938–5942.

Pedrosa C, De Felice FG, Trisciuzzi C, Ferreira ST (2000). Selective neoglycosylation increases the structural stability of vicilin, the 7S storage globulin from pea seeds. Arch Biochem Biophys 382(2):203–210.

Piccirillo CA, Prud'homme GJ (2003). Immune modulation by plasmid DNA-mediated cytokine gene transfer. Curr Pharm Des 9(1):84–94.

Prescott SL (2003). Allergy: the price we pay for cleaner living? Ann Allergy Asthma Immunol 90(6 Suppl 3):64–70.

Prescott SL, Dunstan JA, Hale J, Breckler L, Lehmann H, Weston S, Richmond P (2005). Clinical effects of probiotics are associated with increased interferon-gamma responses in very young children with atopic dermatitis. Clin Exp Allergy 35(12): 1557–1564.

Prescott SL, Jones C (2002). An update of immunotherapy for specific allergies. Curr Drug Targets Inflamm Allergy 1(1):65–75.

Primeau MN, Kagan R, Joseph L, Lim H, Dufresne C, Duffy C, Prhcal D, Clarke A (2000). The psychological burden of peanut allergy as perceived by adults with peanut allergy and the parents of peanut-allergic children. Clin Exp Allergy 30(8):1135–1143.

Prioult G, Nagler-Anderson C (2005). Mucosal immunity and allergic responses: Lack of regulation and/or lack of microbial stimulation? Immunol Rev 206:204–218.

Quirce S, Diez-Gomez ML, Eiras P, Cuevas M, Baz G, Losada E (1998). Inhalant allergy to egg yolk and egg white proteins. Clin Exp Allergy 28(4):478–485.

Rask C, Holmgren J, Fredriksson M, Lindblad M, Nordstrom I, Sun JB, Czerkinsky C (2000). Prolonged oral treatment with low doses of allergen conjugated to cholera toxin B subunit suppresses immunoglobulin E antibody responses in sensitized mice. Clin Exp Allergy 30(7):1024–1032.

Rautava S, Kalliomaki M, Isolauri E (2005). New therapeutic strategy for combating the increasing burden of allergic disease: Probiotics. Nutrition, Allergy, Mucosal Immunology and Intestinal Microbiota (NAMI) Research Group Report. J Allergy Clin Immunol 116(1):31–37.

Raz E, Spiegelberg HL (1999). Deviation of the allergic IgE to an IgG response by gene immunotherapy. Int Rev Immunol 18(3):271–289.

Renz H, Bradley K, Larsen GL, McCall C, Gelfand EW (1993). Comparison of the allergenicity of ovalbumin and ovalbumin peptide 323–339. Differential expansion of V beta-expressing T cell populations. J Immunol 151(12):7206–7213.

Ria F, Gallard A, Gabaglia CR, Guery JC, Sercarz EE, Adorini L (2004). Selection of similar naive T cell repertoires but induction of distinct T cell responses by native and modified antigen. J Immunol 172(6):3447–3453.

Rolinck-Werninghaus C, Staden U, Mehl A, Hamelmann E, Beyer K, Niggemann B (2005). Specific oral tolerance induction with food in children: transient or persistent effect on food allergy? Allergy 60(10):1320–1322.

Romagnani S (1991). Human TH1 and TH2 subsets: Doubt no more. Immunol Today 12(8):256–257.

Roy K, Mao HQ, Huang SK, Leong KW (1999). Oral gene delivery with chitosan—DNA nanoparticles generates immunologic protection in a murine model of peanut allergy. Nat Med 5(4):387–391.

Rupa P, Mine Y (2003a). Immunological comparison of native and recombinant egg allergen, ovalbumin, expressed in *Escherichia coli*. Biotechnol Lett 25(22):1917–1924.

Rupa P, Mine Y (2003b). Structural and immunological characterization of recombinant ovomucoid expressed in *Escherichia coli*. Biotechnol Lett 25(5):427–433.

Rupa P, Mine Y (2006a). Ablation of ovomucoid-induced allergic response by desensitization with recombinant ovomucoid third domain in a murine model. Clin Exp Immunol 145(3):493–501.

Rupa P, Mine Y (2006b). Engineered recombinant ovomucoid third domain can desensitize Balb/c mice of egg allergy. Allergy 61(7):836–842.

Rupa P, Mine Y (2006c). Engineered recombinant ovomucoid third domain can modulate allergenic response in Balb/c mice model. Biochem Biophys Res Commun 342(3):710–717.

Rupa P, Nakamura S, Mine Y (2007). Genetically glycosylated ovomucoid third domain can modulate IgE antibody production and cytokine response in Balb/c mice. Clin Exp Allergy (in press).

Sampson HA (1992). The immunopathogenic role of food hypersensitivity in atopic dermatitis. Acta Derm Venereol Suppl (Stockholm) 176:34–37.

Sampson HA (1997). Food sensitivity and the pathogenesis of atopic dermatitis. J R Soc Med 90(Supl 30):2–8.

Sampson HA (2004). Update on food allergy. J Allergy Clin Immunol 113(5):805–819.

Sampson HA, Bernhisel-Broadbent J, Yang E, Scanlon SM (1991). Safety of casein hydrolysate formula in children with cow milk allergy. J Pediatr 118(4 Pt 1):520–525.

Sampson HA, McCaskill CC (1985). Food hypersensitivity and atopic dermatitis: Evaluation of 113 patients. J Pediatr 107(5):669–675.

Sampson HA, Mendelson L, Rosen JP (1992). Fatal and near-fatal anaphylactic reactions to food in children and adolescents. New Engl J Med 327(6):380–384.

Santeliz JV, Van Nest G, Traquina P, Larsen E, Wills-Karp M (2002). Amb a 1-linked CpG oligodeoxynucleotides reverse established airway hyperresponsiveness in a murine model of asthma. J Allergy Clin Immunol 109(3):455–462.

Sasaki E, Mine Y (2001). IgE binding properties of the recombinant ovomucoid third domain expressed in Escherichia coli. Biochem Biophys Res Commun 282(4):947–951.

Satkauskas S, Bureau MF, Mahfoudi A, Mir LM (2001). Slow accumulation of plasmid in muscle cells: Supporting evidence for a mechanism of DNA uptake by receptor-mediated endocytosis. Mol Ther 4(4):317–323.

Sato R, Takeyama H, Tanaka T, Matsunaga T (2001). Development of high-performance and rapid immunoassay for model food allergen lysozyme using antibody-conjugated bacterial magnetic particles and fully automated system. Appl Biochem Biotechnol 91–93:109–116.

Sato T, Sasahara T, Nakamura Y, Osaki T, Hasegawa T, Tadakuma T, Arata Y, Kumagai Y, Katsuki M, Habu S (1994). Naive T cells can mediate delayed-type hypersensitivity response in T cell receptor transgenic mice. Eur J Immunol 24(7):1512–1516.

Schloss OM (1912). A case of allergy to common foods. Am J Dis Child 3:341–343.

Schuster M, Einhauer A, Wasserbauer E, Sussenbacher F, Ortner C, Paumann M, Werner G, Jungbauer A (2000). Protein expression in yeast; comparison of two expression strategies regarding protein maturation. J Biotechnol 84(3):237–248.

Seo J, Kim J, Lee J, Yoo YC, Kim M, Park KS, Byun MW (2007). Ovalbumin modified by gamma irradiation alters its immunological functions and allergic responses. Int Immunopharmacol 7(4):464–472.

Seo JH, Lee JW, Lee YS, Lee SY, Kim MR, Yook HS, Byun MW (2004). Change of an egg allergen in a white layer cake containing gamma-irradiated egg white. J Food Protect 67(8):1725–1730.

Sette A, Buus S, Colon S, Smith JA, Miles C, Grey HM (1987). Structural characteristics of an antigen required for its interaction with Ia and recognition by T cells. Nature 328(6129):395–399.

Shida K, Takahashi R, Iwadate E, Takamizawa K, Yasui H, Sato T, Habu S, Hachimura S, Kaminogawa S (2002). *Lactobacillus casei* strain Shirota suppresses serum immunoglobulin E and immunoglobulin G1 responses and systemic anaphylaxis in a food allergy model. Clin Exp Allergy 32(4):563–570.

Shimada J, Yano H, Mizumachi K (2005). Trends in food allergy research. Sci Technol Trends 16:26–35.

Shimojo N, Katsuki T, Coligan J, Nishimura Y, Sasazuki T, Tsunoo H, Sakamaki T, Kohno Y, Niimi H (1994). Identification of the disease-related T cell epitope of ovalbumin and epitope-targeted T cell inactivation in egg allergy. Int Arch Allergy Immunol 105(2):155–161.

Shimonkevitz R, Colon S, Kappler JW, Marrack P, Grey HM (1984). Antigen recognition by H-2-restricted T cells. II. A tryptic ovalbumin peptide that substitutes for processed antigen. J Immunol 133(4):2067–2074.

Sicherer SH (2002). Food allergy. Lancet 360(9334):701–710.

Spiegelberg HL, Horner AA, Takabayashi K, Raz E (2002a). Allergen-immunostimulatory oligodeoxynucleotide conjugate: A novel allergoid for immunotherapy. Curr Opin Allergy Clin Immunol 2(6):547–551.

Spiegelberg HL, Takabayashi K, Beck L, Raz E (2002b). DNA-based vaccines for allergic disease. Exp Rev Vaccines 1(2):169–177.

Srivastava KD, Kattan JD, Zou ZM, Li JH, Zhang L, Wallenstein S, Goldfarb J, Sampson HA, Li XM (2005). The Chinese herbal medicine formula FAHF-2 completely blocks anaphylactic reactions in a murine model of peanut allergy. J Allergy Clin Immunol 115(1):171–178.

Strachan DP (1989). Hay fever, hygiene, and household size. Br Med J 299(6710):1259–1260.

Strid J, Thomson M, Hourihane J, Kimber I, Strobel S (2004). A novel model of sensitization and oral tolerance to peanut protein. Immunology 113(3):293–303.

Sun J, Dirden-Kramer B, Ito K, Ernst PB, Van Houten N (1999). Antigen-specific T cell activation and proliferation during oral tolerance induction. J Immunol 162(10):5868–5875.

Suzuki K, Inoue R, Sakaguchi H, Aoki M, Kato Z, Kaneko H, Matsushita S, Kondo N (2002). The correlation between ovomucoid-derived peptides, human leucocyte antigen class II molecules and T cell receptor-complementarity determining region

3 compositions in patients with egg-white allergy. Clin Exp Allergy 32(8):1223–1230.

Swint-Kruse L, Robertson AD (1995). Hydrogen bonds and the pH dependence of ovomucoid third domain stability. Biochemistry 34(14):4724–4732.

Szepfalusi Z, Ebner C, Pandjaitan R, Orlicek F, Scheiner O, Boltz-Nitulescu G, Kraft D, Ebner H (1994). Egg yolk alpha-livetin (chicken serum albumin) is a cross-reactive allergen in the bird-egg syndrome. J Allergy Clin Immunol 93(5):932–942.

Takagi K, Teshima R, Okunuki H, Itoh S, Kawasaki N, Kawanishi T, Hayakawa T, Kohno Y, Urisu A, Sawada J (2005). Kinetic analysis of pepsin digestion of chicken egg white ovomucoid and allergenic potential of pepsin fragments. Int Arch Allergy Immunol 136(1):23–32.

Takeda K, Akira S (2005). Toll-like receptors in innate immunity. Int Immunol 17(1):1–14.

Tanabe S, Tesaki S, Watanabe M (2000). Producing a low ovomucoid egg white preparation by precipitation with aqueous ethanol. Biosci Biotechnol Biochem 64(9): 2005–2007.

Tanabe S, Tesaki S, Yanagihara Y, Mita H, Takahashi K, Arai S, Watanabe M (1996). Inhibition of basophil histamine release by a haptenic peptide mixture prepared by chymotryptic hydrolysis of wheat flour. Biochem Biophys Res Commun 223(3):492–495.

Tariq SM, Matthews SM, Hakim EA, Arshad SH (2000). Egg allergy in infancy predicts respiratory allergic disease by 4 years of age. Pediatr Allergy Immunol 11(3):162–167.

Taylor SL, Hefle SL, Bindslev-Jensen C, Atkins FM, Andre C, Bruijnzeel-Koomen C, Burks AW, Bush RK, Ebisawa M, Eigenmann PA, Host A, Hourihane JO, Isolauri E, Hill DJ, Knulst A, Lack G, Sampson HA, Moneret-Vautrin DA, Rance F, Vadas PA, Yunginger JW, Zeiger RS, Salminen JW, Madsen C, Abbott P (2004). A consensus protocol for the determination of the threshold doses for allergenic foods: How much is too much? Clin Exp Allergy 34(5):689–695.

Teuber SS, Beyer K, Comstock S, Wallowitz M (2006). The big eight foods: Clinical and epidemiological overview. In Maleki SJ, Burks W, Helm RM, eds. Food Allergy. Washington, DC: AMS Press; pp 49–79.

Tsalik EL (2005). DNA-based immunotherapy to treat atopic disease. Ann Allergy Asthma Immunol 95(5):403–410.

Tulic MK, Fiset PO, Christodoulopoulos P, Vaillancourt P, Desrosiers M, Lavigne F, Eiden J, Hamid Q (2004). Amb a 1-immunostimulatory oligodeoxynucleotide conjugate immunotherapy decreases the nasal inflammatory response. J Allergy Clin Immunol 113(2):235–241.

Untersmayr E, Jensen-Jarolim E (2006). Mechanisms of type I food allergy. Pharmacol Ther 112(3):787–798.

Urisu A, Ando H, Morita Y, Wada E, Yasaki T, Yamada K, Komada K, Torii S, Goto M, Wakamatsu T (1997). Allergenic activity of heated and ovomucoid-depleted egg white. J Allergy Clin Immunol 100(2):171–176.

Urisu A, Yamada K, Tokuda R, Ando H, Wada E, Kondo Y, Morita Y (1999). Clinical significance of IgE-binding activity to enzymatic digests of ovomucoid in the diagnosis and the prediction of the outgrowing of egg white hypersensitivity. Int Arch Allergy Immunol 120(3):192–198.

Valenta R (2002). Recombinant allergen-based concepts for diagnosis and therapy of type I allergy. Allergy 57(Suppl 71):66–67.

Valenta R, Kraft D (2002). From allergen structure to new forms of allergen-specific immunotherapy. Curr Opin Immunol 14(6):718–727.

Valenta R, Kraft D (2004). Recombinant allergens: From production and characterization to diagnosis, treatment and prevention of allergy. Methods 32:207–345.

van Halteren AG, van der Cammen MJ, Biewenga J, Savelkoul HF, Kraal G (1997). IgE and mast cell response on intestinal allergen exposure: A murine model to study the onset of food allergy. J Allergy Clin Immunol 99(1 Pt 1):94–99.

van Ree R (2002). Carbohydrate epitopes and their relevance for the diagnosis and treatment of allergic diseases. Int Arch Allergy Immunol 129(3):189–197.

van Ree R, Aalberse RC (1995). Demonstration of carbohydrate-specific immunoglobulin G4 antibodies in sera of patients receiving grass pollen immunotherapy. Int Arch Allergy Immunol 106(2):146–148.

van Toorenenbergen AW, Huijskes-Heins MI, Gerth van Wijk R (1994). Different pattern of IgE binding to chicken egg yolk between patients with inhalant allergy to birds and food-allergic children. Int Arch Allergy Immunol 104(2):199–203.

Vidard L, Rock KL, Benacerraf B (1992a). Diversity in MHC class II ovalbumin T cell epitopes generated by distinct proteases. J Immunol 149(2):498–504.

Vidard L, Rock KL, Benacerraf B (1992b). Heterogeneity in antigen processing by different types of antigen-presenting cells. Effect of cell culture on antigen processing ability. J Immunol 149(6):1905–1911.

Vidard L, Rock KL, Couderc J, Mouton D, Benacerral B (1992c). Processing and presentation of ovalbumin in mice genetically selected for antibody response. Eur J Immunol 22(8):2165–2168.

Vieleuf I, Besler M, Paschke A, Steinhart H, Vieluf D (2002). Practical approach to adverse food reactions. In: Ring J, Behrendt H, eds. New Trends in Allergy V. Heidelberg: Springer-Verlag; pp 190–202.

von Baehr V, Hermes A, von Baehr R, Scherf HP, Volk HD, Fischer Von Weikersthal-Drachenberg KJ, Woroniecki S (2005). Allergoid-specific T-cell reaction as a measure of the immunological response to specific immunotherapy (SIT) with a Th1-adjuvanted allergy vaccine. J Invest Allergol Clin Immunol 15(4):234–241.

Walsh BJ, Barnett D, Burley RW, Elliott C, Hill DJ, Howden ME (1988). New allergens from hen's egg white and egg yolk. In vitro study of ovomucin, apovitellenin I and VI, and phosvitin. Int Arch Allergy Appl Immunol 87(1):81–86.

Walsh BJ, Elliott C, Baker RS, Barnett D, Burley RW, Hill DJ, Howden ME (1987). Allergenic cross-reactivity of egg-white and egg-yolk proteins. An in vitro study. Int Arch Allergy Appl Immunol 84(3):228–232.

Walsh BJ, Hill DJ, Macoun P, Cairns D, Howden ME (2005). Detection of four distinct groups of hen egg allergens binding IgE in the sera of children with egg allergy. Allergol Immunopathol (Madrid) 33(4):183–191.

Watanabe J, Tanabe S, Watanabe M, Shinmoto H, Sonoyama K (2004). The production of hypoallergenic wheat flour and the analysis of its allergy suppressive effects. Biofactors 22(1–4):295–297.

Watanabe M, Watanabe J, Sonoyama K, Tanabe S (2000). Novel method for producing hypoallergenic wheat flour by enzymatic fragmentation of the constituent allergens and its application to food processing. Biosci Biotechnol Biochem 64(12):2663–2667.

Weber P, Raynaud I, Ettouati L, Trescol-Biemont MC, Carrupt PA, Paris J, Rabourdin-Combe C, Gerlier D, Testa B (1998). Molecular modeling of hen egg lysozyme HEL[52–61] peptide binding to I-Ak MHC class II molecule. Int Immunol 10(12):1753–1764.

Weiss R, Hammerl P, Hartl A, Hochreiter R, Leitner WW, Scheiblhofer S, Thalhamer J (2005). Design of protective and therapeutic DNA vaccines for the treatment of allergic diseases. Curr Drug Targets Inflamm Allergy 4(5):585–597.

Wichers H, Matser A, Van Amerongen A, Wickers J, Soler-Rivas C (2003). Monitoring of and technological effects on allergenicity of proteins in the food industry. In: Mills ENC, Shewry PR, eds. Plant Food Allergens. Oxford: Blackwell; pp 196–212.

Williams LW, Bock SA (1999). Skin testing and food challenges in allergy and immunology practice. Clin Rev Allergy Immunol 17(3):323–338.

Yamada K, Urisu A, Kakami M, Koyama H, Tokuda R, Wada E, Kondo Y, Ando H, Morita Y, Torii S (2000). IgE-binding activity to enzyme-digested ovomucoid distinguishes between patients with contact urticaria to egg with and without overt symptoms on ingestion. Allergy 55(6):565–569.

Yamada T, Yamada M, Sasamoto K, Nakamura H, Mishima T, Yasueda H, Shida T, Iikura Y (1993). Specific IgE antibody titers to hen's egg white lysozyme in allergic children to egg. Arerugi 42(2):136–141.

Yano H, Kato Y, Matsuda T (2002). Acute exercise induces gastrointestinal leakage of allergen in lysozyme-sensitized mice. Eur J Appl Physiol 87(4–5):358–364.

Yazdanbakhsh M, Kremsner P, van Ree R (2002). Allergy, parasites, and the hygiene hypothesis. Science 296(5567):490–494.

Yoshino S, Sagai M (1999). Induction of systemic Th1 and Th2 immune responses by oral administration of soluble antigen and diesel exhaust particles. Cell Immunol 192(1):72–78.

Youn CJ, Miller M, Baek KJ, Han JW, Nayar J, Lee SY, McElwain K, McElwain S, Raz E, Broide DH (2004). Immunostimulatory DNA reverses established allergen-induced airway remodeling. J Immunol 173(12):7556–7564.

Yunginger JW, Sweeney KG, Sturner WQ, Giannandrea LA, Teigland JD, Bray M, Benson PA, York JA, Biedrzycki L, Squillace DL, et al (1988). Fatal food-induced anaphylaxis. JAMA 260(10):1450–1452.

Zeiger RS (2002). Current issues with influenza vaccination in egg allergy. J Allergy Clin Immunol 110(6):834–840.

Zemann B, Schwaerzler C, Griot-Wenk M, Nefzger M, Mayer P, Schneider H, de Weck A, Carballido JM, Liehl E (2003). Oral administration of specific antigens to allergy-prone infant dogs induces IL-10 and TGF-beta expression and prevents allergy in adult life. J Allergy Clin Immunol 111(5):1069–1075.

Zhang JW, Mine Y (1998). Characterization of IgE and IgG epitopes on ovomucoid using egg-white-allergic patients' sera. Biochem Biophys Res Commun 253(1):124–127.

Zhang JW, Mine Y (1999). Characterization of residues in human IgE and IgG binding site by chemical modification of ovomucoid third domain. Biochem Biophys Res Commun 261(3):610–613.

Zuercher AW, Fritsché R, Corthésy B, Mercenier A (2006). Food products and allergy development, prevention and treatment. Curr Opin Biotechnol 17(2):198–203.

7

PRODUCTION OF NOVEL PROTEINS IN CHICKEN EGGS

Robert J. Etches
Origen Therapeutics, Burlingame, California

7.1. INTRODUCTION

The potential to produce novel proteins in chicken eggs was identified in the early 1990's, when the introduction of genetic modifications into the chicken genome was first suggested (Shuman 1990; Sang 1994). The concept was further supported by early studies of the ovalbumin and vitellogenin genes demonstrating the potential to direct deposition of the novel products into egg white and egg yolk, respectively (Deeley et al. 1991). While an appropriate technology for the production of transgenic chickens that deposited novel proteins in their eggs was slow to develop, the concept remained attractive throughout the 1990s and into the first decade of the twenty-first century (Mohammed et al. 1998; Zajchowski and Etches 2000; Morrison et al. 2002; Ivarie 2003; Kamihira et al. 2004; Lillico et al. 2005). A major technical hurdle has been the introduction of large DNA constructs that would direct high-level and tissue-specific deposition of proteins into eggs. This hurdle has now been overcome (van de Lavoir et al. 2006a) and high levels of therapeutic monoclonal antibodies have been produced in egg white (Zhu et al. 2005). In this chapter, we present the background and most recent developments in the production of novel proteins in eggs.

Egg Bioscience and Biotechnology Edited by Yoshinori Mine

7.2. PRODUCTION OF NOVEL PROTEINS IN EGG YOLK

There are about 6g of protein in an egg equally divided between the egg yolk and egg white. The egg yolk proteins include transthyretin, vitellogenin, apoprotein B, and immunoglobulins, and each of these molecules is recruited into yolk via a receptor-mediated uptake mechanism (Morrison et al. 2002; Vieira et al. 1995; Deeley et al. 1991). Fusion proteins designed for production in and export from the liver and for selective delivery into the oocyte have been envisioned for some time (Deeley et al. 1991), although to date there are no reports demonstrating that recombinant or transgene-encoded proteins can be manufactured in the liver and recruited into yolk.

Approximately 200mg of immunoglobulin is recruited into the yolk to provide maternal immunity for the neonatal chick (Kowalczyk et al. 1985). Uptake of immunoglobulin into yolk is dependent on the C_H2/C_H3 interface and requires a HEAL motif at positions 429–432. Since this motif is shared by human IgG and IgA, both classes of human immunoglobulin are transported into egg yolk (Morrison et al. 2002). These data support the conclusion that human immunoglobulins can be produced in transgenic chickens by B cells or other tissues and subsequently purified from egg yolk (Etches et al. 2005b). Human antibodies have been produced in mice carrying transgene-encoding sequences of the human heavy- and light-chain immunoglobulin genes, and it has been suggested that this principle could be applied to produce human polyclonal antibodies in chicken egg yolk (Lonberg 2005; Etches et al. 2005a).

7.3. PRODUCTION OF NOVEL PROTEINS IN EGG WHITE

The egg white proteins are produced in tubular gland cells and the secretory epithelial cells in the magnum region of the reproductive tract (Fig. 7.1). It is generally believed that the major egg white proteins and some of the less abundant proteins are produced in the tubular gland cells and that the secretory epithelial cells produce some of the minor egg white components. To date, however, a systematic analysis using immunohistochemical techniques that would precisely identify the sources of each protein in egg white has not been undertaken. The egg white proteins are produced continuously and stored in secretory granules within the tubular gland cells and secretory epithelial cells. As the ovum is transported through the magnum, the egg white proteins are released from the secretory granules and deposited as a gelatinous mass around the yolk. Immediately prior to the passage of an ovum through the magnum, the tubular gland cells and secretory epithelial cells contain a reserve of the egg white proteins that is sufficient to produce more than one egg.

Egg white contains approximately 40 proteins, but more than 50% of the mass of egg white is ovalbumin (Table 7.1). The gene regulating ovalbumin production is located in a cluster of genes encoding serpin family proteins,

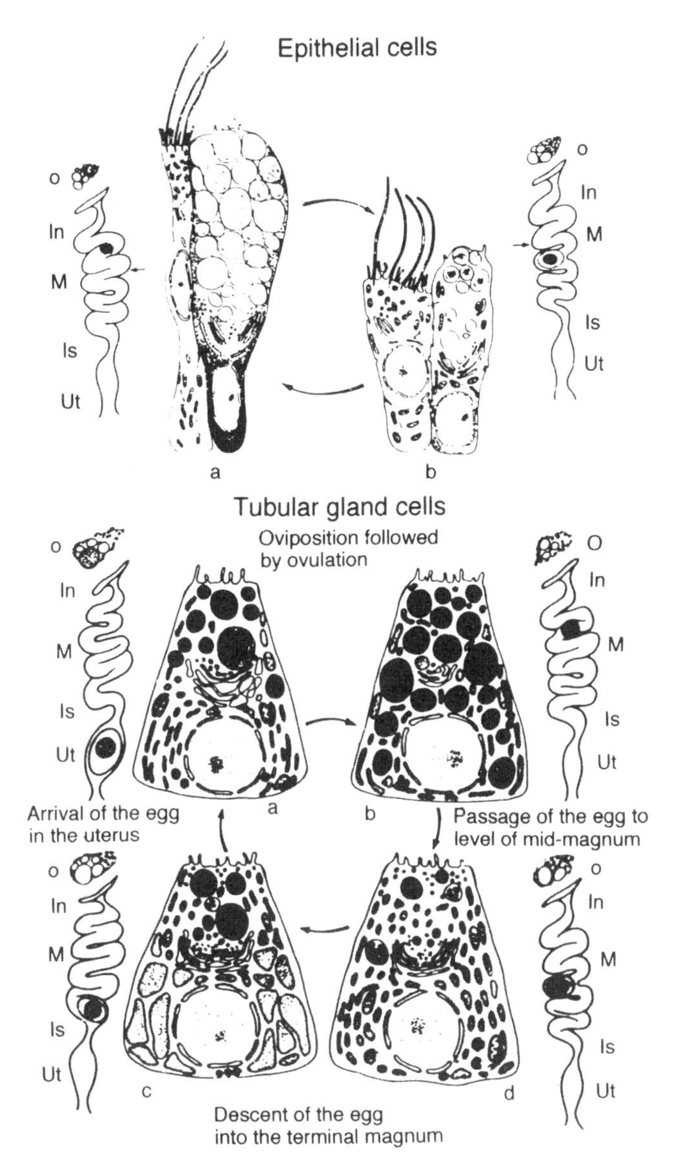

Figure 7.1. A diagrammatic representation of the changes in the secretory cells of the epithelium (upper panel) and the tubular gland cells (lower panel) of the magnum throughout the ovulatory cycle. As the egg enters the magnum, the epithelial cells are 20–30 μm in height (upper panel, a) and the nonsecretory ciliated cells are compressed between the nonciliated cells, which are packed with secretory granules at their apical surface. When an egg has traversed the magnum (upper panel, b), the secretory granules have discharged most of their contents and the epithelium is 13–18 μm in height. The histological characteristics of the cells in the tubular glands indicate that they produce egg white proteins continuously throughout the ovulatory cycle and store the proteins in secretory granules (shown in black in the lower panel, a and b). Most of the preformed protein is released as the egg passes through the magnum (lower panel, d), and the rough endoplasmic reticulum (ER) becomes more conspicuous within an hour after the egg leaves this region (lower panel, c). The proteins produced in the ER are packaged by the Golgi apparatus into secretory granules that are most abundant as the egg enters the magnum. [From Etches (1996).]

which includes ovalbumin, on chromosome 2 (Benarafa and Remold-O'Donnell 2005). The majority of albumin RNA is believed to be expressed by the ovalbumin locus, although the X and Y genes, which are located ovalbumin 5′ of the ovalbumin locus, are very similar and may produce small amounts of ovalbumin message. Given that there is approximately 1.5 g of ovalbumin per egg, each allele at the ovalbumin locus has the capacity to produce 750 mg of ovalbumin. This remarkable level of protein production makes the ovalbumin regulatory regions of interest for the production of proteins in transgenic animals.

The regulatory regions of the ovalbumin locus can be used in at least three strategies for the production of novel proteins in eggs. To accommodate the limitations of retroviral and lentiviral vectors in which only small amounts of DNA can be included, a number of investigators have suggested that the regulatory elements known to be present in regions 5′ of the ovalbumin-coding sequence could be assembled into a minigene that would provide a high level and tissue-specific expression. For example, a 1.4-kb (kilo–base pair) fragment of the chicken ovalbumin promoter has been linked to the reporter gene chloramphenicol transferase and inserted into the chicken genome using a retroviral vector (Harvey et al. 2002b). Similarly, a 1.3-kb fragment (MacArthur 2004) and a 1.4-kb fragment (Ivarie et al. 2004) of the 5′ regulatory region of the ovalbumin locus in combination with a signal sequence derived from the lysozyme locus have been suggested as regulatory sequences that could direct high-level and tissue-specific expression of transgenes (MacArthur 2004). Recently, tissue specific expression has been demonstrated using a 2.8–3.5 kb sequence derived from the ovalbumin locus (Lillico et al. 2007).

A second strategy has been to insert the coding sequence of the protein of interest into the ovalbumin locus by homologous recombination (Pain et al. 2003). This strategy has the advantage of using both endogenous regulatory elements 5′ and 3′ of the ovalbumin locus and should, in theory, yield approximately 750 mg of protein per egg (Ivarie 2006). While yields in this range might be desirable, they may be fraught with perturbations of egg formation that preclude formation of a hard-shelled egg. Evidence has been presented to suggest that erythropoietin and marker genes can be incorporated into the ovalbumin locus of ES cells by homologous recombination, although the capacity of genetically modified cells, to contribute to chimeras, differentiate into tubular gland cells, and subsequently deliver large amounts of protein into egg white has not yet been reported (Pain et al. 2003).

A third strategy to harness the ability of the ovalbumin regulatory regions to direct high-level and tissue-specific expression of a transgene has been executed by randomly integrating transgenes carrying extensive regions of genomic DNA that lie immediately 5′ and 3′ of the ovalbumin-coding sequence (Zhu et al. 2005). Using this approach, tissue-specific expression was achieved with 7.5 kb of 5′ sequence and 15 kb of 3′ sequence. Increasing the amount of 5′ sequence to 15 kb yielded up to 3 mg of a human therapeutic monoclonal antibody per egg, although paradoxically, this transgene was expressed at a low level in the gut as well as at high levels in the tubular gland cells. Extending the amount of genomic DNA to include approximately 125 kb of 5′ and approximately 30 kb of 3′ of the ovalbumin-coding region yielded up to 6 mg of the therapeutic monoclonal antibody per egg (Zhu and Etches 2008). Eggs from hens that contained milligram quantities of the therapeutic monoclonal antibody were fertile, and normal chicks were hatched, indicating that the inclusion of a human monoclonal antibody at these concentrations was compatible with normal development.

Although ovalbumin is the most abundant protein in egg white, the regulatory regions of other egg white proteins have been considered for driving the expression of foreign proteins in the tubular gland cells and their deposition into egg white. Transgenes that included regulatory regions of the chicken lysozyme locus have been prepared, but their function has yet to be demonstrated *in vivo* (Lampard and Verrinder Gibbins 2002). Transgenic chickens carrying the chicken lysozyme enhancer and promoter elements fused to the lacZ gene and flanked by two lysozyme gene 5′ matrix attachment regions have been produced (Love et al. 1994), but expression of the transgene has not been reported. A lysozyme promoter has been described and shown to function in cultured oviduct cells (Rapp 2002), although to date there are no reports of protein production in eggs using this system.

The second and third most abundant egg white proteins are ovotransferrin (also known as *conalbumin*) and ovomucoid, which each account for approximately 12% of the egg white (Table 7.1). In order to achieve high-level and

TABLE 7.1. Characteristics of Major Proteins Found in Egg Albumen

Protein	Proportion of Albumen (%)	Molecular Weight (kDa)	Characteristics
Ovalbumin	54	45	Enzyme inhibitor; binds Fe, Mn, Cu, and other trace metals
Ovotransferrin	12	76	Binds metallic ions
Ovomucoid	11	28	Inhibits trypsin
Ovomucin	3.5	110	Inhibits viral hemagglutination
Lysozyme	3.4	14.3	Lyses some bacteria
Ovoglobulin G2	~4	33–49	Good foaming agent
Ovoglobulin G3	~4	33–49	Good foaming agent
Ovoinhibitor	1.5	49	Inhibits serine proteases
Cystatin	0.05	12.7	Inhibits thioproteases
Ovoglyoprotein	1.0	24.4	Sialoprotein
Ovoflavoprotein	0.8	32	Binds riboflavin
Ovomacroglobulin	0.5	760–790	Strongly antigenic
Avidin	0.05	68.3	Binds biotin
Other proteins	4.2	—	—

Source: Etches (1996).

tissue-specific expression, the regulatory regions for both proteins have been suggested as candidates for inclusion in transgenes to produce therapeutic proteins (MacArthur 2004; Ivarie et al. 2004). Transgenic birds expressing erythropoietin regulated by the MDOT promoter, which contains elements from both the ovomucoid and ovotransferrin promoters, however, produced approximately 70 ng/mL of erythropoietin in blood and eggs (Ivarie et al. 2004). Furthermore, erythropoietin was observed at similar levels in the blood of chicks. It would appear, therefore, that the MDOT promoter failed to provide high-level, tissue-specific, and developmentally regulated gene expression.

The ovoinhibitor locus, which encodes the eighth most abundant egg white protein, and the ovomucoid locus are located in close proximity in the chicken genome. These two genes have been isolated in a bacterial artificial chromosome, and a promoter system incorporating both loci has been prepared (Harvey et al. 2005). Following injection of a transgene comprising the ovomucoid/ovoinhibitor promoter and either the immunoglobulin heavy- or light-chain coding region into newly fertilized eggs, 0.3% of the hens that were reared to sexual maturity expressed between 19 and 150 ng of product per milliliter of egg white. It is unclear whether these birds were transgenic founders or mosaic chickens.

7.4. UBIQUITOUS EXPRESSION OF PROTEINS IN CHICKENS

Transgenes have been introduced into the chicken genome under the regulation of promoters such as CMV (McGrew et al. 2004; Harvey et al. 2002a, 2002b; Rapp et al. 2003), SV40 (Mozdziak et al. 2003), RAV-2 (Thoraval et al. 1995), PGK (Chapman et al. 2005), RSV (Kamihira et al. 2005), and β-actin (van de Lavoir et al. 2006a, 2006b; Zhu et al. 2005), all of which are believed to be expressed ubiquitously. In some cases, the claim to ubiquitous expression has been supported by data, whereas in other cases it has been assumed that the transgenes would be expressed in all tissues. Each of these promoters is very useful in the assessment of techniques for genetic modification and has widespread utility in developmental biology. However, ubiquitous expression suffers from the inherent inability to deliver high-level, tissue-specific, and developmentally regulated gene expression. Conceptually, it seems prudent to restrict the production of large amounts of a foreign protein to the physiological compartments that have the greatest impact on productivity. In the case of protein production in egg white, therefore, transgenes should be designed to yield high-level expression in the tubular gland cells of mature hens in egg production. In addition to potentially affecting the health and welfare of the birds, restricting expression to only one cell type yields proteins with a consistent glycosylation. For example, monoclonal antibodies produced by the tubular gland cells are nonfucosylated (Zhu et al. 2005), whereas antibodies produced by B-cells-contain fucose (Raju et al. 2000). If a transgene-encoding synthesis of a monoclonal antibody were expressed ubiquitously, a portion of the B-cell-derived material would be passively transferred into egg white and the product would contain a mixture of fucosylated and nonfucosylated antibody. In this case, the presence of glycovariants is important because the antibody-dependent cellular cytotxicity (ADCC) of nonfucosylated antibody is 10–100-fold greater than that of fucosylated antibody (Zhu et al. 2005) and ADCC is an important attribute of several monoclonal antibodies used in cancer therapy such as Herceptin (trastuzumab) and Rituxan (rituximab).

7.5. INTRODUCTION OF TRANSGENES INTO THE CHICKEN GENOME

The technical challenges of introducing genetic modifications into the genome of chickens has been reviewed by several authors (Zajchowski and Etches 2000; Petitte 2002; Petitte et al. 2004; Sang 2004). Briefly, the challenges have revolved around the need to insert large amounts of DNA to obtain high-level and tissue-specific expression of exogenous proteins, the large number of birds that must be evaluated to identify founder transgenic animals, and the perceived dangers of using retroviral vectors derived from pathogenic avian and human viruses.

7.5.1. Viral Vectors

The ability of retroviruses to insert DNA into host cells was utilized in the earliest technologies for producing transgenic chickens (Bosselman et al. 1989a, 1989b; Salter et al. 1987). The original technology was modified to produce vectors that were replication-defective to prevent viremia in the transgenic birds (Harvey et al. 2002b; Thoraval et al. 1995). In spite of these advances, the amount of DNA that can be loaded into these vectors is restricted to 8 kb or less. By contrast, transgenes that have been shown to yield high-level expression contain 40–160 kb of sequence. In addition to their inability to incorporate large amounts of DNA, the frequency of incorporation of the transgene into the germline is very low. For example, 395 chicks hatched following injection of 1474 embryos with vectors encoding β-lactamase or chloramphenical transferase (Harvey et al. 2002b). Three mosaic males from the cohort of 395 birds produced a total of seven transgenic founders among 2696 offspring. The low frequency of males that transmitted the transgene and the low rate of transmission of transmitting males requires a burdensome investment in time and resources to identify the founder animals. Similar rates of germline transmission were observed using an SV40-β-galactosidase transgene in a retroviral vector (Mozdziak et al. 2003), a PGK-GFP transgene in a lentiviral vector (Chapman et al. 2005), and a CMV-EGFP transgene using a retroviral vector (Koo et al. 2006). The notable exception to low rates of germline transmission is the CMV-GFP and CMV-β-galactosidase transgenes that were inserted into the genome using a lentiviral vector (McGrew et al. 2004). In this case, 10 of 20 males that were injected with the vector trans-mitted the transgene to their offspring, and the rate of germline transmission of each male varied from 4% to 45%.

7.5.2. Injection into Newly Fertilized Oocyte

The first nonviral system to introduce genetic information into the genome of chickens utilized direct injection of DNA into the newly fertilized egg (Love et al. 1994). While this system has the theoretical advantage of being able to insert transgenes of any size into the genome, it has the disadvantages of requiring a donor chicken to provide a newly fertilized egg for each injection and significant technical expertise to nurture the embryo in surrogate shells throughout development (Rowlett and Simkiss 1987; Perry 1988, 1991; Perry and Sang 1993; Sang et al. 1993; Naito and Perry 1989). In addition, the nuclei and pronuclei of the newly fertilized egg are very difficult to identify, and therefore the injection of DNA cannot be precisely directed into an optimal site for integration into the genome. Consequently, the overall rate of success in the system is very low. Attempts have been made to improve the rate of success using two-photon laser-scanning microscopy to identify the nucleus and to transfer the newly fertilized egg into the oviduct of a surrogate hen following injection (Christmann 2002; Rapp and Christmann 2003). Of 5164 ova that were injected and transferred to recipients, 500 chicks were hatched, and the transgene was identified in blood from 58 of them (Rapp et al. 2004), but germline transmission has not yet been reported. In addition to using two-photon laser-scanning microscopy to increase the accuracy of injection, artificial chromosomes have been suggested as vectors to increase the amount of DNA that can be introduced into the chicken genome by direct injection (Christmann et al. 2005). The transmission of an artificial chromosome through the germline has been reported for one bird, and artificial chromosomes have been constructed carrying the ovomucoid promoter driving expression of EPO and G-CSF (Christmann et al. 2005). However, there are no reports of lines of chickens carrying an artificial chromosome with or without a transgene.

The successful cloning of mammals has inspired similar strategies for the chicken (Christmann et al. 2002). To overcome the anatomic challenges that are posed by the large yolk-filled egg in chickens, two-photon laser-scanning microscopy has been suggested as a tool for identifying and ablating the endogenous nucleus in newly fertilized eggs. As in mammals, the replacement nucleus could be derived from embryonic fibroblasts. The advantageous of the cloning strategy is its potential to insert transgenes of any size and to introduce site-specific changes in the chicken genome. To date, however, there are no reports indicating that chickens have been cloned.

7.5.3. Cell-Based Systems for Modification of the Avian Genome

Genetic modification of ES cells and PGCs is theoretically the most attractive route for introducing DNA into the genome because selection for the genetic modification can be done *in vitro*, large numbers of the cells can be produced for evaluation of transgene integrity, and the site of insertion of the transgene into the genome can be established *in vitro* before the transgenic animals are made. Although these advantages were recognized long ago, the technology evolved during a 15-year period as the technical challenges were resolved.

7.5.3.1. Embryonic Stem Cells

Embryonic stem cell technology grew out of the recognition that blastodermal cells isolated from stage X embryos could contribute to both somatic tissues and the germline following transfer to recipient embryos (Petitte et al. 1990). Subsequently, methods for the culture of ES cells were established (Petitte and Yang 1994; Samarut and Pain 2000, 2002, 2006; Pain et al. 1996; Petitte et al. 2004), although none of these methods supported the proliferation of cells that colonized the germline or provided transgenic cells that colonized the somatic tissues of live chimeras. Nevertheless, the repeated observation that freshly isolated blastodermal cells contributed to both somatic tissues and the germline (Kagami et al. 1995, 1997; Naito et al. 1992; Carsience et al. 1993; Kino et al. 1997; Ono et al. 1994) fueled the search for culture systems that would (1) support the expansion of PGCs (that were known to colonize the germline) and ES cells (that were known to colonize the somatic tissues), (2) provide a population of 10^5–10^6 cells for transfection, and (3) allow the clonal derivation of genetically modified cells.

Transfection of ES cells has been described using small (van de Lavoir et al. 2006b) and large (Zhu et al. 2005) transgenes. Briefly, 10^6–10^7 cells are transfected using standard electroporation or lipofection protocols, and genetically modified cells are produced at rates between 10^{-5} and 10^{-6}. These rare transfected cells are isolated *in vitro* by including a gene encoding resistance to an antibiotic such as neomycin or puromycin as either an integral component of the transgene or a separate transgene that is co-transfected with the transgene of interest. The population of transfected cells is then diluted to a concentration that is anticipated to yield only one transfected colony per well and exposed to the antibiotic for which resistance has been putatively established. Following selection, a few wells will contain clonally derived colonies that carry the transgene of interest and also express the gene encoding resistance to the antibiotic. This powerful protocol yields large populations of cells that are uniformly genetically modified. The presence of the transgene can be assessed by Southern blot analysis in DNA extracted from these cells to confirm that the transgene has been integrated into the genome and that a clonal population of cells has been selected. In cases where the gene of interest is expressed in ES cells or PGCs, further analysis of gene expression can be done by RT-PCR. Information about the site of integration of the transgene can be obtained by Fluorescence *in situ* hybridization (FISH) analysis of the clonally derived populations of cells or by sequencing DNA flanking the transgene and comparing this sequence with the reference sequence of the chicken genome (Hillier et al. 2004).

Transfected ES cells can be used to evaluate transgene function following injection into stage X (EG&K) embryos to produce high-grade chimeras (van de Lavoir et al. 2006b; Zhu et al. 2005). These birds will contain genetically modified cells in most, if not all, of the somatic tissues, and the ES-derived cells can be identified by including a marker gene such as GFP under the control of a ubiquitous promoter in the transgene. Histologic analysis of tissues from these high-grade chimeras will reveal the range of tissues and cells within tissues that express the transgenes and provide important information about the performance of the transgene *in vivo*. For example, tissue specificity and expression of a transgene-encoding production of a human therapeutic monoclonal antibody under the control of the ovalbumin promoter was validated in estrogen-induced 2-week-old chicks (Zhu et al. 2005). The interval from transfection of ES cells to selection and expansion of clonal populations of cells for injection into chimeras is approximately 8 weeks. Although colonization of ES cells in chimeras has to date been restricted to somatic tissues, their use in evaluating transgene functionality is very appealing because the assessment can be done quickly and it occurs *in vivo*.

7.5.3.2. Primordial Germ Cells

The quest for culture systems that would support the proliferation, transfection, and clonal selection of genetically modified PGCs was fueled by the observation that freshly isolated PGCs could colonize the germline following transfer from donor to recipient embryos (Vick et al. 1993; Park et al. 2003; Naito 2003; Naito et al. 1994a, 1994b, 1998; Tajima et al. 1993, 1998; Song et al. 2005). After more than a decade of searching, a culture system for growing, transfecting, and clonally selecting PGCs is now available (van de Lavoir et al. 2006a).

To demonstrate this principle, genes encoding GFP and resistance to puromycin were placed under the control of the ubiquitously expressed β-actin promoter and introduced by electroporation into PGCs. To obtain expression of the transgenes in PGCs, HS4 sequences derived from the chicken β-globin locus were placed in regions 5′ and 3′ of the transgene. The genetically modified cells were injected into recipient stage 14–17 (H&H) embryos, and the genetic modifications were transmitted through the germline to produce chickens expressing GFP. As expected, the transgenes were inherited according to Mendelian principles (van de Lavoir et al. 2006a). A major advantage of the PGC technology is the greatly reduced requirement for *in vivo* screening of putative founders to identify birds carrying the transgene. Selection of the genetically modified cells occurs *in vitro*, and they are characterized by PCR and/or Southern blot analysis prior to injection into the recipient to ensure that the transgene is incorporated into the genome intact. Colonization of the germline is usually efficient. For example, in a cohort of 10 males produced from transgenic PGCs, several birds will have rates of transmission of the PGC genotype in excess of 50%. Because the transgenes are inserted into only one of the homologous pairs of chromosomes, only 50% of the PGC-derived offspring carry the transgene.

In a typical example, a G0 chimeric male will sire 400 offspring, and 50% of these will be derived from PGCs to yield 200 PGC-derived G1 birds. Approximately 50% of these birds will carry the transgene to yield 50 transgenic G1 males and 50 transgenic G1 females. The 50 G1 males can be used to expand the flock to several thousand birds, and/or the 50 G1 females can be used to assess the functionality of the transgene if it is encoding the production of a novel protein in eggs. Production of G1 males and females is highly scalable. If a population of several thousand G1 hens is required, they can be produced quickly from a larger cohort of G0 males or by mating G0 males with high rates of germline transmission to 100–150 hens per week. The latter strategy can produce up to 1300 transgenic hens per month from a single G0 male with 90% fertility, 90% hatchability, and a germline transmission rate of 90%. Additional G0 males producing transgenic offspring at the same rate yield a linear increase in the size of the flock of G1 hens depositing the transgenic product in their eggs.

7.6. PURIFICATION OF PROTEINS FROM EGG WHITE

A number of proteins, including lysozyme and avidin, are routinely isolated from egg white for use in the pharmaceutical industries, biomedical research, and the food-processing industry. The experience gained in processing eggs for these uses and the extensive descriptions of the chemistry of egg proteins (Burley and Vadhera 1989) provide a basis for designing purification protocols for products deposited in egg white or egg yolk by transgenic chickens. To date, the only protocol for transgenic products has been developed to isolate a therapeutic monoclonal antibody from egg white (Zhu et al. 2005). Briefly, the viscosity of the material was reduced by initiating precipitation of ovomucin by diluting raw egg white threefold in water and adjusting the pH to 6. The material was then centrifuged, filtered and purified on a protein A column to yield a pure preparation by Coomassie staining and Western blotting. As the number of products increases and the products enter into clinical trials, considerably more effort will be required to develop purification protocols.

An attractive feature of eggs as a starting material for protein products is the absence of proteases, which frequently contaminate products during their extraction from plant and animal cells. These proteases yield breakdown products that aggregate and are immunogenic in patients. By contrast, a therapeutic monoclonal antibody produced in chickens was free of aggregates, whereas the same product produced in Chinese Hamster Ovary (CHO) cells contained 2.8% aggregates (Zhu et al. 2005).

7.7. ATTRACTIVE ATTRIBUTES OF EGG PROTEIN PRODUCTION

Among the various options for producing proteins in cells, plants, and animals, the deposition of proteins in eggs is attractive from several perspectives. The amount of product is easily scalable by breeding additional birds. The generation interval is short, and the reproductive capacity of a single male is prolific. For example, if a single male that is heterozygous for a transgene is mated to 150 hens per week, a flock of 5000 hens can be produced in 6 months. If each egg contains 50 mg of a novel protein and each hen lays 300 eggs, then 75 kg of crude protein can be produced annually. If smaller or larger amounts are needed, the size of the flock can be adjusted accordingly. For protein therapeutics, the requirements for most products could be accommodated by flocks of 1000–15,000 hens. Using industrial standards, these are small flocks that can be housed in buildings operated under GMP and SPF conditions. Standardization of SPF and GMP conditions has been established for the production of elite breeding stock for poultry production and for the manufacture of some vaccines in eggs.

Although the number of proteins that have been extensively examined is limited, the glycosylation of human proteins produced in the chicken indicated that egg-derived proteins contain sugar residues that closely mimic those found in humans (Zhu et al. 2005; Ivarie et al. 2004). The sugar residues that decorated a human therapeutic monoclonal antibody produced in the tubular gland cells and deposited in eggs contained moieties that are shared by humans and therefore are unlikely to be immunogenic when injected into patients. In addition, the antibodies were nonfucosylated, which improves the ADCC activity of the antibody (Zhu et al. 2005).

The manufacture of proteins by chickens requires systems for handling a large number of eggs. For example, if each egg contains 50 mg of protein and 100 kg of protein is required annually, 2 million eggs are needed if 50% of the crude protein is extracted during purification. Automated egg collecting, handling, and processing systems that have been designed for the egg production industry can be adapted for this purpose. Eggs can be oiled to prevent dehydration, stored at 4°C, and processed in batches to meet manufacturing requirements.

7.8. FUTURE PROSPECTS

The potential attractiveness of producing proteins in eggs has been recognized for at least a decade, but the lack of a suitable technology for introducing transgenes into the avian genome has prevented development of this potential until very recently. During that time, at least seven biotechnology companies dedicated to capturing the value of this technology have been established, and two of the major pharmaceutical companies have invested in the technology. While the initial work demonstrating the potential was done in academic environments, subsequent development has occurred primarily within the biotechnology sector. Accordingly, much of the information about the technology is found in the patent literature, which has been extensively quoted in this chapter.

The advent of a simple technology using PGCs to introduce large transgenes into the germline and the demonstration that high-level, tissue-specific, and developmentally regulated gene expression of transgenes is possible has opened the door for further commercialization of the technology. The attractiveness of both the medium and the product that is derived from eggs provides a unique and potentially profitable platform for the production of novel proteins.

REFERENCES

Benarafa C, Remold-O'Donnell E (2005). The ovalbumin serpins revisited: Perspective from the chicken genome of clade B serpin evolution in vertebrates. Proc Natl Acad Sci USA 102:11367–11372.

Bosselman RA, Hsu R-Y, Boggs T, Hu S, Bruszewski J, Ou S, Kozar L, Martin F, Green C, Jacobsen F, Nicolson M, Schultz JA, Semon KM, Rishell W, Stewart RG (1989a). Germline transmission of exogenous genes in the chicken. Science 243:533–535.

Bosselman RA, Hsu, R-Y, Boggs T, Sylvia H, Bruszewski J, Ou S, Souza L, Kozar L, Martin F, Nicolson M, Rishell W, Schultz JA, Semon KM, Stewart RG (1989b). Replication-defective vectors of reticuloendotheliosis virus transduce exogenous genes into somatic stem cells of the unincubated chicken embryo. J Virol 63:2680–2689.

Burley RW, Vadhera DV (1989). The Avian Egg: Chemistry and Biology. New York: Wiley.

Carsience RS, Clark ME, Verrinder Gibbins AM, Etches RJ (1993). Germline chimeric chickens from dispersed donor blastodermal cells and compromised recipient embryos. Development 117:669–675.

Chapman SC, Lawson A, Macarthur WC, Wiese RJ, Loechel RH, Burgos-Trinidad M, Wakefield JK, Ramabhadran R, Mauch TJ, Schoenwolf GC (2005). Ubiquitous GFP expression in transgenic chickens using a lentiviral vector. Development 132: 935–940.

Christmann L (2002). Microinjection Assembly and Methods for Microinjecting and Reimplanting Avian Eggs. WO Patent 02/064727 A2.

Christmann L, Eberhardt DM, Harvey AJ, Leavitt MC (2005). Genomic Modification. US Patent 2005/0198700 A1.

Christmann L, Pratt SL, Rapp JC (2002). Cloned Cells, Embryos, and Animals and Methods of Producing Them. WO Patent Application 02/20752.

Deeley RG, Burtch-Wright RA, Grant CE, Hoodless PA, Ryan AK, Schrader TJ (1991). Synthesis and deposition of egg proteins. In: Gibbins RJE ed. Manipulation of the Avian Genome. Boca Raton, FL: CRC Press; pp 205–222.

Etches RJ (1996). Reproduction in Poultry. Wallingford, CT: CAB International.

Etches RJ, Kay RM, Leighton P, Zhu L (2005a). Transgenic Aves Producing Human Polyclonal Antibodies. US Patent Application 20050246782.

Etches RJ, Mohammed M, Morrison S, Wims L, Trinh KM, Wildeman AG (2005b). Production of Proteins in Eggs. US Patent 6,861,572.

Harvey AJ, Leavitt MC, Wang Y (2005). Ovomucoid Promoters and Methods of Use. US Patent Application 2005/003414 A1.

Harvey AJ, Speksnijder G, Baugh LR, Morris JA, Ivarie R (2002a). Expression of exogenous protein in the egg white of transgenic chickens. Nat Biotechnol 20:396–399.

Harvey AJ, Speksnijder G, Baugh LR, Morris JA, Ivariet R (2002b). Consistent production of transgenic chickens using replication-deficient retroviral vectors and high-throughput screening procedures. Poultry Sci 81:202–212.

Hillier LW, Miller W, Birney E, Warren W, Hardison RC, Ponting CP, Bork P, Burt DW, Groenen MA, Delany ME, Dodgson JB, Chinwalla AT, Cliften PF, Clifton SW, Delehaunty KD, Fronick C, Fulton RS, Graves TA, Kremitzki C, Layman D, Magrini V, McPherson JD, Miner TL et al. (2004). Sequence and comparative analysis of the chicken genome provide unique perspectives on vertebrate evolution. Nature 432:695–716.

Ivarie R (2003). Avian transgenesis: Progress towards the promise. Trends Biotechnol 21:14–19.

Ivarie R (2006). Competitive bioreactor hens on the horizon. Trends Biotechnol 24:99–101.

Ivarie R, Harvey AJ, Morris JA, Liu G, Rapp JC (2004). Exogenous Proteins Expressed in Avians and Their Eggs. US Patent Application 2004/0019923 A1.

Kagami H, Clark ME, Verrinder Gibbins AM, Etches RJ (1995). Sexual differentiation of chimeric chickens containing ZZ and ZW cells in the germline. Mol Reprod Devel 42:379–387.

Kagami H, Tagami T, Matsubara Y, Harumi T, Hanada H, Maruyama K, Sakurai M, Kuwana T, Naito M (1997). The developmental origin of primordial germ cells and the transmission of the donor-derived gametes in mixed-sex germline chimeras to the offspring in the chicken. Mol Reprod Devel 48:501–510.

Kamihira M, Nishijima K, Iijima S (2004). Transgenic birds for the production of recombinant proteins. Adv Biochem Eng Biotechnol 91:171–189.

Kamihira M, Ono K, Esaka K, Nishijima K, Kigaku R, Komatsu H, Yamashita T, Kyogoku K, Iijima S (2005). High-level expression of single-chain Fv-Fc fusion protein in serum and egg white of genetically manipulated chickens by using a retroviral vector. J Virol 79:10864–10874.

Kino K, Pain B, Leibo SP, Cochran M, Clark ME, Etches RJ (1997). Production of chicken chimeras from injection of frozen-thawed blastodermal cells. Poultry Sci 76:753–760.

Koo BC, Kwon MS, Choi BK, Kim J-H, Cho S-K, Sahn SH, Cho EJ, Lee HT, Chang W, Jeon I, Park J-K, Park JB, Kim T (2006). Production of germ line transgenic chickens expressing enhanced green fluorescent protein using a MOMLV-based retrovirus vector. FASEB J 20:2251–2260.

Kowalczyk K, Diass J, Halpern J, Roth TF (1985). Quantitation of maternal-fetal IgG transport in the chicken. Immunology 54:755–762.

Lampard GR, Verrinder Gibbins AM (2002). Secretion of foreign proteins mediated by chicken lysozyme gene regulatory sequences. Biochem Cell Biol 80:777–788.

Lillico SG, McGrew MJ, Sherman A, Sang HM (2005). Transgenic chickens as bioreactors for protein-based drugs. Drug Discov Today 10:191–196.

Lillico SG, Sherman A, McGrew MJ, Robertson CD, Smith J, Haslam C, Barnard P, Radcliffe PA, Mitrophanaus KA, Elliot EA, Sang HM (2007). Oviduct specific expression of two therapeutic proteins in transgenic lens. Proc Natl Acad Sci USA 104:1771–1776.

Lonberg N (2005). Human antibodies from transgenic animals. Nat Biotechnol 23:1117–1125.

Love J, Gribbin C, Mather C, Sang H (1994). Transgenic birds by DNA microinjection. Bio/Technology 12:60–63.

MacArthur WC (2004). Methods for Tissue Specific Synthesis of Protein in Eggs of Transgenic Hens. USA Patent 6,825,396.

McGrew MJ, Sherman A, Ellard FM, Lillico SG, Gilhooley HJ, Kingsman AJ, Mitrophanous KA, Sang H (2004). Efficient production of germline transgenic chickens using lentiviral vectors. EMBO Rep 5:728–733.

Mohammed SM, Morrison S, Wims L, Trinh KR, Wildeman AG, Bonselaar J, Etches RJ (1998). Deposition of genetically engineered human antibodies into the egg yolk of hens. Immunotechnology 4:115–125.

Morrison SL, Mohammed MS, Wims LA, Trinh R, Etches R (2002). Sequences in antibody molecules important for receptor-mediated transport into the chicken egg yolk. Mol Immunol 38:619–625.

Mozdziak PE, Borwornpinyo S, McCoy DW, Petitte JN (2003). Development of transgenic chickens expressing bacterial beta-galactosidase. Devel Dynam 226:439–445.

Naito M (2003). Cryopreservation of avian germline cells and subsequent production of viable offspring. J Poultry Sci 40:1–12.

Naito M, Nirasawa K, Oishi T (1992). Preservation of quail blastoderm cells in liquid nitrogen. Br Poultry Sci 33:449–453.

Naito M, Perry MM (1989). Development in culture of the chick embryo from cleavage to hatch. Br Poultry Sci 30:251–256.

Naito M, Tajima A, Tagami T, Yasuda Y, Kuwana T (1994a). Preservation of chick primordial germ cells in liquid nitrogen and subsequent production of viable offspring. J Reprod Fertil 102:321–325.

Naito M, Tajima A, Yasuda Y, Kuwana T (1994b). Production of germline chimeric chickens, with high transmission rate of donor-derived gametes produced by transfer of primordial germ cells. Mol Reprod Devel 39:153–161.

Naito M, Tajima A, Yasuda Y, Kuwana T (1998). Donor primordial germ cell-derived offspring from recipient germline chimeric chickens: Absence of long term immune rejection and effects on sex ratios. Br Poultry Sci 39:20–23.

Ono T, Muto, S-I, Agata K, Mochii M, Kino K, Otsuka K, Ohta M, Yoshida M, Eguchi G (1994). Production of quail chimera by transfer of early blastodermal cells and its use for transgenesis. Jpn Poultry Sci 31:119–129.

Pain B, Clark ME, Shen M, Nakazawa H, Sakurai M, Samarut J, Etches RJ (1996). Long-term in vitro culture and characterisation of avian embryonic stem cells with multiple morphogenetic potentialities. Development 122:2339–2348.

Pain B, Samarut J, Valarche I, Champion-Arnaud P, Sobczyk A, Kunita R (2003). Exogenous Protein Expression System in an Avian System. WO Patent Application 03/043415 A1.

Park TS, Jeong DK, Kim JN, Song GH, Hong YH, Lim JM, Han JY (2003). Improved germline transmission in chicken chimeras produced by transplantation of gonadal primordial germ cells into recipient embryos. Biol Reprod 68:1657–1662.

Perry MM (1988). A complete culture system for the chick embryo. Nature 331:70–72.

Perry MM (1991). In vitro Embryo Culture Technique. US Patent 5,011,780.

Perry MM, Sang HM (1993). Transgenesis in chickens. Transgen Res 2:125–133.

Petitte JN (2002). The avian germline and strategies for the production of transgenic chickens. J Poultry Sci 39:205–228.

Petitte JN, Clark ME, Liu G, Verrinder Gibbins AM, Etches RJ (1990). Production of somatic and germline chimeras in the chicken by transfer of early blastodermal cells. Development 108:185–189.

Petitte JN, Liu G, Yang Z (2004). Avian pluripotent stem cells. Mech Devel 121:1159–1168.

Petitte JN, Yang Z (1994). Method for Producing an Avian Embryonic Stem Cell Culture and the Embryonic Stem Cell Culture Produced by the Process. US Patent 5,340,740.

Raju TS, Briggs JB, Borge SM, Jones, AJS (2000). Species-specific variation in glycosylation of IgG: Evidence for the species-specific sialylation and branch-specific galactosylation and importance for engineering recombinant glycoprotein therapeutics. Glycobiology 10:477–486.

Rapp JC (2002). Avian Lysozyme Promoter. US Patent Application 2002/0199214 A1.

Rapp JC, Christmann L (2003). Production of a Transgenic Avian by Cytoplasmic Injection. WO Patent Application 03/025146 A2.

Rapp JC, Christmann L, Harvey AJ, Leavitt MC (2004). Production of Transgenic Avians. US Patent Application 2004/0255345.

Rapp JC, Harvey AJ, Speksnijder GL, Hu W, Ivarie R (2003). Biologically active human interferon alpha-2b produced in the egg white of transgen hens. Transgen Res 12:569–575.

Rowlett K, Simkiss K (1987). Explanted embryo culture: In vitro and in ovo techniques for domestic fowl. Br Poultry Sci 28:91–101.

Salter DW, Smith EJ, Hughes SH, Wright SE, Crittenden LB (1987). Transgenic chickens: Insertion of retroviral genes into the chicken germ line. Virology 157:236–240.

Samarut J, Pain B (2000). Active Retinoic Acid-Free Medium for Chicken Embryonic Stem Cells. US Patent 6,114,168.

Samarut J, Pain B (2002). Culture Medium for Avian Embryonic Cells. US Patent 6,500,668.

Samarut J, Pain B (2006). Active Retinoic Acid-Free Culture Medium for Avian Totipotent Embryonic Stem Cells. US Patent 6,998,266.

Sang H (1994). Transgenic chickens—methods and potential applications. TIBTECh 12:415–419.

Sang H (2004). Prospects for transgenesis in the chick. Mech Devel 121:1179–1186.

Sang H, Gribbin C, Mather C, Morrice D, Perry M (1993). Transfection of Chick Embryos Maintained under in vitro Conditions. New York: CRC Press.

Shuman RM (1990). Genetic engineering. In Crawford: RD, ed. Poultry Breeding and Genetics. Amsterdam: Elsevier; pp 585–598.

Song Y, D'Costa S, Pardue SL, Petitte JN (2005). Production of germline chimeric chickens following the administration of a busulfan emulsion. Mol Reprod Devel 70:438–444.

Tajima A, Naito M, Yasuda Y, Kuwana T (1993). Production of germ line chimera by transfer of primordial germ cells in the domestic chicken (*Gallus domesticus*). Theriogenology 40:509–519.

Tajima A, Naito M, Yasuda Y, Kuwana T (1998). Production of germ line chimeras by transfer of cryopreserved gonadal primordial germ cells (gPGCs) in chicken. J Exp Zool 280:265–267.

Thoraval P, Afanassieff M, Cosset, F-L, Lasserre F, Verdier G, Coudert F, Dambrine G (1995). Germline transmission of exogenous genes in chickens using helper-free ecotropic avian leukosis virus-based vectors. Transgen Res 4:369–376.

van de Lavoir MC, Diamond JH, Leighton P, Mather-Love C, Heyer, BS, Kerchner A, Hooi L, Gessaro T, Swanberg S, Delany ME, Etches RJ (2006a). Germline transmission of genetically modified primordial germ cells. Nature 441:766–769.

van de Lavoir MC, Mather-Love C, Leighton P, Diamond JH, Heyer BS, Roberts R, Zhu L, Winters-Digiacinto P, Kerchner A, Gessaro T, Swanberg S, Delany ME, Etches RJ (2006b). High-grade transgenic somatic chimeras from chicken embryonic stem cells. Mech Devel 123:31–41.

Vick L, Luke G, Simkiss K (1993). Germ-line chimaeras can produce both strains of fowl with high efficiency after partial sterilization. J Reprod Fertil 98:637–641.

Vieira AV, Sanders EJ, Schneider WJ (1995). Transport of serum transthyretin into chicken oocytes. A receptor-mediated mechanism. J Biol Chem 270:2952–2956.

Zajchowski LD, Etches RJ (2000). Transgenic chickens past, present, and future. Avian Poultry Biol Rev 11:63–80.

Zhu L, van de Lavoir MC, Albanese J, Beenhouwer DO, Cardarelli PM, Cuison S, Deng DF, Deshpande S, Diamond JH, Green L, Halk EL, Heyer BS, Kay RM, Kerchner A, Leighton PA, Mather CM, Morrison SL, Nikolov ZL, Passmore DB, Pradas-Monne A, Preston BT, Rangan VS, Shi M et al. (2005). Production of human monoclonal antibody in eggs of chimeric chickens. Nat Biotechnol 23:1159–1169.

Zhu L, Etches RJ (2008). Production of human therapeutic monoclonal antibodies in chicken's eggs. In: Zhigiang A, Stroh WR, eds. Therapeutic Antibodies: From Theory to Practice. Hoboken, NJ: Wiley, in press.

8

EGG PRODUCTS INDUSTRY AND FUTURE PERSPECTIVES

GLENN W. FRONING
Food Science and Technology, Institute of Agriculture and Natural Resources, University of Nebraska, Lincoln

8.1. INTRODUCTION

As future new applications of egg components are pursued, it is important to explore the egg products industry and its application of new technologies through the years and its status today. The egg products industry has seen vast changes since the mid-1950s or 1960s. With the advent of the present-day egg products industry, many advances have been seen including new high-speed egg-breaking machines, improved pasteurization technology, advanced freezing methods, better desugaring techniques, and improved multistage spray driers (Stadelman and Cotterill 1995).

Along with changes in egg-processing technology, there has been a continuing growth of further processed egg products. Today, approximately 30% of the total consumption of eggs is in the form of further processed egg products (Table 8.1). Many of these egg products are used as ingredients in various food applications (e.g., liquid whole egg, yolk, and whites; frozen salted yolk or sugared yolk; dried whole eggs, dried yolks, or dried egg whites). Examples of other egg products may include hard-cooked chopped eggs, precooked scrambled eggs or omelets, quiches, precooked egg patties, scrambled egg mixes, and crepes. As indicated in Table 8.1, egg consumption has increased. Some of this growth is due to the increased egg consumption in the form of new egg products. The Framingham Dietary Study has also shown that egg consumption is unrelated to plasma cholesterol or the risk of

Egg Bioscience and Biotechnology Edited by Yoshinori Mine

TABLE 8.1. Per Capita Consumption

Year	Total Egg Consumption	Egg Products (%)
1996	234.6	28.0
1997	235.7	29.1
1998	239.7	29.3
1999	249.8	30.1
2000	251.7	30.5
2001	252.8	30.2
2002	255.9	30.6
2003	254.7	28.7
2004	257.1	29.8
2005	255.1	30.9

heart disease (Dawber et al. 1982). With these data and other more recent studies, consumers now can eat eggs with no significant concern for the risk for heart disease.

Nutrient composition of egg products is shown in Table 8.2. Eggs are an excellent source of high-quality protein containing all nine essential amino acids and other nutrients required in our diets (Watson 2002). Whole-egg protein has a biological value of 94, which is the highest of any of the major food protein sources. Eggs have a 2:1 ratio of unsaturated to saturated fatty acids and are free of trans–fatty acids (The FDA regulations allow labels to be listed as 0 if transfats are less than 0.5 g per 100 g.) Other nutrients add to the complete nutrition coming from eggs, including vitamins A, D, E, B_{12}, biotin, folic acid, pantothenic acid, riboflavin, and thiamine; choline; and minerals such as calcium, phosphorus, iron, and zinc.

Enrichment of eggs with certain other nutrients has opened new opportunities. Nutrients such as omega-3 fatty acids, lutein, and vitamin E are now being fortified in eggs through the diet of the hen (Sim et al. 2000). The use of eggs as nutraceuticals is discussed by other authors in this book.

8.2. EGG-PROCESSING UNIT OPERATIONS

When converting shell eggs to liquid, frozen, and dehydrated egg products, numerous unit operations are involved. Breaking and processing of egg products is regulated in the United States by the USDA (FDA 1971, 2002a, 2002b; USDA 1980). Also, equipment and their installation must comply with E3A Sanitary Standards and Accepted Practices (2006). As the egg-processing industry considers further ultilization of biologically active components, it is important to consider how such approaches would be implemented into present unit operations.

TABLE 8.2. Nutrient Composition of Egg Products

	Liquid/Frozen					Dried[a]		
Nutrient per 100g	Whole Egg	Yolk	Sugared Yolk	Salted Yolk	White	Whole Egg	Yolk	Stabilized White
Protein (g)	12.0	15.3	13.9	14.9	9.3	48.4	33.7	84.6
Moisture	75.2	56.8	51.2	50.9	89.0	3.7	2.7	6/5
Fat (g)	9.7	23.0	20.8	20.9	0.076[b]	39.2	52.9	0.4[b]
Ash (g)	0.8	1.4	1.1	10.4	0.4	3.4	3.3	3.6
Carbohydrates	2.2	3.6	13.0	3.8	1.3	5.4	7.3	4.8
Calories (cal)	144.0	282.0	294.0	259.3	43.0	568.0	640.0	361.0
Cholesterol (mg)	400.0	991.0	917.0	912.0	3.3[b]	1630.0	2307.0	20.0[b]
Transfat (g)	0.11	0.24	0.18	0.16	0.02[b]	0.35	0.63	<0.004[b]

[a]After rehydration, values would be comparable to those of liquid egg products.
[b]A small amount of fat in white is observed from the breaking operation.

Source: American Egg Board (2006).

8.2.1. Holding and Transfer Rooms

Eggs for breaking operations are generally maintained at 12.8°C when being held for a short time. Most eggs in today's breaking plants are broken 3–4 days after being laid, requiring minimum storage time. Some inline operations will break eggs the same day that the eggs are laid. If eggs are held for long periods of time, it is recommended that they be stored at 7°C. Since most eggs are held a short time, eggs broken and processed today are of high quality. Eggs processed in the 1950s and 1960s were often held for long periods of time. Thus, eggs at that time were much lower in quality and more prone to bacterial contamination. Egg production and egg-handling methods used in the 1960s were inferior to our present procedures. Eggs were produced predominantly by small flocks with less control of washing, handling, and refrigeration practices. In fact, many farms had no refrigeration facilities and eggs were held at the farm for long periods of time after being laid.

Prior to the breaking operation, all eggs must be properly washed and sanitized (Fig. 8.1). The washing facilities must be separate from the breaking room, and air movement must have a positive flow from the breaking room to the washing area. The temperature of the washwater must be maintained at 32°C (89.6°F) or higher and should be at least 11°C (20°F) warmer than the shell temperature. An approved cleaning compound must be used and the washing operation must be continuous. The washwater must be changed at least every 4h or more often to maintain sanitary conditions. Freshwater (preferably with chlorine added) should maintain continuous overflow. Eggs should not be allowed to soak in water at any time. After washing, eggs should be sanitized by spray rinsing with 100–200ppm available chlorine. Prior to breaking, eggs should be sufficiently dried to prevent contamination from free

Figure 8.1. Egg washer. (Courtesy of Diamond Automations, Inc.)

moisture on the shell. Recent research has indicated that proper washing of eggs prevents *Salmonella* contamination of the egg contents (Hutchison et al. 2004; Stadelman and Cotterill 1995).

8.2.2. Breaking Room and Liquid Egg Processing

Modern egg-breaking machines have greatly changed the egg products industry. These machines may break and separate as many as 144,000 eggs per hour (Fig. 8.2). Figure 8.3 illustrates the breaking–separation operation. These machines have improved the sanitation conditions and labor required in plants today. Although these machines are efficient, operators must be observant to ensure that yolk contamination in white for foaming applications is not excessive. Generally, egg processors have an upper limit of 0.05% yolk contamination to ensure optimum foaming properties. Egg-breaking equipment must be cleaned and sanitized approximately every 4h or more often if needed.

After breaking, liquid eggs are filtered and blended and ingredients are added (salt, sugar, foaming agents, etc.). During blending, it is important to avoid excessive blending or homogenization to avoid loss of foaming properties of egg white (Forsythe and Bergquist 1951).

Liquid egg white generally has a solids content of 11–12%. The pH of egg white ranges from 7.6 to 9.4. Albumen pH is affected by the age of the egg. As the egg is stored, the albumen pH will increase. This pH increase is due to carbon dioxide loss through the shell during storage. Eggs coming from inline systems (eggs transferred directly on a conveyor belt from the production facility to the breaking plant) will generally have a very low pH. On the other

Figure 8.2. Automatic egg breaking machine (144,000 eggs/h). (Courtesy of Diamond Automations, Inc.)

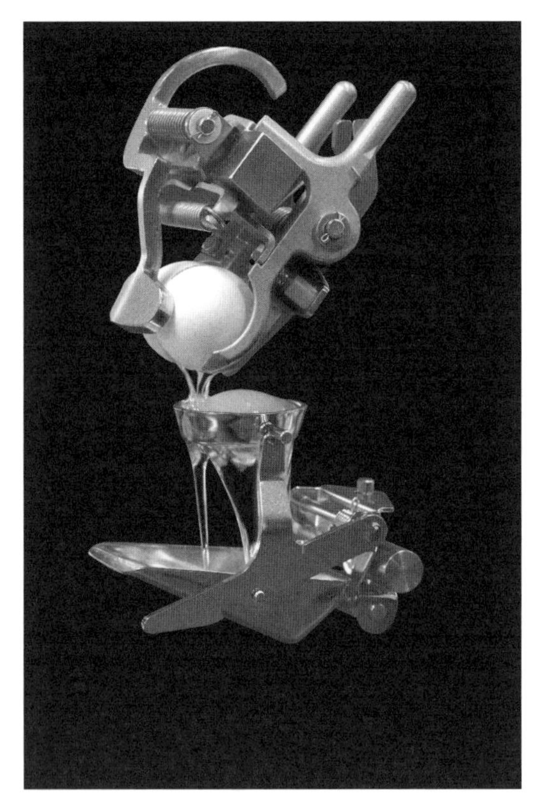

Figure 8.3. Breaking and separation on the automatic breaking machine. (Courtesy of Diamond Automations, Inc.)

hand, eggs arriving from offline facilities will commonly have a higher pH, since they are hauled from the production facility 2–3 times a week. In either case, eggs arrive at the plant much more quickly today than they did years ago, when eggs were shipped to the plant less frequently.

Liquid yolk coming from an egg-breaking operation can vary considerably in solids content. Egg yolk from a breaking machine may contain approximately 20% white, which lowers the present solids (Varadarajulu and Cunningham 1972). The required standard for solids in liquid yolk is 43.0% or better. The pH of egg yolk from shell eggs will be around 6.0–6.2. Yolk pH from breaking operations will be higher due to the presence of egg white.

Liquid whole egg, when considering natural proportions, may vary from 21.5% to 25.0% solids (American Egg Board 2006). According to this survey, inline processed eggs were observed to have lower percent solids. As one would expect, inline natural-break eggs also generally have a lower pH as compared to offline processed whole eggs. For whole egg obtained from blending egg yolk and egg white, the USDA has a requirement of 24.2% solids. To standardize to this level of solids, it is generally necessary to add extra yolk.

The shell egg production practices may influence the composition, physical properties, and functional performance of egg products. For example, it is well known that breed, strain, age of the bird, molting practices, and storage influence egg solids. Cotterill and Geiger (1977) reported changes in egg solids from 1966 to 1976. During that time it was observed that whole-egg solids decreased. Since that time, the industry has seen other changes such as inline and offline operations affecting solids. Also, breaking operations have changed, such as high-speed breaking machines, which may affect the egg solids, yolk contamination of egg white, and functional properties of egg products.

8.2.3. Frozen Egg Products

Egg white is affected minimally by freezing. Some thinning of egg white is commonly encountered, but functional properties are not significantly altered by freezing. Also, research has shown no changes in electropheretic or chromatographic patterns due to freezing (Stadelman and Cotterill 1995).

When freezing and storing of egg yolk below –6°C, the viscosity increases and gelation is encountered (Stadelman and Cotterill 1995). Palmer et al. (1970) found that heating thawed egg yolk at 45–55°C for one hour partially reversed the gelation. Today, the egg industry controls gelation by adding 10% salt or sugar to egg yolk to be frozen. Actually salt increases the emulsifying ability of egg yolk in mayonnaise (Miller and Winter 1951). Wakamatu et al. (1983) presented an excellent review of factors affecting gelation when they indicated that the low-density lipoprotein was the primary yolk component altered by freezing.

Whole egg undergoes less gelation during freezing than that noted from frozen yolk. Also, functional properties are affected minimally during freezing. In fact, Miller and Winter (1951) observed better mayonnaise functionality from frozen whole egg than unfrozen whole egg. Also, freezing of whole egg did not affect the whipping properties (McCready et al. 1971; McCready and Cotterill 1972).

8.2.4. Pasteurization

The early work on egg pasteurization was reviewed in the Egg Pasteurization Manual (USDA 1969). This early research provided the data for development of present-day pasteurization requirements as described in the Egg Products Inspection Act of 1970 (FDA 1971). This act required that all egg products be *Salmonella*-free using approved pasteurization methods. Current pasteurization requirements for various egg products are shown in Table 8.3. Liquid egg products are pasteurized today using high temperature short time (HTST) pasteurizing systems. These are plate or tubular-type heat exchangers (Fig. 8.4).

Other pasteurization methods are also approved for use by the egg industry. Heat plus hydrogen peroxide is used routinely to pasteurize egg white. The

TABLE 8.3. USDA Pasteurization Requirements

Liquid Egg Product	Minimum Temperature (°C)	Minimum Holding Time Requirement (min)
Albumen without chemicals	56.7—55.6	3.5—6.2
Whole egg	60.0	3.5
Whole-egg blends (<2% nonegg ingredients)	61.1—60.0	3.5—6.2
Fortified whole-egg blends (24–36% solids, 2–12% nonegg blends)	62.2—61.1	3.5—6.2
Salted whole egg (with ≥2% salt added)	63.3—62.2	3.5—6.2
Sugared whole egg (with ≥2% salt added)	61.1—60.0	3.5—6.2
Plain yolk	61.1—60.0	3.5—6.2
Sugared yolk (with ≥2% sugar added)	63.3—62.2	3.5—6.2
Salted yolk (with ≥2% salt added)	63.3—62.2	3.5—6.2

Sources: FDA (2002b); USDA (1980).

Figure 8.4. Tubular pasteurizer (7000 lb/h). (Courtesy of Diamond Automations, Inc.)

Armour patented method (Lloyd and Harriman 1957) involves heating to 51.7°C and holding for 1.5 min to inactivate natural catalase. Hydrogen peroxide (10% solution) is metered into the holding tube at a level of 0.5 lb per 100 lb of egg white and held at 51.7°C for 2 min. After pasteurization the egg white is cooled to 7°C and catalase is added to remove residual hydrogen peroxide. Standard Brands (USDA 1969) modified the Armour method. After the egg white leaves the regenerator tube and before it enters the heating section, 10% hydrogen peroxide is injected at a level of 0.875 lb per 100 lb of egg white and heated at 51.7°C for 3.5 min. The egg white is then cooled and catalase is added to remove residual hydrogen peroxide. Heat plus hydrogen peroxide has the advantage of causing less damage to heat-sensitive egg white proteins. Muriana (1997) re-evaluated the hydrogen peroxide procedure and recommended some changes to eliminate *Listeria monocytogenes*.

Dried egg white is pasteurized by the hotroom method to eliminate *Salmonella* (Ayres and Slosberg 1949; Banwart and Ayres 1956). In order to obtain the maximum kill of *Salmonella*, the moisture content should be 6.5–8.0%. The bulk-packed egg white solids are heated at 54.4°C and held at that temperature for 7–10 days until they are *Salmonella*-negative. Research has shown that this process improves whipping properties of egg white (Mine 1995; Handa et al. 1998). Baron et al. (2003) reported that heating dried egg white for 15 days at 67°C and 75°C increased gel strength 2- and 7-fold, respectively, and increased foaming capacities 2.3- and 2.5-fold, respectively. Both heating treatments were effective for eliminating *Salmonella*. Heating at 75°C affected the bacteriostatic ability of reconstituted egg white to a greater extent than did heating at 67°C. They attributed loss of bacteriostatic ability to denaturation of ovotransferrin. Hammershoj et al. (2006) investigated dry pasteurization of egg albumen using a fluidized bed at temperatures of 90°C or 130°C and at high or low air moisture levels for 10 min–3 h. When using the higher temperature and higher moisture levels, higher gel strength and coater holding capacity was observed. Foaming properties were not affected.

There has been a need recently (as of 2007) to reevaluate the present USDA pasteurization requirement for various liquid egg products. Since 1969, there have been many changes in the egg industry. As mentioned earlier, eggs reach the plant very quickly. Therefore, eggs are much fresher and the albumen pH is lower. Cotterill (1968) reported that a better kill of *Salmonella* in egg albumen is achieved at higher pH values. There are other concerns such as thermal resistance of *Salmonella* in salted yolk products. Scientists have concentrated on re-evaluation of present-day pasteurization requirements (Palumbo et al. 1995; Michalski et al. 1999; Froning et al. 2002). Palumbo et al. used the submerged vial technique, while Michalski and Froning used the capillary tube method. These studies indicated that it was necessary to increase the pH of egg white to above 8.8 to achieve a good kill of *Salmonella*. Palumbo reported a much smaller kill of *Salmonella* in salted yolk and egg white than did Michalski and Froning. Schuman et al. (1997) reported that capillary tubes gave more accurate D values as compared to test tubes. Capillary tubes provide an instant come-up time, which likely contributes to the improved accuracy. Froning et al. (2002) investigated a variety of liquid egg products and recommended some changes in the egg pasteurization guidelines. It was recommended that for egg white pasteurized at 56.7°C the pH should be adjusted to around 9 and the holding time increased to 4.5 min. The suggested holding time for salted yolk at 63.3°C was increased to 4.5 min.

The ultrapasteurized and aseptically packaged process was developed by Ball et al. (1987). Pasteurization temperatures ranged from 63.7°C to 72.2°C for 2.7–192.2 s. Swartzel et al. (1989) patented this process and claimed a shelf life of 4–24 weeks at 4°C.

8.2.5. Desugarization

During the 1930s, the United States was having difficulty producing dried egg products with desirable flavor and color. It was known that China was utilizing a fermentation process that yielded dried eggs with a superior flavor and color. With the advent of World War II, there was a need for improved dried egg products for the U.S. Army that would be stable after storage under field conditions. This stimulated research on removal of glucose to improve flavor and color stability. If glucose is removed, the glucose–protein (Maillard) reaction producing undesirable color and the glucose–cephalin reaction causing off-flavors are eliminated (Sebring 1995).

Three methods have been used by the egg industry to remove glucose from liquid egg products prior to drying (Sebring 1995). Controlled bacterial fermentations using special cultures are used most commonly to desugar egg white (Ruangtrakook 1987). Many firms have their own secret bacterial cultures. Yeast fermentation *Saccharomyces cervisiae* is occasionally used to desugar egg white or whole egg. Today, almost all whole-egg and yolk products are desugared using the glucose oxidase–catalase system.

8.2.6. Dehydration and Concentration

Dehydration of eggs was reviewed in depth by Bergquist (1995). Dried egg production in the United States began to develop in the 1930s and has greatly expanded since then. Much growth was stimulated by World War II with the needs of the military.

Most eggs are spray-dried by heating at an inlet temperature between 121°C and 232°C. Atomizing is accomplished by spraying with high-pressure nozzles [500–6000 psi (lb/in.^{2})]into a hot-air stream that evaporates water instantly (Fig. 8.5). Generally, indirect heating such as steam coils are preferred to prevent off-flavors or formation of nitric oxides, which may be formed by direct heat using natural gas or propane. Powder separates from the drying chamber, and the air is removed by an exhaust fan. After drying, the powder is cooled and sifted prior to packaging. Both horizontal and vertical spray driers are used today. A horizontal dryer is shown in Fig. 8.6.

Figure 8.5. High-pressure spray nozzles used in dryer. (Courtesy of Sanova Engineering.)

Figure 8.6. Horizontal spray dryer. (Courtesy of Sanova Engineering.)

The confection trade in some cases still utilizes pan-dried egg white (flake, granules, or powder). Otherwise, other egg products are normally spray-dried. Some freeze-dried scrambled eggs are produced for campers. Freeze-dried eggs are expensive, and commercial use is limited.

The egg industry has utilized reverse osmosis or ultrafiltration to concentrate egg white to reduce energy costs. Froning et al. (1987) compared ultrafiltration and reverse osmosis for concentration of egg white. Ultrafiltration removed water and concentrated egg white from 11.7% solids to 23.2% solids, while reverse osmosis increased solids from 11.2% to 22.8% solids. Ultrafiltration not only removed water but also lowered glucose, sodium, and potassium levels by about 50% in the retentate. Neither of these concentration methods significantly affected foaming properties as measured by angelfood cake volume. Ultrafiltration was observed to improve the gel strength of egg white, which was likely related to the increase in protein concentration.

8.2.7. Processes to Alter Lipid and Cholesterol Content

Various processes have been investigated to alter the lipid and cholesterol content of eggs. Organic solvents can be used remove lipids and cholesterol. Larsen and Froning (1981) extracted an egg oil using hexane-isopropanol which had 40% less cholesterol than that observed in unextracted egg yolk. Warren et al. (1988) extracted lipids from dried egg yolk using a combination of hexane, hexane–isopropanol (2:1), and chloroform–methanol (2:1). Polar solvents were more efficient for extracting cholesterol, but proteins were denatured, adversely affecting functional properties. Environmental problems are a major consideration when handling and disposing of organic solvents. These obstacles likely limit the use of organic solvents to extract lipids and cholesterol from egg yolk.

Supercritical carbon dioxide extraction of lipid components has shown some promise. When a gas is compressed to its critical pressure and heated above the critical temperature, the supercritical fluid becomes an excellent solvent. This process has been utilized commercially to decaffeinate coffee and extract flavor components from foods. Froning et al. (1990) used supercritical carbon dioxide extraction technology to extract cholesterol and lipids from dried egg yolk. Extraction at 306 atm/45°C or 374 atm/55°C removed approximately 67% of the cholesterol and 35% of the lipids. Proteins and phospholipids were concentrated during the extraction process without adversely affecting functional properties. Supercritical carbon dioxide extraction of liquid products has been much less successful. Also, high costs have rendered this process less practical.

β-Cyclodextrin has been used to absorb cholesterol from dairy products and eggs (Haggin 1992). β-Cyclodextrins are cyclic carbohydrates with seven linked D-glucopyranose units forming a central cavity that is hydrophobic. Cholesterol is nonpolar, thereby having an affinity for the interior of the β-cyclodextrin structure. At one time, Michael Foods developed a joint venture with a German firm (SKW Trostberg AG) to remove cholesterol from eggs using β-cyclodextrin (Hammer et al. 1992). When β-cyclodextrin was mixed with egg yolk, the complexed cholesterol was spun out using a centrifuge. This process removed about 80% of the cholesterol from egg yolk. Egg white was mixed with the yolk to form a whole-egg product. This product was marketed for a short period of time.

Since the cholesterol issue is less of a concern with eggs today, there is less incentive to market cholesterol-reduced products. There is still a significant market for egg substitutes that contain about 99% egg white. The yolk replacement is generally created from such ingredients as vegetable oil, nonfat dry milk, gums, food coloring, and artificial flavors. Some egg substitutes are both fat-free and cholesterol-free.

8.3. NEW OPPORTUNITIES

It is apparent that the egg processing industry has made tremendous progress since the 1950s or 1960s. The next 50 years (i.e., until 2057 or so) should provide even more exciting breakthroughs for the egg industry. As we look ahead, there are real opportunities to capitalize on the egg's attributes. The egg-processing industry has a strong base to expand into new applications. Their main task is to establish a vision and goals to make further progress.

Research by food scientists is providing new knowledge related to the egg's components. These components not only provide unique functional properties but also have potential for nutraceuticals and human health applications.

8.3.1. Functionality

The egg is multifunctional, including foaming, coagulative, emulsification, and binding properties in many food applications (bakery products, meringues, mayonnaise, cookies, meat products, etc.). No replacers can match the superior functional attributes in eggs. Nevertheless, there is a continuing need to optimize the egg's functional properties. Food scientists are now finding new technologies that show promise for improving the egg's functional properties.

Mine (1995) reviewed the most recent advances for improving egg white functionality. They described a two-step heating method over a wide range of NaCl concentrations to produce a transparent egg white gel. He also described improved egg white gels formed at pressures of 5000–10,000 kg/cm^2 that were more adhesive and elastic than heated ones. Kato et al. (1994) formed protein–polysaccharide conjugates using ovalbumin and lysozyme along with dextran or galactomannan. These conjugates had excellent emulsifying properties. The lysozyme–polysaccharide conjugates also exhibited antimicrobial activity against Gram-positive and Gram-negative bacteria.

Liang and Kristinsson (2005) investigated low- and high-pH unfolding and refolding for improving foaming properties of egg white. They found that foaming capacity and foaming stability could be improved by unfolding and refolding. The foaming capacity of egg albumen was greatly improved with refolding at pH 6.5, 7.5, or 8.5. Foam stability was improved most with unfolding at pH 12.5. They attributed improved foaming properties to interactions among egg albumen proteins through disulfide and/or hydrophobic groups.

Egg yolk is well known as an excellent emulsifier. The emulsifying ability of egg yolk has been shown to be improved by enzymatic treatment with phospholipase A_2. Phospholipase A_2 converts phosphatidylcholine to lysophosphatidyl–choline, which has high water solubility and emulsifying properties (Hell et al. 1970; Dutilh and Groger 1981). Some egg processors are marketing this enzyme-modified egg yolk with superior emulsifying characteristics.

8.3.2. Nutraceutical and Other Applications

Egg components are unique and offer several nutraceutical applications. To date the industry has not fully capitalized on all of the unique egg fractions found in eggs. Research is just beginning to bring these opportunities to the forefront. Some of these are discussed here. Other authors in this book have expanded in much more detail on many potential avenues for the egg industry to pursue.

Kovaco-Nolan et al. (2005) reviewed many of the egg components contributing to human health. It is well known that egg white has several proteins that offer antibacterial activity. Of these proteins, lysozyme and ovotransferrin are mentioned as likely major contributors to antibacterial properties (Table 8.4). Lysozyme has been shown to lyse Gram-positive organisms, while ovo-

TABLE 8.4. Antimicrobial Proteins in Egg Albumen

Protein	Mode of Action	Significance
Lysozyme	Hydrolysis of β (1–4) glycosidic bonds in bacterial cell wall	Effective against Gram-positive organisms
Ovotransferrin	Chelates metal cations, making them unavailable to microorganisms	Most effective against spoilage bacteria
Avidin	Binds biotin making it unavailable to bacteria	Depends on bacterial requirement for this vitamin
Ovomucoid	Inhibits trypsin	Role unknown
Ovoinhibitor	Inhibits trypsin, chymotrypsin, subtilisin, elastin	Role unknown
Ovomacroglobulin	Inhibits trypsin, papain	Role unknown
Cystatin	Inhibits papain, bromelain, ficin	Role unknown
Flavoprotein	Binds riboflavin, making it unavailable to Bacteria	Role unknown

transferrin chelates iron. The only egg white protein that is routinely commercially removed from egg white is lysozyme (Proctor and Cunningham 1988). Lysozyme, which is isolated using ion exchange resins, is used as a preservative for cheese and in some pharmaceutical applications. When using the ion exchange resins for lysozyme removal, it is also possible to remove avidin at the same time. Avidin binds biotin, making it unavailable to bacteria. Avidin has been used in cancer treatment.

Egg yolk phospholipids, particularly lecithin (phosphatidylcholine), has been associated with memory retention (Kovaco-Nolan et al. 2005). Choline from phosphatidylcholine has been reported as an important nutrient in brain development, liver function, and cancer prevention. Some firms are now separating lecithin from eggs for certain applications.

One underutilized portion of the egg is the shell and shell membranes. Today, some processors dry the shell and shell membranes and utilize them in egg-laying rations as a calcium source. Others may spread them on fields as a fertilizer source. More potential utilization may lie in its utilization as bacterial inhibitors. Mine et al. (2003) characterized the eggshell matrix proteins and found that they were effective inhibitors of bacteria (*Pseudomonas aureginosa, Bacillis cereus*, and *Staphylococcus aureus*). In another study, Poland and Sheldon (2001) found that the shell membranes contain lysozyme and α-*N*-acetylglucsaminidase. Lysozyme is known to inhibit Gram-positive bacteria, while α-*N*-acetylglucsaminidase breaks down the cell wall of Gram-negative bacteria. They found that the shell membrane extract was effective in altering the heat resistance of *Salmonella enteritidis, Salmonella typhimurium, Escherichia coli* O157:H7, *Listeria monocytogenes*, and *Staphylococcus aureus*. It was postulated that the extract may be added to liquid egg prior to pasteurization to improve the pathogen kill. These studies may provide another potential use of the shell and shell membranes.

Enzymatic hydrolysis of egg white to produce various peptides has shown promise. Mine et al. (2004) found that an egg white lysozyme hydrolysate showed bacteriostatic activity against *Staphylococcus aureus* and *E. coli* K-12. Another study by Davalos et al. (2004) observed antioxidant activity of peptides produced from enzymatic hydrolysis of egg white by pepsin. In a more recent investigation, Miguel et al. (2005) reported that an egg white hydrolysate using pepsin produced a peptide fraction (<3000 Da) that effectively lowered blood pressure of hypertensive rats. These studies suggest that egg white peptides may have potential as a functional food. Further research with human subjects may be warranted.

Gennadios et al. (1996) investigated egg albumen films made from desugared spray-dried egg white solids. They concluded that edible and nonedible films could effectively be made from egg albumen. Egg albumen films were observed to be highly hydrophilic and could be used for water-soluble packets (pouches) for ingredients in the food, chemical, and pharmaceutical industries. Egg white edible films could also be used for carriers of antioxidants and flavors for food applications.

Technology employing the immunoglobulin (IgY) lipoprotein has shown promise against several bacteria and viruses. Yolk antibodies have been used to prevent *P. aeruginosa* infections in the lungs of cystic fibrosis patients (Larsson et al. 2006). Use of IgY technology to replace antibiotics has shown promise as an antimicrobial agent (Sunwoo et al. 2006). IgY against *S. mutans* has been effective in preventing dental carries (Smith and Godiska 2006). Kim (2006) utilized IgY as an alternative for antibiotic treatment of *Helicobacter pylori* infection. IgY technology should have some excellent possibilities for usage in the future.

The egg-processing industry needs to look ahead with further emphasis on the health benefits in eggs. In the past the egg industry has been a leader in utilizing new technologies. The industry can gain inspiration from what has happened in the peanut and soybean industry. They have learned to fractionate their products and move forward with new innovative applications. Contributions by other authors in each chapter in this book lay out the opportunities available to the egg-processing industry today.

REFERENCES

American Egg Board (2006; June). Survey of Selected Egg Processors. Park Ridge, IL: American Egg Board.

Ayres JC, Slosberg HM (1949). Destruction of *Salmonella* in egg albumen. Food Technol 3:180–183.

Ball HR Jr, Hamid-Samini M, Foegeding PM, Swartzel KR (1987). Functionality and microbial stability of ultrapasteurized, aseptically-packaged refrigerated whole egg. J Food Sci 52:1212–1218.

Banwart GJ, Ayres JC (1956). The effect of high temperature storage on the content of *Salmonella* and on the functional properties of dried egg white. Food Technol 10:68–73.

Baron F, Nau F, Guerin-Dubiard C, Gonnett FB, Gautier M (2003). Effect of dry heating on microbiological quality, functional properties and natural bacteriostatic ability of egg white after reconstitution. J Food Protect 66:825–832.

Bergquist DH (1995). In: Stadelman WJ, Cotterill OJ, eds. Egg Science and Technology, 4th ed. Binghamton NY: Haworth Press; pp 335–376.

Cotterill OJ (1968). Equivalent pasteurization temperatures to kill *Salmonella* in liquid egg white at various pH levels. Poultry Sci 47:352–365.

Cotterill OJ, Geiger GS (1977). Egg produced yield trends from shell eggs. Poultry Sci 56:1027–1031.

Davalos A, Miguel M, Bartolome B, Lopez-Fandino R (2004). Antioxidant activity of pepties derived from egg white proteins by enzymatic hydrolysis. J Food Protect 67(9):1939–1944.

Dawber TR, Nickerson RJ, Brand FN, Pool J (1982). Eggs, serum cholesterol and coronary heart disease. Am J Clin Nutr 36:617–625.

Dutilh CE, Groger W (1981). Improvement of product attributes of mayonnaise by enzymatic hydrolysis of egg yolk with phospholipase A2. J Sci Food Agric 32:451–458.

E3A Sanitary Standards and Accepted Practices (2006). International Association for Food Protection, 6200 Aurora Avenue, Suite 200 W, Des Moines, IA 50322–2863.

FDA (1971). The Egg Products Inspection Act. Federal Register 36, 9814.

FDA (2002a). Eggs and Egg Products. Code of Federal Regulations CFR 9, 2002.

FDA (2002b). Pasteurization Requirements. Code of Federal Regulations 9:590, 570.

Forysthe RH, Bergquist DH (1951). The effect of physical treatments on some properties of egg white. Poultry Sci 30:302–311.

Froning GW, Wehling RL, Ball HR Jr, Hill RM (1987). Effect of ultrafiltration and reverse osmosis on the composition and functional properties of egg white. Poultry Sci 66:1168–1173.

Froning GW, Wehling RL, Cuppett SL, Pierce MM, Niemann L, Siekmann SL (1990). Extraction of cholesterol and other lipids from dried egg yolk using supercritical carbon dioxide. J Food Sci 55:95–98.

Froning GW, Peters D, Muriana P, Eskeridge K, Travnicek D, Sumner S (2002). International Egg Pasteurization Manual. Park Ridge, IL: American Egg Board.

Gennadios A, Weller CL, Hanna MA, Froning GW (1996). Mechanical and barrier properties of egg-albumen films. J Food Sci 61(3):585–589.

Haggin J (1992). Cyclodextrin research focuses on variety of applications. Chem Eng News (May 18):25–26.

Hammershoj M, Rasmussen HC, Carstens JH, Pedersen H (2006). Dry-pasteurization of egg albumen powder in a fluidized bed. II. Effect on functional properties: Gelation and foaming. Int J Food Sci Technol 41:263–274.

Hammer J, Houston P, Springer K (1992). Betting on guilt-free egg. Newsweek (March 30):56.

Handa A, Takshashi K, Kurvda N, Froning GW (1998). Heat-induced egg white gels as affected by pH. J Food Sci 63:403–407.

Hell R, Menz H, Wieski T (1970). Food Emulsion. UK Patent 1215868.

Hutchison ML, Gittens J, Walker A, Sparks N, Humphrey TJ, Burton C, Moore A (2004). An assessment of the microbiological risks involved with egg washing under commercial conditions. J Food Protect 67:4–11.

Kato A, Ibrahim HR, Nakamura S, Kobayashi K (1994). New methods for improving the functionality of egg white proteins. In: Sim JS, Nakai S, eds. Egg Uses and Processing Technologies—New Developments. Wallingford, CT/Oxon UK: CAB International; pp 250–268.

Kim JW (2006). Use of IGY as an alternative antibiotic treatment control of *Helicobacter pylori* infection. In: Sim JS, Sunwoo HH, eds. The Amazing Egg—Nature's Perfect Functional Food for Health Promotion. Edmonton, Alberta, Canada: Department of Agricultural, Food and Nutritional Science, University of Alberta; pp 354–381.

Kovaco-Nolan J, Phillips M, Mine Y (2005). Advances in the value of eggs and egg components for human health. J Agric Food Chem 53:8421–8431.

Larsen JE, Froning GW (1981). Extraction and processing of lipid components from egg yolk solids. Poultry Sci 60:160–167.

Larsson A, Carlander E, Kollberg H (2006). IGY as a new group of pharmaceutical drugs: Oral immunotherapy with yolk antibodies to prevent *P. aeruginosa* infections in cystic fibrosis patients. In: Sim JS, Sunwoo HH, eds. The Amazing Egg-Nature's Perfect Functional Food for Health Promotion. Edmonton, Alberta, Canada: Department of Agricultural, Food and Nutritional Science, University of Alberta; pp 283–294.

Liang Y, Kristinsson HG (2005). Influence of pH-induced unfolding and refolding of egg albumen on its foaming properties. J Food Sci 70(3):C222–C230.

Lloyd WE, Harriman LA (1957). Method of Treating Egg White. US Patent 2,776,214.

McCready ST, Cotterill OJ (1972). Centrifuged liquid whole egg. 3. Functional performance of frozen supernatant and precipitate fractions. Poultry Sci 51:877–881.

McCready ST, Norris ME, Sebring M, Cotterill OJ (1971). Centrifuged liquid whole egg. 1. Effects of pasteurization on the composition and performance of the supernatant fraction. Poultry Sci 50:1810–1817.

Michalski CB, Brachett RE, Hung YC, Ezeike GOI (1999). Use of capillary tubes and plate heat exchanger to validate USDA pasteurization protocols for elimination of *Salmonella enteritidis* from liquid egg products. J Food Protect 62:112–117.

Miguel M, Lopez-Fandino R, Ramos M, Aleixandre A (2005). Short-term effect of egg-white hydrolyzate products on the arterial blood pressure of hyupertensive rats. Br J Nutr 94:731–737.

Miller C, Winter AR (1951). Pasteurized frozen whole egg and yolk for mayonnaise production. Food Res 16:43–49.

Mine Y (1995). Recent advances in the understanding of egg white functionality. Trends Food Technol 6:225–232.

Mine Y, Oberle C, Kassify Z (2003). Egg matrix proteins as defense mechanism of avian eggs. J Agric Food Chem 51:249–253.

Mine Y, Ma F, Laurian S (2004). Antimicrobial peptides released by enzymatic hydrolysis of hen egg white lysozyme. J Agric Food Chem 52:1088–1094.

Muriana PM (1997). Effect of pH and hydrogen peroxide on heat inactivation of *Salmonella* and *Listeria* in egg white. Food Microbiol 14:11–19.

Palmer HH, Ijichi K, Roff H (1970). Partial thermal reversal of gelation in thawed egg yolk products. J Food Sci 35:403–406.

Palumbo MS, Beers SM, Bhaduri S, Palumbo SA (1995). Thermal resistance of *Salmonella* spp. and *Listeria monocytogenes* in liquid egg yolk and egg yolk products. J Food Protect 58:960–966.

Poland AL, Sheldon BW (2001). Altering the thermal resistance of foodborne bacterial pathogens with an eggshell membrane waste by-product. J Food Protect 64:386–492.

Proctor VA, Cunningham FE (1988). The chemistry of lysozyme and its use as a food preservative and pharmaceutical. CRC Crit Rev Food Sci Nutr 26:359–395.

Ruangtrakook B (1987). Microbial Desugarization of Treated Egg White. PhD dissertation, University of Missouri—Columbia.

Schuman JD, Sheldon BW, Foegeding PM (1997). Thermal resistance of *Aeromonas hydrophila* in liquid whole egg. J Food Protect 60:231–236.

Sebring D (1995). In: Stadelman WJ, Cotterill OJ, eds. Egg Science and Technology, 4th ed. Binghamton, NY: Haworth Press; pp 323–334.

Sim JS, Nakai S, Guenter W, eds (2000). Egg Nutrition and Biotechnology. New York: CAB International.

Smith DJ, Godiska R (2006). Passive immunization approaches for dental caries prevention. In: Sim JS, Sunwoo HH, eds. The Amazing Egg—Nature's Perfect Functional Food for Health Promotion. Edmonton, Alberta, Canada: Department of Agricultural, Food and Nutritional Science, University of Alberta; pp 341–354.

Stadelman WJ, Cotterill OJ, eds (1995). Egg Science and Technology 4th ed. Binghamton, NY: Haworth Press.

Sunwoo HH, Sadeghi G, Karami H (2006). Natural anti-microbial egg antibody. In: Sim JS, Sunwoo HH, eds. The Amazing Egg—Nature's Perfect Functional Food for Health Promotion. Edmonton, Alberta, Canada: Department of Agricultural, Food and Nutritional Science, University of Alberta; pp 295–325.

Swartzel KR, Ball HR Jr, Hamid-Samini MH (1989). Method for the Ultrapasteurization of Liquid Whole Egg. US Patent 4,808,425.

USDA (1969). Egg Pasteurization Manual. ARS 74-48. Albany, CA: USDA Western Utilization Research and Develolpment Division, Agricultural Research Service, U.S. Department of Agriculture.

USDA (1980). Regulations Governing Inspection of Egg Products. 7 CFR, Part 2859. Effective May 8, 1980.

Varadarajulu P, Cunningham FE (1972). A study of selected characteristics of hens yolk. I. Influence of albumen and selected additives. Poultry Sci 51:542–546.

Wakamatu T, Sato Y, Saito Y (1983). On sodium chloride action in the gelation process of low density lipoprotein (LDL) from hen egg yolk. J Food Sci 48:507–516.

Warren HW, Brown HG, Davis DR (1988). Solvent extraction from egg yolk solids. J Am Oil Chem Soc 65:1136–1139.

Watson RR, ed (2002). Eggs and Health Promotion. Ames, IA: Iowa State Press.

INDEX

References followed by t indicate material in tables.

Egg Bioscience and Biotechnology Edited by Yoshinori Mine